PREFACE

The Tongue and Quill was produced by the Air Force to improve writing and speaking skills. Although there are many references to the Air Force in this publication that does not mean the information conveyed only applies to those in the Air Force. This publication is also a helpful resource for those in the Army, Navy, Marines, and Coast Guard, as well as educators, students, and civilian corporations around the United States.

The purpose of this publication is to provide the necessary information to ensure clear communications—written or spoken. Whether you are in the Air Force or not, the knowledge you will acquire from this book will prove to be a valuable tool that you will find yourself referring to often in the future.

D1224567

121 Castle Drive STE F, Madison, AL 35758
Info@MentorEnterprisesInc.com
256.830.8282

ISBN-13: 978-0-9840749-7-6
ISBN-10: 0-9840749-7-x

AIR FORCE HANDBOOK 33-337
1 AUGUST 2004

THE
CONTENTS

PART I:

COMMUNICATION BASICS

CHAPTER 1

A BASIC PHILOSOPHY OF COMMUNICATION

This chapter covers:

- How we define communication.

- Why communication is important in today's Air Force.

- Characteristics of effective communication.

- Common problems encountered when people try to communicate.

This is an exciting time to be in the United States Air Force! Our mission and our operations tempo reflect the larger world around us—a world of rapidly accelerating technology and nearly unlimited access to information. Airmen are successfully accomplishing more missions with fewer people than ever before, and there is a constant battle to cover the bases with limited resources. With many demands on their time, why should the men and women of the Air Force care about effective communication? This chapter answers that question, identifies five principles of effective communication, and describes common speaking and writing problems.

WHAT DO WE MEAN BY COMMUNICATION,
...AND WHY IS IT SO IMPORTANT?

com•mu•ni•ca•tion *n* 1: an act or instance of transmitting information; 2: a verbal or written message; 3: a process by which information is exchanged between individuals through a common system of symbols, signs or behavior.

Communication is defined as the process of sharing ideas, information and messages with others. In the Air Force, most communication involves speaking and writing, but this definition includes nonverbal communication such as body language, graphics, etc.

Any communication can be broken into three parts: the **sender**, the **message** and the **audience**. For communication to be successful, the audience must not only get the message, but must interpret the message in the way the sender intended.

Since communication requires effort, it should always have a purpose. If the purpose isn't clear to the audience, you will have a problem! Most Air Force communication is intended to direct, inform (or educate), persuade or inspire. Often the sender has some combination of these motives in mind.

Chapter 3 describes the process of determining your purpose and audience in detail, but here are a few examples of Air Force communication targeted toward a specific objective:

The headquarters staff (the sender)

    writes a new policy on trip report procedures (the message)

        and sends a copy to all subordinate units (the audience).

    Purpose of this communication: to direct

An aircraft technician (the sender)

    reports the results of an aircraft engine inspection (the message)

        to his supervisor (the audience).

    Purpose of this communication: to inform

A branch chief (the sender)

    requests additional funding for new computers (the message)

        in a meeting with the division chief (the audience).

    Purpose of this communication: to persuade

Most communication outside the Air Force falls in these categories as well. Can you spot the purpose of each of the following sentences?

> "You didn't wash the car like you promised."

> "But Dad! Everyone else is going to the beach. Why can't I go?"

> "Son, I know you're a fine young man, and fine young men keep their promises."

> "Aw, Dad…"

> "Wash the car NOW!"

If you look carefully, you can see the efforts to inform, direct, persuade or inspire in this common conversation.

COMMUNICATION, TEAMWORK AND LEADERSHIP

Communication skills are vitally important in any environment where teamwork is important. Simply put, communication enables us to come together to accomplish things better than we can accomplish as individuals. Communication skills are particularly important for leaders. The ability to communicate a vision and direction, to motivate and inspire others and to persuade our superiors are all essential in bringing people together to achieve a common goal.

The military environment is unique, and much of its uniqueness requires extraordinary communication skills. We operate highly technical equipment in a lethal environment and we are held to very high standards by the country we serve. Miscommunication can cause expensive mistakes, embarrass our organization and in some cases cause accidents or even death.

A CALL TO ARMS…

This book is designed to give you tools and ideas that will help you learn to communicate better … and to teach others as well.

Both the Air Force and the large culture we live in are drowning in a sea of information. Around-the-clock media coverage, universal electronic mail (e-mail), and the expansion of the Internet and other electronic information sources make it difficult for us to sift out the valuable information we need to accomplish our mission. Now, more than ever, it's important to communicate with clarity and focus.

The only way to become a better writer and speaker is to work at it—there are no short cuts. The good news is that service in the Air Force will provide plenty of opportunities for you to improve. Your communication skills will become stronger with practice, regardless of your initial ability, and this book is designed to help you on your journey.

> "Perseverance is a great element of success. If you only knock long enough and loud enough at the gate, you are sure to wake somebody up."
>
> – Henry Wadsworth Longfellow

PRINCIPLES OF EFFECTIVE COMMUNICATION

Once you accept that communication is important, it's important to understand what makes communication succeed and what makes it fail. Most mistakes are caused by forgetting one of five principles of good communication. This section addresses these core principles, which we've organized to spell out the acronym FOCUS. It also describes some of the most common mistakes that occur when you miss the mark.

① FOCUSED: ADDRESS THE ISSUE, THE WHOLE ISSUE AND NOTHING BUT THE ISSUE

The first hallmark of good communication is that it is focused—the sender has a clear idea of purpose and objective, locks on target and stays on track.

In staff or academic environment, writing and speaking often attempts to answer a question provided by either a boss or an instructor. In such situations, the principle may also be stated as the following:

Answer the question, the whole question and nothing but the question.

Failure to focus comes in three forms:

> **fo·cus** *n*
> 1. a state or condition permitting clear perception or understanding: direction; 2. a center of activity, attraction or attention; a point of concentration; directed attention: emphasis.

> **FOCUS Principles**
> **Strong Writing and Speaking:**
>
> **Focused**
>
> Address the issue, the whole issue and nothing but the issue.
>
> **Organized**
>
> Systematically present your information and ideas.
>
> **Clear**
>
> Communicate with clarity and make each word count.
>
> **Understanding**
>
> Understand your audience and its expectations.
>
> **Supported**
>
> Use logic and support to make your point.

1. Answering the wrong question. This happens when we don't understand the assignment or what the audience really wants. Have you ever written what you thought was an excellent paper, only to be told you answered the wrong question or you missed the point? Have you ever asked someone a question and received a long answer that had nothing to do with what you asked?

2. Answering only part of the question. If a problem or question has multiple parts, sometimes we work out the easiest or most interesting part of the solution and forget the unpleasant remainder.

3. Adding irrelevant information. Here the communicator answers the question, but mixes in information that is interesting but unnecessary. Though the answer is complete, it's hard to understand—it's like finding that needle in the haystack.

Failure to focus can really hurt staff communication. Time and time again our efforts crash and burn because we never carefully read the words or really listen to the speaker for the real message ... for the specific question! Most executive officers will tell you that failing to answer the question is one of the primary reasons staff packages are dumped back into the laps of hapless action officers.

Chapter 3 provides suggestions on how to be clear on your purpose and avoid these problems.

② ORGANIZED: SYSTEMATICALLY PRESENT YOUR INFORMATION AND IDEAS

Good organization means your material is presented in a logical, systematic manner. This helps your audience understand you without reading your words over and over, trying to sort out what you're really trying to say.

When writing or speaking is not well organized, audiences become easily confused or impatient, and may stop reading or listening. Even if you're providing useful, relevant information, your audience may underestimate its value and your own credibility.

Chapter 6 is full of suggestions on how to organize well. Problems with organization are relatively easy to fix, and the payoffs are enormous. In these high-tempo environments, a little effort on your part will save your audience a lot of time and pain.

③ CLEAR: COMMUNICATE WITH CLARITY AND MAKE EACH WORD COUNT

This principle covers two interrelated ideas. First, to communicate clearly, we need to understand the rules of language—how to spell and pronounce words, and how to assemble and punctuate sentences. Second, we should get to the point, not hide our ideas in a jungle of words.

People are quick to judge, and mangled, incorrect language can cripple your credibility and limit acceptance of your ideas. Acceptable English is part of the job, so commit to improving any problems you may have. Developing strong language skills is a lot like developing strong muscles—steady commitment produces steady improvement. Always remember that progress, not perfection, is the goal.

Grammar scares most of us, but the good news is that many common mistakes can be corrected by understanding a few rules. Start by scanning our section on editing sentences, phrases and words on pages 95-101. If you want to dig deeper, then check out some of the books and Internet sites that address grammar and writing—contact your local librarian or our *References* section for some suggestions.

Using language correctly is only half of the battle, though—many Air Force writers and speakers cripple themselves with bureaucratic jargon, big words and lots of passive voice. These bad habits make it hard to understand the message. See Chapter 7 pages 73-88 for some suggested cures to these problems.

④ UNDERSTANDING: UNDERSTAND YOUR AUDIENCE AND ITS EXPECTATIONS

If you want to share an idea with others, it helps to understand their current knowledge, views and level of interest in the topic. If you've been asked to write a report, it helps to understand the expected format and length of the response, the due date, the level of formality and any staffing requirements. It's easy to see how mistakes in understanding your audience can lead to communication problems, and I'm sure you've watched others make this mistake. Check out Chapter 3 for some helpful hints on audience analysis.

⑤ SUPPORTED: USE LOGIC AND SUPPORT TO MAKE YOUR POINT

> **sup·port** *n* information that substantiates a position.
> *v* to furnish evidence for a position.

Most writers and speakers try to inform or persuade their audience. Part of the communicator's challenge is to assemble and organize information to help build his or her case. Support and logic are the tools used to build credibility and trust with our audience.

Nothing cripples a clearly written, properly punctuated paper quicker than a fractured fact or a distorted argument. Avoiding this pitfall is most difficult, even for good writers and speakers. Logic is tough to teach and learn because it challenges the highest levels of human intellect—the ability to think in the abstract. We slip into bad habits at an early age, and it takes effort to break them. Chapter 4 provides practical advice on how to use support and logic to enhance your effectiveness as a speaker and how to avoid common mistakes.

SUMMARY

In this chapter, we defined communication as the process of sharing ideas, information, and messages with others, and described how effective communication enables military personnel to work together. To help writers and speakers stay on target, we introduced five FOCUS principles of effective communication. In the next chapter, we'll describe a systematic approach to help you attain these principles and meet your communication goals.

"Jargon allows us to camouflage intellectual poverty with verbal extravagance."

— David Pratt

CHAPTER 2

SEVEN STEPS TO EFFECTIVE COMMUNICATION: AN OVERVIEW

This chapter covers:

- A systematic process to help achieve communication success.

- How this process helps attain FOCUS principles from Chapter 1.

- Where to find detailed information on each step in later chapters.

Chapter 1 introduced five FOCUS principles of good communication. In this chapter, we'll introduce a seven-step approach to hitting the target. Here you'll get the big picture introduction, but later chapters will describe each of the seven steps in greater detail.

You can tailor the steps to your own style and approach, but completing each of them will increase your chances of speaking and writing success. These steps are not always used in sequence, and for long and complicated assignments you may find yourself moving back and forth between steps. That's OK—it's better to deviate from a plan, than to have no plan at all.

PREPARING TO WRITE AND SPEAK: THE FIRST FOUR STEPS

Like many things, good communication requires preparation, and the first four steps lay the groundwork for the drafting process. Though much of this seems like common sense, you'd be surprised at how many people skip the preparation and launch into writing sentences and paragraphs (or speaking "off the cuff"). DON'T DO IT! Good speaking or writing is like building a house—you need a good plan and a firm foundation.

> **SEVEN STEPS FOR EFFECTIVE COMMUNICATION**
>
> 1. Analyze Purpose and Audience
> 2. Research Your Topic
> 3. Support Your Ideas
> 4. Organize and Outline
> 5. Draft
> 6. Edit
> 7. Fight for Feedback and Get Approval

① ANALYZE PURPOSE AND AUDIENCE

"Where there is no vision, the people perish."
– Proverbs 29:18

Too many writers launch into their project without a clear understanding of their purpose or audience. This is a shame—a few minutes spent on this step can save hours of frustration later, and help determine whether you end up looking like an eagle or a turkey. You're much more likely to hit the target if you know what and who you're aiming at.

Carefully analyzing your purpose helps with FOCUS Principle #1: "Focused—answer the question, the whole question and nothing but the question." In some cases, if you take a hard look at the purpose you might find that a formal paper or briefing might not be needed. You'd be startled at how many briefings, paper documents and electronic messages are processed in a typical day in a MAJCOM or wing. Formal communication takes effort and costs money—make sure you don't unnecessarily add to everybody's workload.

If you take the time to "understand your audience" (FOCUS Principle #4) and think about their current knowledge, interest and motives, you'll be better able to tailor your message so that you'll accomplish your purpose, regardless of what it is. Instructing a hostile audience about changes in medical benefits will be different than inspiring a friendly audience at a Veteran's Day celebration, and writing for the general's signature will be different than writing for the base newspaper. Chapter 3 has lots of helpful suggestions about analyzing purpose and audience.

② RESEARCH YOUR TOPIC

"Truth is generally the best vindication against slander."
– Abraham Lincoln

Remember that FOCUS Principle #5 states good communication should be *supported* with information relevant to your point. Step Two—"Research your topic"—gives you the raw material to build your case.

For many of us, "research" sounds intimidating—it brings back memories of painful school projects and hostile librarians who wouldn't let us sneak coffee into the building. Don't let the idea of research scare you. In the context of the seven-step approach, research is the process of digging up information that supports your communication goals. Think of it as "doing your homework" to get smart on your communication topic. Chapter 4 is full of helpful advice on how to approach the challenge. For those of you interested in academic research, Appendix 2 has additional information on the topic.

③ SUPPORT YOUR IDEAS

> "If you can't dazzle them with brilliance, baffle them with bull."
> — Anonymous

Often our communication goal involves persuasion. In such cases, throwing information at our audiences isn't enough—we have to assemble and arrange our facts to support our position. Different kinds of information gathered during the research process can be used to form a *logical argument*. A logical argument is not a disagreement or a fight—it's how we assemble information to make decisions and solve problems.

At the same time we are trying to persuade others, others are trying to persuade us and not all their arguments are airtight. A *logical fallacy* is a weakness or failure in the logic of an argument. Chapter 5 describes logical arguments and several common logical fallacies—allowing you to recognize mistakes in other's arguments and avoid them in your own.

Building logical arguments are part of everyday life. We build arguments when we decide which new car to buy, who to nominate for a quarterly award or how we should spend our training budget. You'll find that many of the ideas described in Chapter 5 are part of the way you think, even if you didn't know the formal terminology.

④ ORGANIZE AND OUTLINE

> "Organizing is what you do before you do something, so that when you do it, it's not all mixed up."
> — Christopher Robin in A.A. Milne's *Winnie the Pooh*

You know your purpose and audience, you've done your homework—it's time to deliver your message, right? Not so fast! Before starting to write sentences and paragraphs (or deliver your speech), you'll save time and frustration by organizing your thoughts and developing an outline of how you are going to present your information.

Successful communicators organize their material logically and in a sequence that leads their audience from one point to the next. Audiences often "tune out" a speaker or writer who rambles on without a logical pattern. Poorly organized essays are a common complaint in both civilian and military schools. Save yourself and your audience a lot of pain—read Chapter 6 to learn different patterns and techniques to organize and outline your material.

FOCUS Principle #2 states that good communication should be *organized* so that the audience can efficiently understand your point. You've taken the first steps towards accomplishing this principle when you take the time to organize and outline your work before starting to write ... but how you actually draft and edit paragraphs will take you the rest of the way.

DRAFTING AND EDITING:
WHEN THE FINGERS HIT THE KEYBOARD

The first four steps are identical for both writing and speaking assignments, but the drafting and editing processes are somewhat different for the two forms of communication. In this section we'll describe the steps from a writing perspective, and Chapters 9 and 10 will describe how the steps are adapted for Air Force speaking.

⑤ DRAFT

> "Writing is easy. All you do is stare at a blank sheet of paper until drops of blood form on your forehead."
>
> – Gene Fowler

When we think about the writing process, we immediately think of drafting sentences and paragraphs. If you're uncomfortable with your writing skills, this step usually causes the most anxiety. The good news is that your work on Steps 1-4 will make the drafting process less painful and more efficient.

Once you've completed the preliminaries and are ready to write, there are several practical ways to ensure you connect with your readers.

• First, get to the point quickly—use one or more introductory paragraphs to state your purpose up front. Most Air Force readers don't have the time or patience to read a staff paper written like a mystery novel with a surprise ending.

• Second, organize your paragraphs so the readers know where you're leading them, and use transitions to guide them along.

• Third, make sure your sentences are clear and direct. Cut through the jargon and passive voice, use the right word for the job and don't make them wade through an overgrown jungle of flowery words.

• Finally, summarize your message in a concluding paragraph that connects all the dots and makes the message feel complete.

Chapter 7 is full of practical advice on drafting, and it takes a top-down approach. It begins with preliminaries such as writing tone and formats, transitions to paragraph construction, provides practical tips on writing clear, vigorous sentences, then concludes with advice on overcoming writer's block.

⑥ EDIT

Experienced writers know that editing should be a separate, distinct process from drafting. When you draft, you create something new. When you edit, you shift from creator to critic. This change in roles can be tough, and no one wants to admit that his baby is ugly. Remember that criticism and judgment are inevitable in communication. The better you are at critically evaluating and correcting your own writing, the fewer people will be doing it for you.

There are two important aspects of the editing process—WHAT you are editing for, and HOW to edit efficiently. What to edit for is simple—remember those FOCUS principles from Chapter 1? How to edit is a little more complicated, but we recommend starting with the big picture and working down to details like spelling and punctuation. Ironically, many people do just the opposite; they focus on details first. Some even think that editing is all about the details. Nothing could be farther from the truth. Though details are part of editing, they're only part of the puzzle.

⑦ FIGHT FOR FEEDBACK AND GET APPROVAL

> "Wisdom is the reward you get for a lifetime of listening when you'd have preferred to talk."
>
> – Doug Larson

When you've completed the editing process and done what you can to improve your communication, it's time to move outside yourself to get feedback. We are all limited in our ability to criticize our own work, and sometimes an outside opinion can help us see how to improve or strengthen our communication. Your objective is to produce the best possible product; don't let pride of authorship and fear of criticism close your mind to suggestions from other people. Also, what we write or say at work often must be approved by our chain of command through a formal coordination process. Your supervisor needs to see it, the executive officer needs to see it, then the big boss, and so on…. Chapter 9 provides tips on how to give and receive feedback and how to manage the coordination process.

SUMMARY

In this chapter, we summarized a systematic process—**Seven Steps to Effective Communication**—that will help you achieve the five FOCUS principles. These steps will help you improve your writing and speaking products. Each step is described in greater detail in subsequent chapters.

THE BASIC STEPS...	FOR MORE DETAILS, REFER TO:
1. Analyze Purpose and Audience	Chapter 3
2. Research Your Topic	Chapter 4
3. Support Your Ideas	Chapter 5
4. Organize and Outline	Chapter 6
5. Draft	Chapter 7 (Writing); Chapter 11 (Speaking)
6. Edit	Chapter 8 (Writing); Chapter 11 (Speaking)
7. Fight for Feedback and Approval	Chapter 9

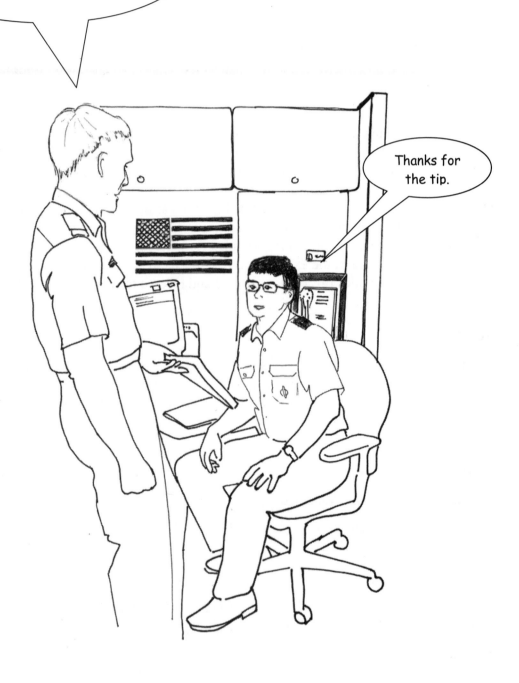

PART II:

PREPARING TO WRITE AND SPEAK

CHAPTER 3

STEP 1: ANALYZING PURPOSE AND AUDIENCE

This chapter covers:

- Evaluating purpose and emphasis in your communication.
- Drafting a purpose statement.
- Audience analysis—know yourself, your unit, and the audience.
- Tips for success with your audience.

In Chapter 2, we introduced the Seven Steps for Effective Communication. Now we're going to discuss the first of these steps in further detail. First and most importantly, you can save yourself a lot of work by asking yourself if the PowerPoint briefing, memo, meeting, e-mail, you name it, is going to help get the mission done or a task accomplished. In today's fast-past work environment, we don't need to create unnecessary work for anyone.

Once you're clear on the need for communication, Step 1 requires you to get clear on your purpose and audience. As you'll see in this chapter, these are not two distinct categories. The characteristics of your audience will influence all parts of your message, and your purpose often involves influencing your audience.

KEY QUESTIONS

Masters in the art of communication stay focused on their objective and approach audience analysis seriously. The more you know about your audience, the more comfortable you will feel writing that memo or delivering that briefing to the commander. Where do you start in this? Here are some questions to help you begin to analyze your purpose and audience and get you on the right track. We'll discuss many of them in more detail later in the chapter.

- What is the overall purpose of the communication? Are you trying to make a change in your audience? Are you writing just to inform your audience?

- If you had one sentence or 30 seconds to explain your specific objective, what would you write or say?

- What format are you using to communicate? How much time do you have to prepare?

- Is there anything unusual about the time and place your audience will receive your communication? (i.e., is it 1600 on a Friday before a holiday weekend?) A lengthy informative e-mail sent out late on a Friday afternoon may not be appreciated or even worse—not read!

- Who will read this communication? Your boss? Your subordinates? Civilians? The answers will have a direct bearing on the tone and formality of your message.

- What are the education levels, career fields and areas of expertise of your readers/listeners?

- Do you need to supply any background information, explanation of terms, or other information to your audience? Does your audience have experience with the ideas and concepts you are presenting?

- What does the audience think of you? Are you known and trusted?

- Is your audience motivated to hear/read your communication?

- Do you need to coordinate your communication?

- Are you making promises your organization will have to keep?

WHAT IS MY PURPOSE?

Most Air Force writing or speaking falls under one of the following purposes: **to direct, inform, persuade or inspire.** Your task is to think about the message you want to send (the "what") and make some sort of determination what your purpose is (the "why"). Some communication has primary and secondary purposes, so don't kill yourself trying to make sure your message fits neatly in one of these categories. Once you decide the purpose, you'll know where to place the emphasis and what the tone of your communication should be. Here's a quick synopsis of these purposes and how they might work for you.

- **To Direct.** *Directive communication* is generally used to pass on information describing actions you expect to be carried out by your audience. The emphasis in directive communication is clear, concise directions and expectations of your audience.

- **To Inform.** The goal of *informative communication* is to pass on information to the audience. The communication is successful if the audience understands the message exactly the way the speaker or writer intended. The emphasis in informative communication is clear, direct communication with accurate and adequate information tailored to the education and skill levels of the audience. Audience feedback and interaction may be appropriate in some situations to make sure they "got the message."

- **To Persuade.** *Persuasive communication* is typically used when you are trying to "sell" your audience on a new idea, new policy, new product or change in current operations. Though emotions are one tool of persuasion, most persuasive communication in the workplace requires convincing evidence put together in a logical way. Audience analysis is critical, because different audiences have different views on what evidence is convincing. Since the purpose is to guide your audience to a specific course of action, you cannot overlook tone and delivery. Chapter 5, *Supporting Your Ideas*, describes how to build your persuasive skills.

- **To Inspire.** One final purpose for writing or speaking that doesn't get much attention but is frequently used in the military is to *inspire*. As you climb the leadership ladder you will increasingly be requested to perform retirements, promotions, commander's calls—opportunities where you will want to inspire the audience with your profound insight on someone's career or possibly your philosophy on leadership. Although protocol drives portions of these events, the opportunity to send a personal message and inspire the audience should not be overlooked. The emphasis in inspirational communication is delivery, a thorough knowledge of your topic and likewise your audience.

Regardless of whether your mission is to inspire, direct, persuade or inform, there are general principles that apply to almost all communication. See "Tips for Success" at the end of this chapter to round out your portfolio for analyzing purpose and audience.

GET CLEAR ON YOUR BOTTOM LINE: DRAFTING A PURPOSE STATEMENT

OK, we're making progress. You have a feel for the general purpose of your communication (to direct, inform, persuade or inspire), but what is your "bottom line" you need to communicate to your audience? If you have difficulty nailing down your objective, your audience will be equally confused.

One way to make sure you're clear on your objective is to write a *purpose statement,* which is one sentence that captures the essence of what you're trying to do—your "bottom line." It's the one sentence you'd keep if you were allowed only one.

Developing a clear purpose statement will help you in two ways:

- It will help you FOCUS as you develop your communication.

- It will help your audience FOCUS when you deliver your message.

Think about it; your audience wants a clear statement of your position and where you are going. This is especially true when your audience consists of higher-ranking individuals with many demands on their time and issues requiring their attention.

Let's look at some examples of draft purpose statements:

Purpose Statement: To inform individuals in the Civil Engineering Squadron about new policies on hazardous waste disposal.

Purpose Statement: To encourage (inspire) at-risk high school students to work hard, stay in school and have hope in the future.

Purpose Statement: To persuade the division chief to buy three laptops for use during official travel.

As we'll mention in Chapter 5, you may update your purpose statement after researching your topic. (For example, you might find out you need four laptops, not three.) Even if it isn't locked down, a draft statement will help guide your research and support efforts.

ANALYZING PURPOSE: OTHER ISSUES

Knowing your "bottom line" is not the end of analyzing purpose. Here are some questions you want to ask to make sure you look at all parts of the equation. The answers will help refine your purpose and shape your entire project.

- **What format will I use to communicate?** Today's commanders are increasingly more vigilant of the value of their troops' time—and their own. Make sure that the communication is required and that you select the most appropriate format for delivery. Think about how much time you will have to deliver your message before you go any further. Will you have 2 minutes in a staff meeting with the general that starts in 1 hour? One hour at your commander's call next week?

- **How much time do I have to prepare my communication?** The breadth and scope of a report your boss needs the next day will be different than a staff project due by the end of the fiscal year. What's the suspense? How long is it going to take to write the report? Be sure to budget adequate time for all "Seven Steps," especially "Researching your Topic" (Step 2),

"Drafting" (Step 5), and "Editing" (Step 6). If coordination is part of the master plan, it will also affect your timelines.

There's more to successful communication than getting clear on purpose. There's always that human factor, and in the communication game that translates to *analyzing your audience.*

AUDIENCE ANALYSIS: THE HUMAN FACTOR

As stated in Chapter 1, all communication involves a sender, a message and an audience. "A" sends the message (either verbally or in writing) with a specific intended meaning to "B." "B" receives the message, processes the message and attaches perceived meaning to the message. This is where it gets interesting. Did the intended message actually get to "B" (or not to "B"— that is the question)? Was the perceived meaning what "A" intended? If not, why not? Ah, thus begins our journey into audience analysis.

KNOW YOURSELF: Before you look around, you need to look in the mirror. Knowing your strengths and weaknesses will help you meet your communication goals.

- Do you do better with certain communication formats than others? Be aware of your personal strengths. If you know that you'd rather be buried in a pile of fire ants than speak in public, you may choose to send your message in writing—if that's an option. If not, spend some time in Chapter 10 and improve your speaking skills!

- Are you an inexperienced briefer that needs notes? If so, make sure they are written in a format that is easy to refer to while the general is listening intently and watching the beads of sweat form on your forehead.

- Do you have expertise in the area? If so … great! But don't lose your audience with lots of lingo and unfamiliar jargon. You may think it's cool; others may not. On the other hand, if you lack expertise in the area, you will need to focus your research to beef up on unfamiliar territory. Remember, there's always someone in the crowd that knows as much or more—no pressure!

- What is your relationship with the audience? Are you personally familiar with them? You may be able to present a more informal briefing or written document if you know this is acceptable to the audience. See the section on tone for more guidance.

KNOW YOUR ORGANIZATION: Once you've taken a hard look at yourself, you need to take a look at your work environment and your organization. In the military we rarely act or speak in a vacuum. Often we represent our organization, unit or functional area and must understand them and accommodate their views, capabilities or concerns in our communications. The following questions may help bring things into focus.

- Am I promising something my organization can deliver? (You can substitute boss or personnel for organization.) If not, why are you bothering?

- Is what I'm saying consistent with previous policy or operating philosophy? If not, you need to shift to a persuasive tone and explain why your approach warrants a change or breach in policy.

- Who needs to coordinate on this? Who else owns a piece of this action? The coordination game can be a mind maze, but if you leave a key player out, you will undoubtedly hear about it.

KNOW YOUR AUDIENCE: The receiving audience falls into one of four sub-categories. Depending on the type of communication and coordination necessary, you may or may not deal with each and every one of these.

- **Primary receiver:** The person you directly communicate with either verbally or in writing.

- **Secondary receiver:** People you indirectly communicate with through the primary receivers. Let's say, for example, that you're a group commander. You send a written memo to the unit's first sergeants (primary receivers) identifying establishments near your base that are now designated "off limits" to the troops. The first sergeants post these areas in the squadron orderly rooms for "widest dissemination." The secondary receivers would be the troops that read (and hopefully heed!) the commander's directive.

- **Key decision makers:** These are the most powerful members of the audience … the ones that really make the decisions. Knowing who they are will help focus your attention and potentially your delivery in larger briefings and certain written communication.

- **Gatekeepers:** These are people in the chain that typically review the communication before it reaches your intended audience. Knowing who they are and what their expectations are can save you embarrassment and help ensure your success in the long run. We all know that administrative assistants are keenly aware of their bosses' preferences. Listen to their inputs!

> "Therefore I say: Know the enemy and know yourself; in a hundred battles you will never be in peril. When you are ignorant of the enemy but know yourself, your chances of winning or losing are equal. If ignorant of your enemy and of yourself, you are certain in every battle to be in peril."
>
> — Sun Tzu

Sun Tzu was ahead of his time. He knew the importance of audience analysis! We're not recommending you view your audience as "the enemy," but Sun Tzu had the right idea. Read on for some final tips for connecting with your audience.

TIPS FOR SUCCESS WITH YOUR AUDIENCE

- **RANK—don't be afraid of it.** Differences in military rank can be a real barrier to communication in the Air Force. Too many of us become tongue-tied when communicating with those senior in rank, and cursory or impatient with those who are junior in rank. We must constantly remind ourselves we are all communicative equals and should strive to be candid, direct and respectful with everyone.

- **JARGON—tailor to your audience.** Don't overestimate the knowledge/expertise of your readers, but don't talk down to them either. Be careful with excessive use of career-field specific jargon and acronyms. Yes, they are second nature to most of us in the military, but you can lose your audience with unfamiliar terminology.

- **Be INCLUSIVE—remember our diverse force.** Sometimes we inadvertently exclude members of our audience by falling into communication traps involving references to race, religion, ethnicity or sex. Remember this concept when designing your visual support. Your visual aids should show a range of people who represent our Air Force. Avoid traditional stereotyping of jobs based on sex or race. Inclusiveness also applies to humor. Humor is not universal, and joke telling is the biggest area where otherwise sensitive people unknowingly get themselves into trouble. Knowing your audience and adhering to good taste and sensitivity will keep you in check.

- **TONE—it's not just what you say, it's how you say it.** Closely tied to the purpose of your communication is the tone you take with your audience. Speakers have gestures, voice and movements to help them communicate. Writers only have words on paper. How many times have you seen colleagues get bent out of shape over a misunderstood e-mail? Why? Because the nonverbal signals available during face-to-face communication are absent. Recognize this disadvantage in written communication and pay close attention to it. Words that carry uncomplimentary insinuations (ignorant, opinionated) make negative suggestions (failure, impossible) or call up unpleasant thoughts (liable, unsuccessful) can potentially defeat your purpose.

- **COURTESY—be polite (please!).** The first rule of writing is to be polite. Forego anger, criticism and sarcasm—strive to be reasonable and persuasive. Try not to deliberately embarrass someone if it can be avoided with a more tactful choice of words. Rudeness is a weak person's imitation of strength.

- **Make it PERSONAL—but it's not all about you!** When appropriate, use pronouns to create instant rapport, show concern and keep your reader involved. Using pronouns also keeps your writing from being monotonous, dry and abstract. The pronouns you'll probably use the most are *you, yours, we, us* and *our.* Use *I, me,* and *my* sparingly. One rule of business writing is "put your audience first," so when possible, avoid using *I* as the first word of an opening sentence and avoid starting two sentences in a row with *we* or *I* unless you're trying to hammer home a point. These guidelines will help you to avoid sounding self-centered and repetitive.

- **Be POSITIVE.** To cultivate a positive tone, give praise where praise is due; acknowledge acceptance before focusing on additional improvements; and express criticism in the form of helpful questions, suggestions, requests, recommendations, or clear directives rather than accusations. When having to give bad news, lead with a neutral comment before jumping in with the bad news. Save the positive for the closing by offering alternatives, etc. Stay away from using clichés, restating the refusal, hiding bad news in a fog of wordiness and inappropriate apologizing. Your audience always appreciates sincerity and honesty. To get you started thinking "positive," listen for the tone of the following sentences:

Commanders *will recommend only qualified* persons for training. [Constructive]

Commanders *may not recommend* for training any person who *is not qualified.* [Destructive]

Positive	*Negative*
reception area	waiting room
established policy	old policy
change of schedule	postponement
confirm meeting	reminder
competition is keen	opportunity is limited
start writing well	stop writing badly
use the big hoist	don't use the small hoist
the cup is half full	the cup is half empty

- **FORMAL ("To be, or not to be") versus INFORMAL ("hey dude").** Different communication situations require different levels of formality. The informal tone is more like a conversation between you and your reader, and it is characterized by clear, direct, active language. In today's Air Force, most of your writing will be informal, though ceremonies and awards may require more *elaborate* (formal) language. Whether your tone is formal or informal, you still need to follow the accepted rules of grammar. In any case, the best advice is to keep your writing clear, concise and simple. We'll cover this in more detail under "drafting clear and concise sentences" in Chapter 7.

SUMMARY

This chapter covered the key concepts of analyzing your purpose and audience—the first step towards developing effective communication. Getting clear on your purpose early in the process helps you focus your preparation. Taking the time to understand your audience will help you tailor your message to their knowledge, interests and motives. Once you've determined your purpose, nailed down your purpose statement, and carefully analyzed your audience, you need to do some homework on your topic...

...it's called R-E-S-E-A-R-C-H.

CHAPTER 4

STEP 2: RESEARCHING YOUR TOPIC

This chapter covers:

- Things to think about before getting lost in the library.

- Where to go to get information.

- Internet searches.

- Tips for evaluating sources.

- A list of useful references for Air Force personnel.

Whether your communication goal is to persuade or inform, you'll need more than fancy words to win the day—you'll need substance as well as style. Once you're clear on your purpose and audience (Step 1), you'll need to research your topic to uncover information that will support your communication goals.

In some ways, research has never been easier—electronic databases and the Internet give us access to quantities of information unthinkable 20 years ago. But new opportunities bring new challenges. With so much information, how do we find the data we need to meet our purpose? And how do we know a source should be trusted? This chapter gives some basic suggestions that will be useful in nearly any assignment.

BEFORE YOU HIT THE LIBRARY

Okay—you need to research your topic … how do you approach the task? If you do a little early planning, you'll be more focused and effective when searching for data.

For simple projects, planning means spend a few quiet moments thinking about your task. For longer projects, you may write out a detailed research plan. Regardless of the scope, think through these issues:

- **Review the purpose and scope of the overall project.**

After completing Step 1, Analyzing Purpose and Audience, you should have a good idea of what you need, but sometimes your purpose

> **Chapter scope and related sections**
>
> The word "research" is often used to describe a multistep investigation process used to either answer a question or solve a problem. Academic research expands knowledge by finding answers to questions, while nearly all military staff research revolves around identifying and solving problems. This chapter emphasizes information retrieval and evaluation of sources, and will not describe the end-to-end process associated with academic research (which is covered in Appendix 2), or details of systematic problem solving (which is described in the *Military Staff Study* chapter).

and scope may evolve as you learn more about the topic. You may also need to do some preliminary research just to get smart enough to scope out the effort. If you've been handed a vague topic, try to get some feel for how far you should go in your research, what you can realistically do, and where you should stop.

- **Assign yourself a deadline for the research effort.**

It's easy to get lost in the research process. Don't do an outstanding job of data retrieval, then a marginal job on the presentation because you ran out of time. For larger projects, assign yourself a time budget for the data-gathering process.

- **Ask the boss—Are there unusual sources or knowledgeable individuals you should seek out?**

Your boss gave you this research problem for a reason—he or she thought that you were capable of finding the answer. Even if you can eventually find the answer on your own, you might save some time by asking for suggestions on where to start. An early vector could be particularly helpful if you're working on a practical problem that's "local," specialized, or requires information that isn't available to the general public.

- **How much do I know about this topic and what are my biases?**

Before you look for answers outside yourself, look in the mirror first. You may have valuable knowledge about an assigned research project, but you need to acknowledge and guard against your own biases in working a research problem. It's tough to keep an unbiased attitude; in fact, it's probably impossible if you know anything about the subject in question. The good news is once you realize you may be biased, you'll be less likely to automatically dismiss data that's not consistent with your personal philosophy.

WHERE SHOULD I LOOK FOR INFORMATION, AND WHERE DO I START?

Three things will probably influence how you approach the data gathering process:

- Your research topic.

- Your experience as a researcher, and your expertise in this research area.

- Your experience with Internet and electronic database searches.

If you are seeking information that's publicly available and are comfortable with electronic search tools, you'd probably start with the Internet. If you are dealing with a local problem, a sensitive topic, or feel uncomfortable with the research process or search technology, you'll probably want to start by talking to another person. Regardless of the order you approach them, here are four major categories of information:

① **COWORKERS AND BASE PERSONNEL** that you can easily meet face-to-face may be subject matter experts on your topic. They may also know where you can get valuable information on local projects, even if they are not subject matter experts themselves. The person who gave you the research assignment may also be a resource. Get clear on the big picture and be specific when you ask others for information—you'll save everybody's time.

② **YOUR OFFICE FILES AND REFERENCES**, both in paper form and on your computer network, may be valuable sources of information. We haven't totally evolved to a paperless workplace, but in most offices, a great deal of current information on policies and procedures is contained on the office computer network. Paper files are still used for correspondence, sensitive information and older archives. Each office has its own policies—check them out.

In addition to office files, larger units typically produce unit histories that can be very useful for staff research. They tell what happened, when it happened, why it happened and where it happened. A good history also shows how past experiences relate to current plans and how recent experiences relate to future plans.

③ **THE INTERNET**—Become an instant expert on anything without leaving your chair! The amount of information available from your desktop computer has exploded over the last decade. The web can be intimidating for those who didn't grow up with computers, but set your fears aside—the payoffs are enormous. The two biggest challenges in using the Internet are (1) finding the information you need, and (2) sorting out what you can believe and what you can't. Remember—anyone with an ax to grind can build a web site, and there's no one out there checking to see if the facts are even remotely correct. It's an interesting environment—information ranges from official, credible sources to the lunatic fringe, and web sites may appear and disappear without notice. Later in this chapter, we'll give you details on searching the net and evaluating what you find.

④ **THE LIBRARY**—Libraries have unique benefits for the researcher:

- Librarians (real, honest-to-goodness humans!) who can help you find information and give basic research advice.

- Free access to books and periodicals—most of which aren't available on the Internet.

- Free access to the Internet and electronic databases.

- Interlibrary loans that let you get at nearly any book in print—even at small libraries

Though the Internet is a convenient source of information, a great deal of information needed for serious research is not on the Internet—it's still found in books and periodicals. Libraries purchase these sources, and then give you free access to them. Isn't America great?

Someone other than the author has critically reviewed most information you find in a library. The end result is less trash and outright errors than you find out on the Internet.

Public libraries are available in most bases and cities … and don't forget about your local college or university! Even if you can't borrow books, many of them allow nonstudents to visit and read books.

"Virtual libraries" are another important resource—they're web sites that give you access to several library resources. Though these can be reached through an Internet browser, the information meets the same quality standards as the material in the physical library.

Air University Library's World Wide Web page—PortAUL—is an example of a virtual library that contains several on-line research tools and traditional library resources. They include access to search engines, full-text bibliographies, online library assistance, full-text periodicals, relevant research links and assignment assistance for the Professional Military Education (PME) schools located at Maxwell Air Force Base. PortAUL is available online at http://www.au.af.mil/au/aul/aul.htm.

HOW CAN I ACCESS THE INTERNET TO GATHER DATA FOR MY RESEARCH?

- **For the computer phobic: the absolute minimum you need to know.**

There is nothing to fear but fear itself. But seriously, accessing the Internet is easy … even if you don't want to spend any time learning about it. If someone gives you a web site address, just open your Internet browser software and type it in exactly as it is given. If you DON'T know exactly what you want, you'll need to use a search engine to find web sites that have useful information. Search engines use key words and phrases to search the Internet. Some allow you to type in questions, and most have catalogues that sort a limited number of sites by topic.

One popular and user-friendly search tool worth mentioning is Google. To get to Google, simply type www.google.com into your Internet browser. Once you are at the site, type in a few words or phrases that describe the main concepts of a topic. Google only returns web pages that contain all the words you type in, making it easy to refine or narrow your search.

- **Refining your keyword searches to improve search efficiency.**

Once you're comfortable with the Internet, you may want to learn some tricks to improve your search efficiency.

Use "and" or the **plus symbol "+"** to make sure that your search engine gives you pages that have all the words you enter. For example, either of these searches will keep you away from sites that describe military aviation, but not military aviation accidents.

> Military AND aviation AND accidents
>
> Military+aviation+accidents

Use "not" or the **minus symbol "-"** to eliminate sites with unrelated words that clutter your search. For example, if you're looking for Windows 95 information but keep getting Windows 98 or Windows 3.1 sites, you can eliminate them by using the following search:

> Windows -98 -3.1
>
> Windows NOT 98 NOT 3.1

The Google search engine requires a space before the minus sign.

Using Quotation Marks (" ") to get web sites that have your search words in the order you specify. For example,

> "Operation Desert Storm"
>
> gives a much tighter search than
>
> Operation+Desert+Storm

> **Fun fact**: The words AND, OR and NOT, when used between keywords in a database search, are called *Boolean operators*. Boolean operators were developed in the 1800s by George Boole, an English mathematician. Boolean operators are heavily used in the design of electronic circuits.

Combining Operations: Once you've got the basics down, you can combine operations. For example:

> "Operation Desert Storm" AND "Air Operations"

gives you relevant sites on the war in the air. If you don't want to hear about Gulf War Syndrome, you can subtract it out:

> "Operation Desert Storm" + "Air Operations" – "gulf war syndrome"

Internet searches require some judgment. If you don't use enough keywords to narrow your topic, you'll end up spending a lot of time scanning sites and trying to find the ones that are most relevant. On the other hand, a tightly focused search might overlook a relevant citation. There are no easy answers, but through trial and error you'll probably find the balance that works for your particular topic.

Search Engine Options. If you're having trouble finding material on your topic, you might try more than one search engine. Different search engines may yield different results, and some specialize in certain fields of study. To find out more details, do an Internet search for current rankings and recommendations.

Though anything on the web is subject to change, here are some of the top rated search engines at the time of publication: Google, Alltheweb.com (fast), Yahoo, MSN Search, Lycos, Ask Jeeves, AOL Search, Teoma, Wisenut, Altavista and Netscape Search.

Specialized Military Search Engines. Several search engines located on the Internet are tailored to search for military and government information. **FirstGov, Jane's Defence Discovery Search, Searchmil.com** and **Google!Unclesam** may be useful for such topics. **Call's Military Domain Search Engine** (http://call-search.leavenworth.army.mil/) also searches multiple military related sites, including CALL, Army, Army Field Manuals, Army regulations, Corps of Engineers, Early Bird, Joint Service, Mitre Corp., NATO, Navy, Rand, USAF, USCG, and USMC sites.

Metasearch Engines. Meta search engines search other search engines. Since they often search smaller, less well-known search engines and specialized sites, they are a good choice if you are struggling to find relevant materials on your topic. Again, consider an Internet search to find out if one of these engines specializes in your field of interest. Flipper, Iboogie, Infogrid, Infonetware Real term search, Ithaki, Ixquick, Kartoo, Profusion, Qbsearch, Query Srver, Searchonline, Vivisimo, Dogpile, Excite, Metacrawler and Webcrawler.

EVALUATING THE QUALITY OF INTERNET SOURCES

Once you've found web sites on your topic, you need to decide which ones to take seriously. Some feel that a positive attitude is key to success in life, but in the case of the Internet, your attitude should be deep suspicion unless the source is official and you can confirm the site is what it pretends to be. There are lots of credible web sources with a known pedigree (the Air University Library, for example), but there are also many sites that are either factually incorrect or deliberately misleading.

When looking at a web site, ask yourself some questions. Who is responsible for the web site, and can you confirm that? How distant are the authors from the "event" they are writing about? (See page 32 for a description of primary, secondary and tertiary sources.) What are the authors' motives? Are they part of a group whose goal is to influence public opinion or to sell you something? Are there things about the site that make you question its accuracy, objectivity or currency? Take a look at the checklist on the next page for more specifics to add to your paranoia.

Another way to build confidence in the information you gather is to seek confirmation of the facts from multiple sources. The weaker the source, the more important it is to get a second opinion before believing it. Most of us instinctively do this in the workplace—we have a very short list of people we absolutely believe every time they open their mouths (these folks usually don't talk much!), we have a longer list of people who are right most of the time, and then there's usually one or two who have no credibility whatsoever. If the issue is important and you want to be sure, try to get the answer in stereo.

> The less you know about a topic, the more authoritative the sources sound.
>
> – Quill Law of Research

How to Recognize an Informational Web Page

An informational web page is one whose purpose is to present factual information. The URL Address frequently ends in .edu or .gov, as many of these pages are sponsored by educational institutions or government agencies.

Examples: Dictionaries, thesauri, directories, transportation schedules, calendars of events, statistical data, and other factual information such as reports, presentations of research or information about a topic.

Questions to Ask About the Web Page

NOTE: The greater number of questions listed below answered "yes," the more likely it is you can determine whether the source is of high information quality.

Criterion #1: AUTHORITY

Is it clear who is responsible for the contents of the page?
Is there a link to a page describing the purpose of the sponsoring organization?
Is there a way of verifying the legitimacy of the page's sponsor? That is, is there a phone number or postal address to contact for more information? (Simply an e-mail address is not enough.)
Is it clear who wrote the material and are the author's qualifications for writing on this topic clearly stated?
If the material is protected by copyright, is the name of the copyright holder given?

Criterion #2: ACCURACY

Are the sources for any factual information clearly listed so they can be verified in another source?
Is the information free of grammatical, spelling, and typographical errors? (These kinds of errors not only indicate a lack of quality control, but can actually produce inaccuracies in information.)
Is it clear who has the ultimate responsibility for the accuracy of the content of the material?
If there are charts and graphs containing statistical data, are they clearly labeled and easy to read?

Criterion #3: OBJECTIVITY

Is the information provided as a public service?
Is the information free of advertising?
If there is any advertising on the page, is it clearly differentiated from the informational content?

Criterion #4: CURRENCY

Are there dates on the page to indicate the following:
    When the page was written?
    When the page was first placed on the web?
    When the page was last revised?
Are there any other indications that the material is kept current?
If material is presented in graphs and/or charts, is it clearly stated when the data was gathered?
If the information is published in different editions, is it clearly labeled what edition the page is from?

Criterion #5: COVERAGE

Is there an indication that the page has been completed, and is not still under construction?
If there is a print equivalent to the web page, is there a clear indication of whether the entire work is available on the Web or only parts of it?
If the material is from a work which is out of copyright (as is often the case with a dictionary or thesaurus) has there been an effort to update the material to make it more current?

NOTE: This checklist is the original Web version. The authors' book *Web Wisdom: How to Evaluate and Create Information Quality on the Web* contains a revised and expanded version. Copyrighted by Jan Alexander and Marsha Ann Tate 1996-1999.

PRIMARY, SECONDARY AND TERTIARY SOURCES: HOW CLOSE IS A SOURCE TO WHAT IT REPORTS?

When evaluating a source, one factor to consider is the distance between the writer and his or her subject. Since people and their research are often misquoted, it's better to refer back to original material than rely on someone else's interpretation of existing work. This is true for research published in books and print journals, as well as Internet sites. The material you find can be classified as either a primary, secondary or tertiary source.

A **primary source** is a first-hand account of an historical event, a physical artifact or record of that event or a description of research written by the people that actually performed it.

A **secondary source** is one step removed from the event or research. It documents the findings of someone else who took the time to review primary sources.

A **tertiary source** summarizes findings published in secondary sources.

Let's look at some examples. If you were doing research on a friendly fire incident, primary sources would include interviews of the parties involved, radio recordings, gun camera footage and black box recordings. The *Summary of Official Findings* published by the investigation board would be a secondary source. A tertiary source might be a magazine article that quoted the *Summary of Official Findings* as part of a larger discussion on the topic.

If you wanted to learn about the foundations of logic and persuasion in Western culture, a primary source would be essays on the subject by the Greek philosopher Aristotle, a secondary source would be an academic textbook that refers back to these writings, and a tertiary source would be a lecture given by an instructor that used the academic textbook as a reference.

In general, primary and secondary sources are considered more reliable than tertiary sources. Each level of interpretation can introduce potential errors or bias, and ideas can be misquoted or quoted out of context. On the other hand, sometimes a tertiary source might be useful to get a "big picture view" of a topic before you start slogging through primary sources and secondary sources.

A List of Useful Resource Sources

In some situations, you may want to start your research by referring to one authoritative and relevant source instead of searching the library or Internet for all possible sources of information. This might be particularly useful for staff research, or for situations where you want a quick and official answer to a relatively noncontroversial topic. We'll conclude this chapter with a list of sources that are either "official" in nature or involve publications that include at least one layer of critical review.

AIR FORCE SOURCES

AIR FORCE LINK. http://www.af.mil/

The official web site of the US Air Force. It provides the latest news, information about careers, a library (of public web sites) and an image gallery. It provides information about the Department of the Air Force, the Pentagon, the Unified Combatant Commands, the Air Force Reserve and the Air National Guard. There are links to other related sites including the full text of the report *Air Force Vision 2020* and the electronic journal *Airman Magazine*.

AIR FORCE PORTAL. https://www.my.af.mil/gcss-af/afp40/USAF/ep/home.do

The Air Force portal is a new initiative put forth that combines a number of Air Force web sites and computer systems into one common interface. Individuals must register and establish an account.

AIR FORCE PUBLICATIONS. http://www.e-publishing.af.mil

The full-text of most standard Air Force publications. See Part VI, Air Force Publications.

AIR FORCE HISTORICAL RESEARCH AGENCY. http://www.au.af.mil/au/afhra/

This agency's holdings consist of over 70,000,000 pages devoted to the history of the United States Air Force. Research requests can be submitted at: http://www.maxwell.af.mil/au/afhra/contact.html

AIR FORCE RESOURCE CENTER. http://www.petersons.com/airforce

Educational and career information for both current Air Force personnel and potential recruits. Search for educational information, search for career guidance information, view news and articles and access a collection of Air Force links.

AIR UNIVERSITY LIBRARY'S WORLD WIDE WEB PAGE—PORTAUL.
http://www.au.af.mil/au/aul/aul.htm

Air University Library's World Wide Web page contains several on-line research tools and traditional library resources. Resources on the web page include access to search engines, full-text bibliographies, online library assistance, full-text periodicals, relevant research links and assignment assistance for AWC, SOS, SNCOA, ACSC and other PME schools located at Maxwell AFB. Research requests can be submitted at: http://www.maxwell.af.mil/au/aul/forms/refform.htm

AIR UNIVERSITY PRESS. http://www.maxwell.af.mil/au/aul/aupress

The AU Press publication program is designed primarily to help war fighters and policy makers understand and apply air and space power in peacetime and conflict. Air University Press publishes books, monographs and research papers by military authors and civilian scholars.

AIR UNIVERSITY STUDENT RESEARCH PAPERS. https://www.research.au.af.mil

Includes over 2,000 full-text publications by faculty and students, summaries of AU research and prospective topics for research. Includes ACSC, AFFP, AFIT, AWC, CADRE and SAAAS papers.

AIR WAR COLLEGE (AWC) GATEWAY TO INTERNET RESOURCES.
http://www.au.af.mil/au/awc/awcgate/awcgate.htm

This site includes links organized for topics such as USAF Counterproliferation Center, Center for Strategy and Technology, International/Regional Studies, Issues, Forces, and Capabilities, Information Operations, Special Operations, Space Operations and Systems, Military and Space Doctrine, Strategy and Military Theory, Military Education On-line, Education and Technology, Future Studies, Visioning and Strategic Planning, Military Medicine and Health and Fitness.

INTERNET SEARCHING TOOLS (AND TIPS). http://www.au.af.mil/au/awc/awcgate/awc-srch.htm

This site includes links helpful for anyone wishing to improve their Internet research skills.

OTHER SERVICES

US ARMY. http://www.army.mil/
ArmyLink is the official web site of the US Army. It provides the latest news, as well as information about careers, a library (of public web sites) and an image gallery.

US ARMY PUBLICATIONS. http://www.usapa.army.mil/
The full-text of most standard Army publications.

US ARMY MILITARY HISTORY INSTITUTE. http://carlisle-www.army.mil/usamhi/1collectionoverviewsubheaders.html
USAMHI collects, organizes, preserves, and makes available source materials on American military history to the defense community, academic researchers and the public. Research requests can be submitted at: http://carlisle-www.army.mil/usamhi/1inquiries.html

US COAST GUARD. http://www.uscg.mil/
The official web site of the US Coast Guard. It provides the latest news, as well as information about careers, a library (of public web sites) and an image gallery.

US COAST GUARD PUBLICATIONS. http://www.uscg.mil/hq/g-s/g-si/g-sii/sii-3/sii-3.htm
The full-text of most standard Coast Guard publications.

US MARINE CORPS. http://www.uscg.mil/
MarineLink is the official web site of the US Marine Corps. It provides the latest news, as well as information about careers, a library (of public web sites) and an image gallery.

US MARINE CORPS PUBLICATIONS. http://www.hqmc.usmc.mil/pubs.nsf
The full-text of most standard Marine publications.

US MARINE CORPS HISTORY AND MUSEUMS DIVISION. http://www.usmc.mil
The official United States Marine Corps History and Museums Division web site. This site provides a variety of information about the division and most importantly on the histories and traditions of the United States Marine Corps.

US NAVY. http://www.navy.mil/
The official web site of the US Navy. It provides the latest news, as well as information about careers, a library (of public web sites) and an image gallery.

US NAVY DIRECTIVES. http://www.navy.mil/
The full-text of most standard Navy directives.

US NAVAL HISTORICAL CENTER. http://www.history.navy.mil/
The Naval Historical Center is the official history program of the Department of the Navy. It includes a museum, art gallery, research library, archives and curator as well as research and writing programs.

NATIONAL GUARD. http://www.defenselink.mil/faq/pis/grdres.html
GuardLink is the official web site of the National Guard. It provides the latest news, as well as information about careers, a library (of public web sites) and an image gallery.

RESERVE AFFAIRS. http://www.defenselink.mil/ra/
The official web site for reserve affairs. It provides the latest news, as well as information about careers, a library (of public web sites) and an image gallery.

DOD SOURCES

DEFENSE ACQUISITION DESKBOOK. http://deskbook.dau.mil/jsp/default.jsp
The Defense Acquisition Deskbook is an electronic reference library that provides the most current acquisition policy for the Military Services and Agencies. The Deskbook is free-of-charge and is available as a CD-ROM subscription and as a web edition.

DEFENSELINK. http://www.defenselink.mil/
DefenseLINK is the official web site for the Department of Defense and the starting point for finding US military information online. It contains links to publications, statements, etc.

DEFENSE TECHNICAL INFORMATION CENTER (DTIC). http://stinet.dtic.mil/
DTIC is the central point for the Defense Department's collections of research and development in virtually all fields of science and technology. DTIC documents defense-related research, development, test and evaluation (RDT&E) activities. The DTIC Scientific and Technical Information Network provides access to all unclassified, unlimited document citations added into DTIC from December 1974 to the present. Descriptive summaries and full text information are also available for some documents.

EARLYBIRD. http://ebird.afis.osd.mil
A daily, concise compilation of the most current published news articles and commentary concerning the most significant defense and defense-related national security issues. Access restricted to DoD personnel.

JOINT ELECTRONIC LIBRARY. http://www.dtic.mil/doctrine/index.html
Joint Electronic Library (JEL) provides access to full text copies of Joint Publications, the Department of Defense Dictionary, Service Publications, History Publications, CJCS Directives, research papers and other publications. It also includes the *Joint Force Quarterly* and *A Common Perspective.*

JOINT WARFIGHTING CENTER ELECTRONIC RESEARCH LIBRARY.
http://elib1.jwfc.js.mil/main/index.htm
Provides access to documents on Peace Operations, Future Joint Operations, Joint Publications, Consequence Management, as well as research papers. Individuals must register and establish an account.

MERLN. http://merln.ndu.edu/
On the Internet since 1990, MERLN provides indexing to significant articles, news items, editorials and book reviews appearing in military and aeronautical periodicals, many of which are not indexed elsewhere. MERLN also provides access to the major library catalogs of several PME schools.

THE NATIONAL ARCHIVES AND RECORDS ADMINISTRATION (NARA).
http://www.archives.gov/
An independent federal agency that preserves our nation's history and defines us as a people by overseeing the management of all federal records.

MILITARY BIBLIOGRAPHIES

MILITARY BIBLIOGRAPHIES. A bibliography is a great starting point for research. A librarian has examined many relevant sources and compiled a listing of Internet sites, books, documents, magazine articles and videos. Many bibliographies are done on various topics by different military libraries or groups. Some of the most useful are:

- Air University Library Bibliographies. http://www.au.af.mil/au/aul/bibs/bib97.htm

- Joint Forces Staff College Library. http://www.jfsc.ndu.edu/library/bibliography/bibs.htm

- Marine Corps University Research Center. http://www.mcu.usmc.mil/MCRCweb/library/library.htm

- National Defense University Library. http://www.ndu.edu/library/pubs.html#bibliographies

- Naval Postgraduate School. http://library.nps.navy.mil/home/bibliogs.htm

- US Army Military History Institute. http://carlisle-www.army.mil/usamhi/FindingAids.html

- US Army War College Library. http://www.carlisle.army.mil/library/bibliographies.htm

MILITARY NEWSPAPER AND MAGAZINE INDEXES AND LISTS

Some indexes listed below require a paid subscription. Contact your local or base library for availability.

AIR UNIVERSITY LIBRARY INDEX TO MILITARY PERIODICALS (AULIMP).
http://www.au.af.mil/au/aul/muir1/aulimp1.htm
AULIMP contains citations to articles in English language military journals. Air University Library (AUL) has been producing AULIMP since 1949, and material after 1989 is available on the web.

DOD AND MILITARY ELECTRONIC JOURNALS.
http://www.au.af.mil/au/aul/periodicals/dodelecj.htm
An electronic list compiled and maintained by the staff of the AUL.

READ MILITARY NEWS. http://www.au.af.mil/au/aul/periodicals/milnews.htm
A listing of defense and military service news including base newspapers compiled and maintained by AUL.

STAFF COLLEGE AUTOMATED MILITARY PERIODICALS INDEX (SCAMPI).
http://merln.ndu.edu/
Developed to support the curriculum of the Armed Forces Staff College; provides selected citations to journal articles and selected documents and reports. Subject, author and keyword indexing is provided.

OTHER MAGAZINE AND NEWS INDEXES

Some indexes listed below require a paid subscription. Contact your local or base library for availability.

ACCESS NEWS. http://infoweb.newsbank.com/
Provides news articles covering social, economic, environmental, government, sports, health, science and military issues and events from more than 500 US regional and national newspapers and wire services, and 140 full-text general news, subject-specific and military magazines. Paid subscription required.

ACCESS SCIENCE. http://www.accessscience.com/server-java/Arknoid/science/AS
Provides on-line access to the McGraw-Hill Encyclopedia of Science and Technology, including 7,100+ articles, 115,000 dictionary terms, and hundreds of Research Updates—in all areas of science and technology—updated daily, plus these great features: Over 2,000 in-depth biographies of leading scientists through history; weekly updates on breakthroughs and discoveries in the world of science and technology; resources to guide your research; and links to web sites for further research. Paid subscription required.

AEROSPACE and HIGH TECHNOLOGY DATABASE. http://alt3.csa3.com/
An abstracting and indexing service covering the world's publishing literature in the field of aeronautics and space science and technology. Paid subscription required.

AMERICA HISTORY AND LIFE. http://serials.abc-clio.com/
A comprehensive bibliography of articles on the history and culture of the United States and Canada from prehistory to the present. Paid subscription required.

AIR FORCE TIMES AND OTHER TIMES PUBLICATIONS. Provides full-text access to Army Times, Air Force Times, Navy Times, and the Marine Corps Times and includes current issues and archives from 2000-2003. Paid subscription required for full access.

- Army Times. http://www.armytimes.com/

- Air Force Times. http://www.airforcetimes.com/

- Navy Times. http://www.navytimes.com/

- Marine Corps Times. http://www.marinetimes.com/

COUNTRYWATCH. http://www.countrywatch.com/ip/default.asp
Provides the political, economic, corporate and environmental trends for each of the 191 countries around the world. In addition, Country Wire provides a global news service. Paid subscription required.

CQ LIBRARY. *CQ Weekly*: http://library2.cqpress.com/cqweekly/

CQ RESEARCHER. http://library.cqpress.com/cqresearcher/
Provides access to *CQ Weekly* and the *CQ Researcher*. *CQ Weekly* on the Web includes access to the full text of all articles published since 1983. *CQ Researcher* is a weekly publication that covers the most current and controversial issues of the day with complete summaries, insight into all sides of the issues, bibliographies and more. Paid subscription required.

EBSCOHOST. http://search.epnet.com/
A web-based periodical index that provides abstracts and indexing for over 4,800 scholarly journals covering the social sciences, military science, humanities, education and more. Also included is the full text for over 1,000 journals—with many dating back to 1990—and over 1,700 peer-reviewed journals. Paid subscription required.

FBIS: FOREIGN BROADCAST INFORMATION SYSTEM. https://portal.rccb.osis.gov/
An electronic information service that collects and translates current political, economic, technical and military information from the media worldwide for the US Government. Individual must register and establish an account.

FIND ARTICLES.COM. http://www.findarticles.com/
Search for quality articles in more than 300 reputable magazines and journals.

HISTORICAL ABSTRACTS. http://serials.abc-clio.com/
A comprehensive bibliography of articles on the history and culture of the world from 1450 to the present excluding the United States and Canada. Paid subscription required.

LEXIS-NEXIS ACADEMIC UNIVERSE. http://www.lexis-nexis.com/universe
Provides access to a wide range of full-text news, business, legal, military and reference information. Specifically, this service includes: Top News, Biographical Information, General News Topics, Reference and Directories, Company News, General Medical and Health Topics, Industry and Market News, Medical Abstracts, Government and Political News, Accounting, Auditing, and Tax Legal News, Law Reviews, Company Financial Information, Federal Case Law, Country Profiles, US Code, Constitution, and Court Rules, State Profiles, and State Legal Research. Paid subscription required.

PROQUEST. http://www.proquest.com/pqdauto
Magazine articles from over 8,000 publications, with many in full-text, full image format. Paid subscription required.

STRATFOR.BIZ. http://www.stratfor.biz/
Combines intelligence analysis and news archives on various countries. Paid subscription required.

WILEY INTERSCIENCE. http://www3.interscience.wiley.com/cgi-bin/simplesearch
Provides the full text of over 300 leading scientific, technical, medical and professional journals. Paid subscription required.

OTHER INDEXES

Some indexes listed below require a paid subscription. Contact your local or base library for availability.

AP MULTIMEDIA ARCHIVE. http://accuweather.ap.org/cgi-bin/aplaunch.pl
An electronic library containing the AP's current photos from their 50 million-image print and negative library, as well as charts, graphs, tables, and maps from the AP's graphics portfolio. This database will be useful for your class instruction, briefings or publications. Paid subscription required.

BRITANNICA ON-LINE. http://search.eb.com/
A full text encyclopedia that is accessible via the World Wide Web. It provides articles as well as related resources on the Internet. Paid subscription required for full access.

CIAO (COLUMBIA INTERNATIONAL AFFAIRS ONLINE). http://www.ciaonet.org/
A web-based database designed to be the most comprehensive source for theory and research in international affairs. It publishes a wide range of scholarship from 1991 on, that includes working papers from university research institutes, occasional papers series from NGOs, foundation-funded research projects, and proceedings from conferences. Paid subscription required.

FEDERAL R&D PROJECT SUMMARIES. http://www.osti.gov/fedrnd/
Includes more than 240,000 research summaries and awards by three of the major sponsors of research in the Federal government. The federal databases available via this source are the Department of Energy R&D Project Summaries, the National Institutes of Health (NIH), CRISP (Computer Retrieval of Information on Scientific Projects), Current Awards and the National Science Foundation (NSF) Award Data.

FIRSTSEARCH. http://newfirstsearch.oclc.org

An electronic information service that is accessible over the Internet. This service provides access to over 70 databases covering a wide variety of subjects. Paid subscription required.

GALENET. http://infotrac.galegroup.com

An electronic information service that is accessible over the Internet. This service will give you access to the following databases. Paid subscription required:

● Associations Unlimited. Contains information for approximately 460,000 international and US national, regional, state and local nonprofit membership organizations in all fields, including IRS data on US 501(c) nonprofit organizations.

● Biography and Genealogy Master Index. Focuses on biographical material on people from all time periods, geographical locations and fields of endeavor.

● Contemporary Authors. Provides complete biographical and bibliographical information and references on more than 120,000 US and international authors.

GRAYLIT NETWORK. http://www.osti.gov/graylit

Provides a portal for over 100,000 unclassified full-text technical reports located at the Department of Energy (DOE), Department of Defense, Environmental Protection Agency (EPA), and National Aeronautics and Space Administration (NASA). Collections in the GrayLit Network include the DOE Information Bridge; the Defense Technical Information Center (DTIC) Technical Report Collection, the EPA National Environment al Publications Internet Site (NEPIS), the NASA Jet Propulsion Lab Reports, and the NASA Langley Technical Reports.

GPO ACCESS. http://www.access.gpo.gov/multidb.html

An electronic information service that is accessible over the Internet. This service will give you access to over 60 databases covering a wide variety of government and congressional subjects.

JANE'S ONLINE. http://www.janesonline.com

Provides access to select Jane's resources including Jane's Geopolitical Library, Launched Weapons Image Library, All The World's Aircraft, Land Systems Image Library, Military Aircraft Image Library, Military Biographies and the Warships Image Library. Paid subscription required for full access.

JOINT COMBAT CAMERA CENTER. http://dodimagery.afis.osd.mil/dodimagery/home.html

Database that contains official US Military images from around the world. The DoD Imagery Server contains more than 35,000 still images and over 175 QuickTime previews of videotapes. Individuals must register and establish an account.

LEXIS-NEXIS CONGRESSIONAL UNIVERSE. http://web.lexis-nexis.com/congcomp

Provides access to a wealth of congressional publications—including hearings, reports, prints, documents, hot topics, bills, laws, regulations and CIS Legislative Histories. This site also allows you to locate specific information about members and committees. Paid subscription required.

NETLIBRARY. http://www.netlibrary.com

Access the full-text of books electronically. Paid subscription required for full access.

PERISCOPE. http://www.periscope.ucg.com/

Provides open source intelligence on orders of battle, equipment inventories, and procurement plans and programs for 150 nations. It also provides technical descriptions and characteristics for more than 4,500 global weapon systems. Paid subscription required.

POLITICS IN AMERICA. http://www.library.cqpress.com/pia
Provides access to biographical data, committee assignments, election results and key votes as well as background on members' districts for the 106th and 107th Congresses. Paid subscription required.

STAT-USA. http://www.stat-usa.gov/
Provides access to State of the Nation, GLOBUS and NTDB. These resources provide access to current and historical economic data, financial releases, information on the US Economy, current and historical trade-related releases, international market research, trade opportunities, country analysis, and the STAT-USA trade library. Paid subscription required.

SUMMARY

When you're trying to inform or persuade others, you need to do your homework. In this chapter, we covered the basics of how to research your topic and gather the data you need. The Internet, the library, your office files, and your coworkers are all potential sources of information. Modern technology and the Internet make it easier to gather data than ever before, but remember that you need to critically evaluate your sources. Research can be enjoyable and rewarding, but make sure you review your purpose, scope and schedule before you begin—it helps you stay focused and keeps you from getting lost in the data.

CHAPTER 5

STEP 3:
SUPPORTING
YOUR IDEAS

This chapter covers:

- Logical arguments—what they are and how to build them.

- Types of evidence and characteristics of "good evidence."

- Common problems with logic.

- Ethical issues with arguments.

Once you've researched your topic and collected information, you need to figure out how to use what you've found to meet your communication goals. If you're dealing with a controversial question or problem, throwing facts at your audience won't be enough—you'll need to assemble it into a logical argument that can stand up to critical attack. This chapter will give you some helpful pointers on how to build an argument and support your ideas.

Logical arguments are instruments of power. They're how you make things happen. It's worth the effort to understand some basics, even if some of this chapter makes your head hurt.

"A moment's thought would have shown him. But a moment is a long time, and thought is a painful process."

– A.E. Housman

WHAT ARE LOGICAL ARGUMENTS, AND WHY ARE THEY IMPORTANT?

When you present a solution to a problem or answer a controversial question, persuasion is part of the assignment. There are different approaches to persuade members of your audience—you can appeal to their emotions, their ability to reason or even your own credibility on the topic being discussed. In the Air Force environment, your best approach to support your ideas and persuade others is by building a solid logical argument.

Though the word "argument" is commonly used to describe a quarrel or disagreement, it also has a more positive meaning—it's a series of statements intended to persuade others. In this chapter, when we use the term *logical argument,* we're referring to a coherent set of statements that provide a position and support for that position based on information and facts, not just emotions.

Why should you care about all this stuff? First, you build logical arguments every day—when you talk to a child about chores and allowance; when you talk to your boss about your workload and schedules; and when you sort out how best to accomplish the mission. If you build strong arguments, things are more likely to work out the way you think they should. Second, others are aiming arguments at YOU every day—and some of them are pretty shaky. If you understand how arguments are constructed and where they go wrong, you're less likely to buy into somebody else's muddy thinking.

ELEMENTS OF A LOGICAL ARGUMENT

Logical arguments contain four elements:

- a claim

- evidence that supports the claim

- warrants linking pieces of evidence to the claim

- qualifications that limit the claim

First we'll describe each of these terms, and then we'll illustrate them in a real-life example of an argument in the next section. Don't panic if it doesn't seem clear at first—the example will help clarify each point.

> Different textbooks have different terms and approaches to describe logical arguments. This chapter uses terminology found in *The Craft of Research,* by Booth, Comb and Williams, which has been used in the Air Command and Staff College curriculum.

THE CLAIM. Your claim is simply your position on an issue, your answer to a controversial question or your recommendation for resolving a problem. In some academic writing, a claim is also called a *thesis*.

EVIDENCE THAT SUPPORTS THE CLAIM. By definition, every argument has *evidence* intended to give reasons for your claim. A less formal word for evidence is *support*. (The words "support" and "evidence" are used interchangeably in this chapter.)

If a piece of evidence is questionable, it may be attacked as a *subclaim*. At that point, you either have to provide additional evidence to prove your subclaim is true, or eliminate it from your argument.

WARRANTS THAT LINK EVIDENCE TO THE CLAIM. With every piece of evidence, there are often assumptions, either stated or unstated, that link the evidence to the claim and explain why the evidence is relevant to the argument. These linking statements or concepts are called *warrants*. Warrants are important because they can be potential weaknesses in an argument.

QUALIFICATIONS. Sometimes the argument will have *qualifications*—conditions that limit the claim. You can think of a qualification as a statement you attach to the claim with a big IF statement. We often notice these qualifications as we critically look at the evidence we have and realize its limitations.

ELEMENTS OF A LOGICAL ARGUMENT—A REAL LIFE EXAMPLE

Let's use a real life example of a logical argument to show how the different elements work together. Suppose you're responsible for selecting a guest speaker to teach topic XYZ at a Professional Military Education (PME) school. Ms. Jane Doe spoke last year, and you've decided to invite her back. Your boss wants to know your recommendation and your rationale. Guess what? You've just been asked to produce a logical argument.

> **ev·i·dence:** *n* the data by which proof or probability may be based ... admissible as testimony in a court of law.
>
> **sup·port:** *n* information that substantiates a position; *v* To furnish evidence for a position.

CLAIM: We should invite Ms. Jane Doe to teach topic XYZ at this year's class.

EVIDENCE, item #1: Ms. Doe has spent 26 years working with XYZ and is an expert in this field.

WARRANT, item #1: Spending 26 years of working with XYZ makes her an expert. (Another implied warrant is that we want an expert to teach topic XYZ.)

If someone wanted to attack this bit of evidence, he might ask you to prove the fact that she's spent 26 years in the field—let's see a resume!

If someone wanted to attack the underlying warrant, he may argue that she isn't really an expert—maybe she's been doing an entry-level job for 26 years.

But let's suppose that Ms. Doe is indeed an expert in the field and this is solid evidence.

EVIDENCE, item #2: Last year's course directors all thought she did an excellent job.

WARRANT, item #2: These people know what they're talking about.

If someone wanted to challenge this evidence, he might ask you to produce letters of recommendation. How enthusiastic are the directors about the job she did?

If someone wanted to attack the underlying warrant, he might question the course directors' judgment. Maybe they were new to the job and didn't know much about the topic. Maybe they were TDY during the presentation, and were basing their recommendation on what they heard from others.

In this case, let's assume that the course directors are both credible and enthusiastic.

EVIDENCE, item #3: Ms. Doe is a very dynamic lecturer.

WARRANT, item #3: It's good to have a dynamic lecturer.

Recall that evidence you provide to support your claim can be attacked as a subclaim … and this last bit of evidence looks vulnerable. How do we know that Ms. Doe is a dynamic lecturer? To back it up, you'd have to "support your support" on item #3 with something like this:

SUBCLAIM: Ms. Doe is a very dynamic lecturer.

SUBCLAIM EVIDENCE #1: Students provided five times the amount of feedback than is typical for a lecture.

SUBCLAIM WARRANT #1: Student interest is proportional to volume of feedback.

SUBCLAIM EVIDENCE #2: Ninety-two percent of feedback was very favorable, and 8 percent was very unfavorable.

SUBCLAIM WARRANT #2: Polarized feedback implies a dynamic lecture.

Well, this additional information really does back up the fact that Ms. Doe is a dynamic lecturer, but it also indicates her views are controversial—8 percent of the student population really didn't like her presentation. You may believe that your school's goal is education and not to make every student happy, but you might qualify your claim with the following "IF" statement:

QUALIFICATION:

Ms. Jane Doe should be invited back to teach topic XYZ

IF

it is acceptable to have a controversial speaker at the school.

TYPES OF EVIDENCE

As you see, individual pieces of evidence are used to build your argument. In this section, we'll identify some common types of evidence as well as approaches to help explain your ideas to your audience.

A *definition* is a precise meaning or significance of a word or phrase. In an argument, it can be helpful to establish a common frame of reference for important or ambiguous words, so don't underestimate the importance of definitions.

An *example* is a specific instance chosen to represent a larger fact in order to clarify an abstract idea or support a claim. Good examples must be appropriate, brief and attention arresting. Quite often they are presented in groups of two or three for impact.

Testimony uses the comments of recognized authorities to support your claim. These comments can be direct quotations or paraphrases, but direct quotations tend to carry more weight with listeners or readers. When using testimony as support, make sure the individuals being quoted are both generally credible—no unknown relatives or convicted felons, please—and knowledgeable in the field under discussion.

Statistics provide a summary of data that allows your audience to better interpret quantitative information. Statistics can be very persuasive and provide excellent support if handled competently. Keep them simple and easy to read and understand. Also, remember to round off your statistics whenever possible and document the exact source of your statistics.

The persuasive power of statistics means that you need to be particularly careful to use them properly. Many people will put blind trust in numbers and fall prey to people or papers that spout numbers or statistical proof. (Ironically, people who work with numbers for a living are the most cautious about trusting someone else's statistics!)

Always, *always* examine the basic assumption(s) on which the analysis rests. Some of the most compelling statistical arguments turn out to be intricate sand castles built on a foundation of shaky assumptions. The math may be technically correct, but the assumptions can't stand up to scrutiny.

In their book *Writing Arguments*, John Ramage, John Bean and June Johnson define a *fact* as a noncontroversial piece of data that can be confirmed by observation or by talking to communally accepted authorities. The authors distinguish a fact from an *inference*, which is an interpretation or explanation of the facts that may be reasonably doubted. They recommend that writers distinguish facts from inferences and handle inferences as testimony.

Definitions, testimony, statistics and facts provide data that you can use to construct an argument. This next category—*explanation*—can also be helpful in supporting your ideas.

Explanation makes a point plain or understandable or gives the cause of some effect. It can be used to clarify your position or provide additional evidence to help make your case. The following four techniques can be used as part of an explanation:

- **Analysis:** The separation of a whole into smaller pieces for further study; clarifying a complex issue by examining one piece at a time.

- **Comparison and Contrast:** *Comparison* and *Contrast* are birds of similar feather. Use comparison to dramatize similarities between two objects or situations and contrast to emphasize differences.

- **Description:** To tell about in detail, to paint a picture with words—typically more personal and subjective than definition.

CHARACTERISTICS OF GOOD SUPPORTING EVIDENCE

> "There are two kinds of truth. There are real truths and there are made-up truths."
>
> — Marion Barry, mayor of Washington DC,
> on his arrest for drug use

Though different professions and academic fields have their own standards of what is "good" evidence, there are some common characteristics to consider.

Is the information from an authoritative source?

Will your audience trust this source? And should you? In the previous chapter we stressed the importance of being cautious with Internet sites, but you should be wary of any source's credibility. Also, remember that it's better to refer back to original material than rely on someone else's interpretation of existing work since people and their research are often misquoted. Refer back pages 30-32 for questions to consider when evaluating the credibility of your sources.

Is the information accurate and free from error?

Check and recheck your facts—errors can seriously damage your credibility. Critically evaluate your sources, and if you're uncertain about your facts, be honest with your audience. You can increase your confidence in the accuracy of your information by using multiple sources to confirm key facts.

Is your information appropriately precise?

When we talk about "precision," we mean the information should be specified within appropriately narrow limits. The level of required precision will vary with the topic being discussed. Describing regulations for uniform wear may require a precision of fractions of an inch, and telling someone that his operational specialty badge should be in the middle of their shirt or within a meter of his belt buckle is not adequately precise. On the other hand, when reporting on the designated mean point of impact for munitions, a measurement in meters or feet would be an appropriate level of precision.

When talking about some subset of a group, explain how many or what percentage of the total you're talking about. If you find yourself constantly using qualifiers like "some, most, many, almost, usually, frequently, rarely…" you probably need to find some convincing statistics to help you make your case.

Is your evidence relevant?

Evidence can be authoritative, accurate and precise, yet still be totally irrelevant. Don't shove in interesting facts that have nothing to do with the claim; help the reader understand the relevance of your material by explaining its significance. Explain charts, graphs and figures and use transitions in your writing (pages 70-73) to "connect the dots" for the reader.

Is your evidence sufficient to support your claim and representative of the whole situation or group?

If you are trying to form some conclusions about a situation or group, you need data that represents the complete situation. For example, if you were trying to form conclusions about the overall military population, you would want to gather evidence from all services, not just one career field in one service.

If you find that your evidence is either not representative or not sufficient, you need to either find more evidence, limit the claim to what you can prove, or qualify your claim. You may have to let go of evidence that doesn't fit or data that is no longer current.

COMMON MISTAKES IN ARGUMENTS: INFORMAL LOGICAL FALLACIES

"Many people would sooner die than think—in fact, they do so."
– Bertrand Russell

Some of you may have studied formal logic in school. These classes used a lot of complex language and theory to describe what makes an argument "good" or "bad." Unfortunately, many real life arguments outside of math and engineering are more "squishy" … and sometimes it's hard to draw a diagram or write an equation to explain exactly what's wrong.

Common errors in reasoning are called *informal fallacies*. They are called "informal" fallacies because they're harder to pin down than some of the "formal" errors in logic. Still, you see them around you every day—especially in advertising, talk radio or political debates. Keep them out of your staff work and learn to identify them in others.

The *informal fallacies* below have been clumped into major categories that make sense to the editors, but there's no universally accepted approach to categorizing them. Also note that labeling something as a fallacy requires some judgment—after all, many of these are "grey areas."

1. An **Asserted conclusion** is the practice of slipping in an assertion and passing it off as a fact.

Two variations include circular reasoning and loaded questions.

> **1a. Circular reasoning** (also known as **begging the question**) involves rewording your claim and trying to use it as evidence, usually with a lot of other "filler sentences" designed to confuse the other person. This is popular in advertising where different versions of the claim are repeated over and over again. After a while, it's easy to forget there's absolutely no support there at all.

CLAIM:

> "Hey guys! Drink Brand X beer, and you'll be popular!"

SUPPORT:

> "Beautiful women in bikinis will compete for your attention!"
>
> "You'll have many good looking and physically fit friends!"
>
> "Women will want you, and men will want to be like you…"
>
> If the advertisers have their way, you may not notice that the "support" merely restates the claim using different words—a textbook case of circular reasoning.

1b. A **Loaded question** has an assertion embedded in it—it's another form of an asserted conclusion. One example of a loaded question is "Do you think John Smith is going to improve his rude behavior?" The phrasing of the question itself implies that John has behaved poorly in the past—regardless of how you answer the question. "When are we going to stop sinking money on this expensive program?" has an embedded assumption: the money we've spent to date hasn't been effective.

Sometimes an arguer will assert a conclusion, then challenge someone else to disprove it. The best defense is to ask him or her to prove their claim. "How do you know these programs are effective?" puts the listener on the defensive. The proper response would be, "How do you know the programs are not effective?" Remember, those who assert should have the burden of proof.

2. Character attack involves an assault on your opponent as an individual, instead of his or her position. It's very common in political advertisements, but you see it in the workplace as well. Here are some examples:

> The classic name for a character attack is the **Ad Hominem** fallacy (in Latin, *Ad Hominem* means "to the man"). Character attacks are also sometimes called **Poisoning the Well**.

"Mr. Smith is a tax and spend liberal who voted himself a pay raise last year." (If the topic being discussed is environmental legislation, this is irrelevant to the core of the debate).

"That guy is a egotistical windbag—what would he know about A-76 contract transitions?" (He may know a lot—his personality is irrelevant to the issue).

3. Emotional appeals try to persuade the heart, not the head. Though emotion plays a role in persuasion, when emotion replaces reasoning in an argument, you've committed a foul. Often arguers attempt to appeal to our emotions in an argument through biased language, vivid language and stirring symbols. They may try to persuade us using "character" issues such as glowing testimonials from popular but noncredible sources. Here are some examples of logical fallacies in this area:

3a. Emotional appeal (to force): An argument that targets the audience's fear of punishment. What characterizes these examples as fallacies is that they make no attempt to persuade using anything other than pressure.

"Keep this quiet, or I'll implicate you in my wrongdoing."

"Give me your lunch money, or I'll give you a busted lip."

3b. Emotional appeal (to pity): An argument that targets the audience's compassion and concern for others. Though most people would agree that ethics and values should be part of the decision-making process, an appeal solely to emotion, even a positive one, can be dangerous and misguided.

"You can't give me an D on this paper—I'll lose my tuition assistance!"

"We've got to stop the warlords—look at the poor, starving people on the news!"

3c. Emotional appeal (to popularity/tradition)

1. *Appeal using Stirring Symbols*: Using a powerful symbol or attractive label to build support.

> "I stand before our nation's flag to announce my run for President...."

> "Good management principles demand we take this course of action."

2. *Bandwagon Appeal:* Using peer pressure to build support.

> "It must be right—everybody else thinks so."

> "Buy the Ford Escort; it's the world's #1 best seller."

> "Every good fighter pilot knows...."

3. *Precedent as sole support*: Using custom as the only justification for a decision.

> "It must be right—we've always done it that way."

> "The RAF has found the procedure very useful and we should try it."

> "The last three commanders supported this policy and that's good enough for me."

4. False authority is a fallacy tied to accepting facts based on the opinion of an unqualified authority. The Air Force is chock-full of people who, because of their position or authority in one field, are quoted on subjects in other fields for which they have limited or no expertise. Don't be swayed (or try to sway someone else).

A false authority variant is called the **primacy-of-print** fallacy, where facts are believed because they are published in a book, periodical or on a web site. Be as skeptical and thoughtfully critical of the printed word as you are of the spoken word and check out the suggestions on evaluating sources on pages 30-32.

5. False cause (also known as the **Post Hoc fallacy**) occurs when you assume one event causes a second event merely because it precedes the second event. Many people observe that Event B occurred after Event A and conclude that A caused B. This is not necessarily true—maybe a third factor, Event C, caused both A and B.

Consider the following example:

> Event A = At Base X, "Retreat" plays over the intercom at 1635 each day.

> Event B = At Base X, outbound traffic increases at the gate at 1640 each day.

There is a statistical *correlation* between these two events: if Event A happens, Event B is more likely to happen and visa versa. Does that mean A causes B? Not necessarily—possibly a third event may "cause" both A and B:

> Event C = At Base X the official duty day ends at 1630 for much of the workforce.

6. A **Single cause** fallacy occurs when you assume there is a single cause for an outcome, when in fact multiple causes exist.

Let's consider a real life example of a single cause fallacy. Suppose you're very physically fit, and in a few months you'll take a fitness test that measures your heart rate on a stationary bike. You've set a goal to score in the top 10 percent for your age group—an "outstanding" rating. You know that a disciplined exercise program will cause you to improve your score, but is it this simple?

> Event A = disciplined, intense exercise program
>
> CAUSES
>
> Event B = outstanding score on the fitness test

People who've had trouble with similar fitness tests would be quick to point out that cause and effect may be a little more complicated in this case:

> Event A = disciplined, intense exercise program
>
> Event B = genetically low resting heart rate
>
> Event C = no caffeine or nervousness about the test
>
> Event D = planets all in alignment and a strong tail wind…
>
> CAUSES
>
> Event E = outstanding score on the fitness test

On the other hand, people who have the genetically low heart rate and nerves of steel may think an outstanding rating has a single cause because they've never had to deal with the other ones.

7. Faulty analogy. Though we often make analogies to make a point, sometimes they go astray—there's something about the comparison that isn't relevant. A faulty analogy implies that because two things are alike in one way, they are alike in all the ways that matter. It can be thought of as one example of a **non sequitur** fallacy (see item 11).

> "Leading a coalition is just like leading a fighter squadron."

Well, not exactly. Leadership is required in both situations, but leading a coalition requires the ability to work with people from other services and countries, and requires greater communication, tact and diplomacy. Leading a fighter squadron may require a higher level of current technical proficiency in the cockpit. There are people who would excel in one situation and fail in the other.

8. Faulty dilemma. A *faulty dilemma* is that implication that no middle ground exists between two options. Typically one option presented is what the speaker prefers, and the other option is clearly unacceptable.

> "Spend one hour a day reading *Tongue and Quill* to improve your writing skills … or stay an incoherent loser. It's your choice."

Clearly this is a faulty dilemma—it falsely suggests you only have two choices, when you really have many options. Maybe you can read *The Tongue and Quill* once a week or once a month. Maybe you'll find some other way to improve your writing skills—take a class, find a grammar website, get feedback from your boss, etc. Though sometimes life really does give us an "either-or" choice, in most cases we find a considerable range of options between two positions.

9. Hasty generalization results when we "jump to conclusions" without enough evidence. A few examples used as proof may not represent the whole.

> "I asked three student pilots what they thought of the program, and it's obvious that Undergraduate Pilot Training needs an overhaul."

One of the challenges with this fallacy is it's hard to determine how much evidence is "enough" to form a reasonable conclusion. The rules will vary with the situation; more evidence is needed to form a conclusion if the stakes are high. The Food and Drug Administration may require a great deal of evidence before deciding a drug is safe for human use, while SSgt Snuffy may require very little evidence before forming a generalized conclusion about which candy bars should be sold at the snack bar.

10. Non sequitur is a generic term for a conclusion that does not necessarily follow from the facts presented—*non sequitur* means "it does not follow" in Latin. The facts may not be relevant, or there may be some sort of illogical leap made. Several of the fallacies listed here, including faulty analogy and hasty generalization, can be thought of as different types of non sequitur.

> "John Doe will make a great squadron commander because he was a hot stick in the F-15."

This is a non sequitur because it implies that because he has strong technical skills, he also has the skills needed to command. Another fairly common non sequitur in military circles assumes that athletic prowess translates into leadership ability.

11. Slippery slope implies that if we take one small step in an unpleasant or dangerous direction, we'll have to go all the way—like slipping down a hill. Here's an example from the book of *Writing Arguments* by Ramage, Bean and Johnson:

> "We don't dare send weapons to Country X. If we do so, next we will send in military advisors, then a Special Forces battalion and then large numbers of troops. Finally, we will be in an all-out war."

Though not every slippery slope argument is false, in some cases we can identify lines that we will not cross.

12. Red herring fallacies occur when an arguer deliberately brings up irrelevant information to get the audience off track. The origins of the "red herring" name are debatable, but imagine someone dragging a large smelly fish across a scent trail that a bloodhound is following—the dog would probably lose the scent.

13. Stacked evidence is the tendency to withhold facts or manipulate support so that the evidence points in only one direction. It also happens when you gather only the data or opinions that support your position. Though sometimes this is done deliberately, in other cases it occurs due to unconscious bias or carelessness. We don't even see counterarguments or alternative ways of interpreting the information because of our firm belief in our own position, or we just stop gathering information once we we've found enough data to make our case.

Even if you decide to push for your favorite interpretation of the data, never stack evidence by misrepresenting or manipulating the basic information. Also, even if you decide that you don't want to discuss the opposing viewpoint, you should at least be aware of it, so you can prepare a counterpunch if needed.

14. Straw man is a fallacy where you fail to attack your opponent's argument directly. Instead, you make up a weaker, grossly simplified version of the opponent's argument, and then attack it. In effect, you are attacking a "straw man"—the argument that you wished your opponent made, not the one he actually did.

Straw man fallacies are popular in political campaigns. Suppose a candidate believed that a major goal of prisons should be rehabilitation—breaking the cycle of violence—not just punishment. An opponent could exploit that with a straw man attack:

> "My opponent coddles convicted felons and wants to make life easier behind bars than on the street. I believe prison should be a deterrent, not a reward for bad behavior!"

This list of fallacies captures most of the common errors we hear and see daily. Our challenge is to sharpen our professional sense of smell so we can quickly sniff out the rational from the ridiculous, and avoid adding to the epidemic of poor reasoning and weak support we encounter around us.

ETHICS AND ARGUMENTS: PERSUASION, THE QUEST FOR TRUTH, AND ISSUES OF CREDIBILITY

> "We believe that argument is a matter not of fist banging or of win-lose debate but of finding, through a process of rational inquiry, the best solution of a problem."
>
> — from *Writing Arguments*, by John D. Ramage, John C. Bean, and June Johnson

We've all had experience with using logical arguments to persuade someone else. In the middle of such a discussion, you may have asked yourself, "What's my goal—to persuade the other guy and make my case? Or to find out the truth and the best answer to the problem?" (This usually comes up when your opponent comes up with a valid point you hadn't considered before.) Ever since the ancient Greeks were walking around in togas, people have struggled with this issue.

There will always be reasons to use argument as a tool of persuasion—you want your subordinate to win that award, you really need additional funding for your branch, and you want your wife to visit her in-laws over the Thanksgiving weekend.

Sometimes in the heat of verbal battle, it's tempting to focus on persuasion and forget about truth. Don't do it. Integrity is one of our Air Force Core Values, and you have to look at yourself in the mirror every morning. You don't have to be a doormat, but if you find out about some new information that may change your position, keep an open mind. In most situations, you don't only want YOUR WAY, you want THE BEST WAY. Besides, if you pull a fast one and get your way through deception, you've won a battle, but your credibility is shot and you've crippled yourself for future skirmishes.

Other ways to build credibility with your audience include being knowledgeable and fair. Research your topic carefully, and take the time to get the facts right. Don't bluff if you don't have an answer, or mislead others about the strength of your support. Consider your audience's values and assumptions when selecting evidence. Make sure you get the easy things right—the spelling of names, significant dates, and other details like grammar and punctuation. Demonstrate goodwill in your writing tone—don't be condescending or act superior. If you make a mistake, acknowledge it and move on. Credibility takes a long time to build, but it is invaluable when trying to support your ideas and persuade others.

SUMMARY

In this chapter we covered several topics that should help you support your ideas.

We defined a *logical argument* as a set of statements designed to persuade others. Logical arguments have four components:

- a *claim*—your position on a controversial topic
- *evidence* that supports your claim
- *warrants* that identify why the evidence is relevant (these are often implied)
- *qualifications* that limit the claim

Your argument is built upon evidence, and it should be authoritative, accurate, precise, relevant, and adequate to support your claim. As you build or listen to logical arguments, watch out for *logical fallacies*—common mistakes many people make when building an argument.

Arguments are everywhere. To write and speak persuasively, it helps to understand how arguments are constructed and where they go wrong. These insights will be helpful as you start to organize and outline your thoughts—the next step of the process.

An outline helps you see the big picture and how individual ideas flow together. We are here, on Step 4...

Step-by-Step Outline:
Seven Steps to Effective Communication

1. **Analyze Purpose and Audience**
   a. **Analyze purpose**
   b. **Analyze audience**
2. **Conduct Research**
   a. **Search for relevant information**
   b. **Critically evaluate your sources**
3. **Support your Ideas**
   a. **Build your argument**
   b. **Check for mistakes in reasoning**
4. **Organize and Outline**
   a. **Finalize your purpose statement**
   b. **Pick an organizational pattern**
   c. **Write your outline**
5. **Draft**
   a. **Draft flow of paragraphs**
   b. **Write clear sentences**
   c. **Pick the right word for the job**
6. **Edit**
   a. **Edit for "the big picture"**
   b. **Check your paragraph structure**
   c. **Edit sentences and phrases**
7. **Fight for Feedback and win Approval**
   a. **Seek informal feedback**
   b. **Coordinate to obtain formal approval**

CHAPTER 6

STEP 4:
ORGANIZING
AND OUTLINING
YOUR THOUGHTS

This chapter covers:

- How to develop a purpose statement.

- The importance of a three-part structure.

- Different formats for organizing your outline.

After completing the first three steps, you are well on your way to an outstanding spoken or written product. Now it's time to talk about *organizing* and *outlining*, the final step in prewriting. A detailed outline helps you arrange your material logically, see relationships between ideas, and serves as a reference point to keep you on target as you write your draft. Think of your outline as the blueprint for your communication product, and realize that the time you spend preparing it will pay off when you start writing sentences and paragraphs.

ORGANIZING: FINALIZING YOUR PURPOSE STATEMENT AND "BOTTOM LINE"

Why are we talking about a purpose statement again? Didn't we determine our "bottom line" back in Step 1? Or did we? Sometimes information uncovered during the research process (Step 2) may point you in an unexpected direction. So do you tweak the data to match your original purpose? No! Now is the time to adjust the vector of your purpose statement to something you can reasonably support and live with.

> A *thesis statement* is a specialized form of purpose statement used in academic or persuasive writing.
>
> The thesis statement captures the author's point of view on a controversial topic, which he or she defends throughout the paper. A thesis statement is usually finalized after the research process.

You're less likely to go astray during the outlining process if you write down your purpose statement and refer to it often. Every main point and supporting idea in your outline should support that purpose statement—irrelevant facts or opinions should be eliminated. Discipline at this stage will save pain later.

ORGANIZING: GET YOUR BOTTOM LINE UP FRONT (MOST OF THE TIME)

> "In the future, authors will take a long time to get to the point. That way the book looks thicker."
>
> – Scott Adams in *The Dilbert Future: Thriving on Stupidity in the 21st Century*

In nearly every communication situation, you need to state your bottom line early in the message. In a direct or *deductive* approach, state your position, main point or purpose up front, then go into the details that support your main point. When you take a direct approach to communication, your audience is better prepared to digest the details of the message and logically make the connections in its own mind.

There is an exception to every rule, and you might want to be less direct when trying to persuade a hostile audience. In such a situation, if you state your bottom line up front, you risk turning them off before you build your argument—regardless of how well it is supported. In this case you might consider using an indirect or *inductive* approach: you may present your support and end with your bottom line. Sometimes this successfully "softens the blow" and gives your audience time to warm up to your views.

In the inductive approach, you still need an introduction, but it would be less direct. Here's an example of two purpose statements:

Direct: Women should be allowed in combat because….

Indirect: The issue of women in combat has been hotly debated, and both sides have valid points….

Use the inductive approach with caution; it's an advanced technique and difficult to execute without confusing your audience. In an academic setting, seek your instructor's advice before applying this method to your assignments.

THE OUTLINE: WHY DO I NEED ONE?

To some people, preparing an outline looks like a painful chore. Though an outline does take some effort, it's a time-saver, not a time-waster. An outline contains your main points and supporting ideas arranged in a logical order. It allows you to see and test the flow of your ideas on paper without having to write out complete sentences and paragraphs. If some ideas don't fit together or flow naturally, you can rearrange them without a lot of effort. Like the blueprint of a house, an outline makes the "construction process" more efficient and often results in a better quality product.

Does all writing require you to take the time to write a detailed outline with several layers of detail? No. If you plan to write a short letter, message or report, a list of main points may be all you need. For longer papers, Air Force publications, reports, staff studies, etc., you'll find a detailed outline is usually an indispensable aid.

OUTLINING: THREE-PART STRUCTURE

Chapter 7 will describe how most writing and speaking is organized in a three-part structure consisting of an introduction, a body and a conclusion. Most of the work in developing an outline involves organizing the body of your communication, but if you are building a formal or detailed outline on a lengthy written product, you should probably include the introduction and conclusion in the outline. A skilled communicator writing an informal outline on a short assignment may just outline the body and work out the introduction and conclusion during the drafting process.

OUTLINING FORMATS:
HEADINGS AND STRUCTURE USED IN FORMAL OUTLINES

Though most outlines you produce will never be seen by anyone else, in some cases you might be asked to produce a formal outline for "public consumption." Here are some possible scenarios:

- Your boss wants to review what you plan to cover before you start drafting.

- Your document will have numbered headings and subheadings.

- You're organizing the efforts of multiple writers who must work together.

In these situations, it's helpful to have a consistent approach to numbering or lettering the different components of your outline. Be consistent once you pick your approach.

One option for an outline format is to use a numerical structure to identify different levels of the outline. In complicated documents like military publications, these levels may also be used as headings in the finished document to help readability. Another classic option is to use a mix of Roman numerals, Arabic numerals and uppercase and lowercase letters to identify the different levels. (For a refresher on Roman numerals, check out on page 336. A third option is to use the same headings as the paragraph labels in an official memo; see page 185 for details).

Note that some sections of your outline may be more detailed than others. In the below outline it is perfectly acceptable to divide section 1.2 into smaller elements and leave section 1.1 and 1.3 undivided.

```
NUMERICAL OUTLINE FORMAT
1.  Section 1
      1.1 First subheading to Section 1
      1.2 Second subheading to Section 1
            1.2.1 First subheading to 1.2
            1.2.2 Second subheading to 1.2
      1.3 Third subheading to Section 1

2.  Section 2
      2.1 First subheading to Section 2
            2.1.1 First subheading to 2.1
            2.1.2 Second subheading to 2.1
      2.2 Second subheading to Section 2

3.  Section 3…
```

```
CLASSICAL OUTLINE FORMAT
  I. Section I…
      A. First Subheading to Section I
          1. First subheading to I.A
             a.
             b.
          2. Second subheading to I.A
             a. …
             b. …
      B. Second Subheading to Section I
          1. First subheading to I.B
             a.
             b.
          2. Second subheading to "I.B"
             a.
 II. Section II….
```

These two examples illustrate a cardinal rule of outlining: **any topic that is divided must have at least two parts.** Never create a Part 1 without a Part 2, or a Section A without a Section B.

These are only two of many possible formats for numbering different levels of a detailed outline. If your final product requires a particular format for headings or organization (for example, if you are writing an Air Force publication), you might save time later by building your outline with the format specified for the finished product. If not, any consistent approach will work fine.

Some people get tied up in knots over the mechanics of a formal outline. Remember that the primary purpose of an outline is to help you arrange your thoughts into main points and subordinate ideas, so relax and use a format that works for you.

✪ **Tip:** Each item in an outline should begin at the left margin. Second and following lines should be in block format or indented to align with the first word in the line above.

OUTLINING THE BODY: PICK A PATTERN

Your next step is to select a pattern that enables you and your readers to move systematically and logically through your ideas from a beginning to a conclusion. Some of the most common organizational patterns are listed below. Your purpose, the needs of your audience, and the nature of your material will influence your choice of pattern.

1. TOPICAL/CLASSIFICATION PATTERN

Use this format to present groups of ideas, objects or events by categories. This is a commonly used pattern to present general statements followed by numbered listings of subtopics to support, explain, or expand the statements.

A topical pattern usually follows some logical order that reflects the nature of the material and the purpose of the communication. For example, if you are giving a briefing on helicopters, you might separate them into light, medium, and heavy lift capabilities and briefly describe the weight limits for each category. You could begin with the lightest capability and move to the heaviest or begin with the heaviest and move to the lightest.

✪ **Tip:** To help your readers absorb complex or unfamiliar material, consider organizing your material to move from the most familiar to the unfamiliar or from the simplest category to the most complex. When using this pattern, experiment to find the arrangement that will be most comfortable for your audience.

> **Outline: Comparison of
> F-15 and F-16 Performance**
>
> A. General Description
> 1. F-15
> 2. F-16
> B. Detailed Comparison
> <u>1. Spec 1</u>
> - F-15
> - F-16
> <u>2. Spec 2</u>
> - F-15
> - F-16
> <u>3. Spec 3</u>
> - F-15
> - F-16
> <u>4. Spec 4</u>
> - F-15
> - F-16

2. COMPARISON/CONTRAST PATTERN

Use this style when you need to discuss similarities and/or differences between topics, concepts, or ideas.

When you are describing similarities and differences, it often helps the reader to see a point-by-point comparison of the two items. For example, if you were writing a document that compares and contrasts certain characteristics of the F-15 and the F-16, you might go item by item, discussing similarities and differences between the two as you go.

3. CHRONOLOGICAL PATTERN

When you use this pattern, you discuss events, problems or processes in the sequence of time in which they take place or should take place (past to present or present to future). This commonly used pattern is used in writing histories, tracing the evolution of processes, recording problem conditions, and documenting situations that evolve over time.

This approach is also used in official biographies, which are written in chronological order because they serve as a history of the member's professional career.

This pattern is simple to use, but judgment is required when deciding what events to leave in and what events to leave out. For example, if you were preparing a short biography to introduce a distinguished guest speaker, you may decide to emphasize experiences that demonstrate his subject matter expertise and leave out other important but less relevant details. When unsure what to include, think back to your purpose and audience.

✪ **Tip:** You may want to consider a chronological approach to your topic when it is known to be controversial. Many writers and speakers will announce, "First let's take a look at the history of the problem." This starts the sender and audience out on neutral ground instead of just launching into the issue at hand. This is a type of inductive approach, and again, should be used with caution.

4. SEQUENTIAL PATTERN

The *sequential* or *step-by-step* approach is similar to the chronological pattern. Use this approach to describe a sequence of steps necessary to complete a technical procedure or process. Usually the timing of steps is not as important as the specific order in which they are performed. The outline on the first page of this chapter ("Seven Steps to Effective Communication") is an example of a sequential approach.

The sequential approach is often used in manuals and other instruction books. For example, a Security Forces NCO in charge of small arms training might use this pattern when rewriting the teaching manual on how to safely inspect, load, fire, disassemble, and clean weapons. Since safety is paramount, the process must be written in a precise, stepwise fashion to ensure that nothing is overlooked.

✪ **Tip:** When describing a procedure, explain the importance of *sequence* so your audience is mentally prepared to pay close attention to the order, not just the content, of the information.

5. SPATIAL/GEOGRAPHICAL PATTERN

When using this pattern, you'll start at some point in space and proceed in sequence to other points. The pattern is based on a directional strategy—north to south, east to west, clockwise or counterclockwise, bottom to top, above and below, etc. Let's say you are a weather officer briefing pilots about current and anticipated conditions in the geographic region where they will be flying a mission. You would most likely describe conditions in reference to the terrain and describe weather systems that will affect their mission on a map.

✪ **Tip: CAUTION!** Make sure to use appropriate transitions to indicate spatial relationships— to the left, farther to the left, still farthermost to the left; adjacent to, a short distance away, etc. These signal the flow of the communication; if missing, your audience can easily become confused or disoriented. (We'll talk more about transitions in the next chapter, on pages 70-73).

6. PROBLEM/SOLUTION PATTERN

You can use this pattern to identify and describe a problem and one or more possible solutions, or an issue and possible techniques for resolving the issue. Discuss all facets of the problem—its origin, its characteristics, and its impact. When describing the proposed solution, include enough support to convince your readers the solution is practical and cost effective. After presenting your solution, you may want to identify immediate actions required to implement the solution.

The problem/solution pattern may be used in several variations:

• One Solution: Discuss the problem and follow with the single, most logical solution.

• Multiple Solutions: Discuss the problem, several possible solutions, the effects of each and your recommendation.

• Multiple Solutions, Pro-Con: This popular format includes a discussion of the advantages ("Pros") and disadvantages ("Cons") of each solution.

Remember that a problem-solution pattern is not a format for a personal attack on an adversary; it's simply a systematic approach to use in persuading people either to accept your ideas or to modify their own ideas.

**Problem Solution Example:
The Staff Study**

The Staff Study format described on page 203 is a classic example of a problem/solution pattern. Within this format, you can present several possible solutions or just the one you recommend. A staff study with three options might have an outline that looks like this:

1. PROBLEM

2. FACTORS BEARING ON THE PROBLEM

  a. Facts
  b. Assumptions
  c. Evaluation criteria for solutions

3. DISCUSSION OF POSSIBLE SOLUTIONS

  a. Option 1- pros and cons
  b. Option 2- pros and cons
  c. Option 3- pros and cons

4. CONCLUSION

  "Option 3 is recommended…"

5. ACTION RECOMMENDED

  "Take the following steps…"

Though you can list your options in any order, skilled writers often "save the best for last" and put their recommended option last on the list to help readability.

---

7. REASONING/LOGIC PATTERN

In this pattern, you state an opinion and then make your case by providing support for your position. This is the classic "logical argument" described in Chapter 5. This approach works well when your goal is more than just discussion of problems and possible solutions. Use this pattern when your mission is to present research that will lead your audience down the path to your point of view!

✪ **Tip:** Remember your audience analysis? If members of your audience are hostile to your position, try to look at this issue through their eyes. Start out with the support they are most likely to accept, and then move into the less popular issues that support your main point.

8. CAUSE/EFFECT

You can use this pattern to show how one or more ideas, actions or conditions lead to other ideas, actions or conditions. Two variations of this pattern are possible: (1) begin with the effect, then identify the causes; or (2) begin with the causes, then identify the effects. The technique you use depends on the context of your discussion.

Sometimes an effect-to-cause approach is used when your purpose is to identify WHY something happened. When might you use this approach? Let's say you are the president of the Safety Investigation Board

---

Causes, Effects and Faulty Logic

Be careful to avoid faulty logic traps when writing about cause and effect. You're guilty of a *false cause fallacy* when you assume one event causes a second event merely because it precedes the second event. You're guilty of a *single cause fallacy* when you assume only one factor caused an outcome, when in fact there are multiple causes. For more details on these and other fallacies, refer back to Chapter 5.

---

following a fatal aircraft mishap (*the effect*). Your report might begin by describing the mishap itself, and then explain the factors that led up to the mishap and conclude with your determination of one or more *causes* for the effect.

Sometimes a cause-to-effect pattern is used when your purpose is to explain how current actions or conditions (causes) may produce future consequences (effects). For example, someone might use this pattern to present how a series of causes—larger automobiles, reduced financial incentives for energy conservation and reduced research funding for alternative energy technologies—might result in an undesirable effect—a US shortage of fossil fuels.

SUMMARY

A well-planned outline can ease the pain of writing your first draft. Remember, building a house is much easier with a blueprint! This invaluable tool will help you remain focused on your purpose statement and help ensure your support is organized, relevant and tailored to your mission and audience. The outline will also help in the editing process. Take a break after working on your outline and start fresh before you begin your draft. Good luck!

PART III:

WRITING WITH FOCUS

PREWRITING PROCESS SUMMARY: STEPS 1-4

BEFORE STARTING YOUR DRAFT, YOU SHOULD KNOW...

STEP 1. ANALYZE PURPOSE AND AUDIENCE

Your purpose: to direct, inform, persuade or inspire (reference page 19).
Your purpose statement: one sentence that captures your bottom line (reference page 20).
Your schedule for the assignment.
Your communication format: Point paper? Staff study? Academic essay?
Your audience: experience, education, attitudes about topic, etc. (pages 21-22).
Your unit's position: Could you create problems for others? Should you coordinate this?
The appropriate tone will depend on purpose and audience: usually polite, personal,
    positive, and inclusive; often informal and direct (pages 23-24).

STEP 2. RESEARCH YOUR TOPIC

Relevant information from boss, coworkers, office files, Internet and the library.

STEP 3. SUPPORT YOUR IDEAS

How to use relevant information to support your ideas and meet your purpose.
How to "build a case" (a logical argument) for your position, if needed.
How to use facts, definitions, statistics and testimony as evidence for your position.
How to avoid mistakes in your logic and notice problems with your evidence.

STEP 4. ORGANIZE AND OUTLINE

Your chosen organizational pattern (topical, chronological, problem/solution, etc.).
Your outline, which graphically shows the flow of your main points.

CHAPTER 7

STEP 5: WRITING YOUR DRAFT

This chapter covers:

- Drafting using a three-part structure (introduction, body and conclusion).

- Drafting effective paragraphs.

- Drafting clear and concise sentences.

- Overcoming writer's block.

After completing the prewriting process, you've got what you need to produce a first rate communication product. Congratulations! You're ready to write your first draft! In this chapter we'll take a "top down" approach to writing a draft. We'll start with the big picture: a three-part structure consisting of the introduction, the body, and the conclusion. Next, we'll describe how to write effective paragraphs within the body. Finally, we'll dig down deeper into the sentences, phrases, and words that make up the paragraphs of your draft.

> "The basic rule every military writer ought to live by is this:
> 'I will write only when I must.'"
>
> — Colonel William A. McPeak

DRAFTING BASIC PHILOSOPHY

Keep a few things in mind as you start the drafting process. A draft is not the finished product, and each sentence does not have to be polished and perfect. Your focus should be to get your ideas on paper. Don't obsess about grammar, punctuation, spelling, and word choice at this point—that comes later. You don't have to fix every mistake as you see it—you can catch these during the editing process.

On the other hand, it's helpful to keep an eye on your outline when drafting your masterpiece, especially when you're writing something longer than a page or two. By periodically checking your outline, you are less likely to lose focus and include irrelevant information.

Okay, this sounds great, but have you ever sat down to start your first draft and found yourself just staring at the blank computer screen or paper? If you suffer from writer's block, we'll cover strategies for overcoming this fairly common problem at the end of this chapter.

THREE PART STRUCTURE: AN INTRODUCTION, BODY AND CONCLUSION

What is your draft going to look like? Is it going to be one huge paragraph? No, in most cases, you'll organize your draft in a three-part structure—introduction, body and conclusion.

- The **introduction** must capture your audience's attention, establish rapport and announce your purpose.

- The **body** must be an effective sequence of ideas that flow logically in a series of paragraphs.

- The **conclusion** must summarize the main points stated in the body and close smoothly.

Let's take a closer look at this structure. We'll examine these parts out of order—first, the introduction, then the conclusion and lastly the body where we'll spend most of our "time."

DRAFTING THE INTRODUCTION

The *introduction* sets the stage and tone for your message. Although the content and length of your introduction may vary with the assignment, the introduction should, at a minimum, clearly state your purpose ("bottom line") and the direction you plan to take the audience.

A typical introduction has three components: *stage setting remarks*, a *purpose statement*, and an *overview*.

- *Stage-setting remarks* set the tone of the communication, capture the audience's attention and encourage them to read further. Stage setting remarks are *optional*, so you can omit them in very short messages or in messages where you don't want to waste words.

- The *purpose statement* is the one sentence you'd keep if you had only one. It specifically states your purpose, thesis or main point. For some examples and more details, refer back to Chapter 3 (page 20).

> **Stage Setting Remarks:**
> **Use them properly!**
>
> Stage setting remarks are optional. Though they add polish to an introduction, your reader has to be able to pick which sentences are "setting the stage" and which sentence is the "bottom line."
>
> If you've received feedback that readers are sometimes confused about the purpose of your writing, get to the point quickly and don't overdo stage-setting remarks. Too many preliminaries can backfire and actually confuse the reader.

- The *overview* is like a good roadmap—it clearly presents your main points, previews your paragraph sequence and ties your main points to your purpose.

Here's an example of a short introduction that contains all three components:

> Communication is essential to mission accomplishment, and all Air Force personnel should be able to write effectively. (*Stage Setting Remarks*) This handbook provides general guidelines and specific formats for use in both staff environments and Professional Military Education schools. *(Purpose Statement)* It begins with an overarching philosophy on military communication, then describes processes and techniques to improve writing and speaking products, and summarizes the most common formats used in Air Force communication. (*Overview)*

Even though readers read the introduction first, you don't have to write it first. If the introduction doesn't come easily or naturally, you can work on another part of the communication and then return to it. Some writers backpedal and don't want to work on the introduction until the rest of the communication is written. Others insist it guides them in shaping the content or body of their message. Regardless of when you write the introduction, make sure that it captures your purpose and make sure it prepares your audience for what is to come.

Here's the bottom line on your introduction: It must be an appropriate length for your specific communication and it should contain a clear statement of your purpose and direction.

DRAFTING THE CONCLUSION

The *conclusion* is the last and often neglected part of a well-arranged communication. Sometimes inexperienced writers stop writing as soon as they finish discussing their last main idea. That's not an effective conclusion. The conclusion is your last chance to summarize your communication and give your audience a sense of closure.

An effective conclusion often summarizes the overall theme and main points discussed in the body. If you have a simple, straightforward purpose, you might want to emphasize it by restating it in slightly different words in the conclusion. If you have a complicated purpose or a long, involved communication, you'll probably need to emphasize your main ideas and state your proposals or recommendations.

Introductions and conclusions: How long?

The length of your introduction and conclusion will be proportional to the length of your overall writing assignment. On a one-page assignment, they may be very short, while lengthy staff studies or publications may contain introductions and conclusions that are several paragraphs long. Introductions and conclusions to books are often an entire chapter!

Remember that introductions and conclusions are designed to help your readers; use good judgment in determining the appropriate length for your assignment.

For effective endings, restate the main ideas or observations or emphasize the main thrusts of arguments. Under no circumstances apologize for real or perceived inadequacies or inject weak afterthoughts. Conclude your communication with positive statements based on your preceding discussion. Generally, avoid bringing up new ideas in the conclusion; these belong in the body of your communication. Opening up new "cans of worms" will just confuse your reader.

Your introduction and conclusion should balance each other without being identical. To check this, read your introduction and then immediately read your conclusion to determine if your conclusion flows logically from your introduction and whether it fulfills your purpose. An effective conclusion leaves you with a sense you're justified in ending your communication. You're ready to call it a day only when you assure your audience you've accomplished the purpose stated in your introduction.

Recall our sample introduction on page 67; here's a short conclusion derived from that introduction and the body (which we don't have right now):

> As Air Force personnel, we can't accomplish our mission without effective communication. Hopefully, this handbook has provided you with some practical tools to improve your communication skills, specifically speaking and writing. Keep it handy and refer to it often as you prepare and review a variety of spoken and written products throughout your career.

Even without the "body" available, you can see how the introduction and conclusion complement each other.

DRAFTING THE BODY

The *body* of your communication is the heart of your message. It includes your main ideas about your subject and supporting details under each main idea.

The body typically consists of several paragraphs. The total number of paragraphs (and overall length of the body) will depend on your purpose and subject. As a general rule, write a separate paragraph for each main idea—you might confuse your reader if you have two or more main ideas in a single paragraph. In a longer communication, you may find it necessary to use more than one paragraph to cover one main point or idea.

So much for a quick review of introductions, conclusions and bodies. Now, let's dig down a little deeper into the paragraphs that make up the body of your communication.

DRAFTING EFFECTIVE PARAGRAPHS
PARAGRAPHS SHOULD CONTAIN ONE MAIN POINT

Paragraphs are the primary vehicles to develop ideas in your writing. They serve three purposes:

① To group related ideas into single units of thought.

② To separate one unit of thought from another unit.

③ To alert your readers you're shifting to another phase of your subject.

An effective paragraph is a functional unit with clusters of ideas built around a single main point or idea and linked with other clusters preceding and following it. It's not an arbitrary collection designed for physical convenience. It performs a definite, planned function—it presents a single major idea or point, describes an event, or creates an impression.

Most staff writing depends on relatively short paragraphs of three to seven sentences. If you follow this practice, you'll be more likely to develop clear, easy-to-read paragraphs. The length of individual paragraphs will vary because some main points need more supporting details than others.

In general, the flow of your paragraphs will follow the organizational pattern or format you selected in Step 4: "Organizing and Outlining" (Chapter 6). That is, you build your paragraphs to meet the structural requirements of your overall communication. But you can use analogy, examples, definition, and comparison and contrast to develop single paragraphs within your overall pattern. The guiding principle is to develop one main idea or point in each paragraph.

TOPIC SENTENCES

Capturing the main point of each PARAGRAPH

In staff writing, it's helpful to start off each paragraph with a *topic sentence* that captures the subject or controlling idea of the paragraph. The topic sentence prepares the reader for the rest of the paragraph and provides a point of focus for supporting details, facts, figures and examples.

In the body, don't make your reader search for the topic sentences of your paragraphs. (As stated earlier, the rules are different for introductions and conclusions.) Since the topic sentence is the subject and main idea of the paragraph, the best place for it is up front—the first sentence. This helps with clarity and makes things convenient for your readers. Many people need only general information about the content of certain letters, reports and directives. Scanning topic sentences at the beginning of paragraphs for the most important ideas saves a lot of time. If your readers need more details, they can always read beyond your topic sentences.

> **If your readers are confused, check your topic sentences!**
>
> A *topic sentence* announces your intent for a single paragraph in the same way a *purpose statement* announces your intent for the entire writing assignment.
>
> Most readers are better able to understand how ideas relate to each other if they know what's coming.
>
> If you've received feedback that readers have trouble understanding the "flow" of your writing, check your topic sentences. Does one exist for each paragraph? Can you underline it? Do they start off the paragraph? Do they tie back to your purpose statement?

Once you've written a topic sentence, the rest of the paragraph should fall neatly in place. Other sentences between the topic sentence and the last sentence must be closely related to expand, emphasize, and support the topic sentence. In some paragraphs, the last sentence is used to summarize key points, clinch the main idea in the reader's mind, or serve as a transition to the next topic sentence. (We'll talk more about transitions in the next section.) Eliminate any "extra" sentences that don't perform one of these functions!

Though most writers will draft an entire paragraph at a time, **an alternate drafting strategy is to first write all the topic sentences in your body.** Once the topic sentences are completed, go back and write the rest of the paragraphs, one at a time. Drafting the topic sentences first requires the writer to stay focused on the "big picture" and can help produce a clear and well-organized draft. This technique can be very useful for longer writing assignments and is recommended for writers who struggle to organize their writing.

Here's the bottom line on paragraphs in the body: Each paragraph should have one main point/idea captured in a topic sentence, preferably at the beginning of the paragraph. Use supporting ideas to prove, clarify, illustrate and develop your main point. Your objective is to help your readers see your paragraphs as integrated units rather than mere collections of sentences.

THE BULLETIN BOARD

TO CONTRAST IDEAS

but
yet
nevertheless
however
still
conversely
on the one hand
instead of
neither of these
(to) (on) the contrary
rather than
no matter what
much less as
in contrast
otherwise
on the other hand
in the (first) (second) place
nor
according to

TO SHOW TIME

immediately
presently
nearly a … later
meantime
meanwhile
afterward
next
as of today
this year, however
a little later
then last year
next week
tomorrow
as of now
finally

TO RELATE THOUGHTS

indeed
anyway, anyhow
elsewhere
nearby
above all
even these
beyond
in other words
for instance
of course
in short
in sum
yet
in reality
that is
by consequence
notwithstanding
nonetheless
as a general rule
understandably
traditionally
the reason, of course
the lesson here is
from all information
at best
naturally
in the broader sense
to this end
in fact

TO COMPARE IDEAS

like
just as
similar
this

TO SHOW RESULTS

therefore
as a result
thus
consequently
hence

TO ADD IDEAS

first, second, next, last, etc.
in addition
additionally
moreover
furthermore
another
besides
clear, too, is
the answer does not only lie
to all that
more than anything else
here are some … facts
now, of course, there are
now however

TRANSITIONS: BRIDGES BETWEEN DIFFERENT IDEAS

One way to make sure your paragraphs flow together, both internally and externally, is by using transitions in the form of words, phrases, and sentences. *Internal transitions* improve the flow of sentences within a paragraph, while *external transitions* link separate paragraphs together within the body of your communication. Though some inexperienced writers are intimidated by the idea of transitions, a few examples usually make the point.

INTERNAL TRANSITIONS

Internal transitions are one or more related words that show the relationship between ideas *within a paragraph*. Woven skillfully into your writing, internal transitions help your reader follow your line of thought.

Some internal transitions show a relationship between two ideas inside a single sentence:

> "**First** go home and **then** clean your room."

Other internal transitions show a relationship between two or more sentences within a single paragraph:

> "Our plan for Saturday afternoon involves both business and pleasure. **First,** all the kids will come home at noon and we'll eat lunch. **Next,** we'll get the house cleaned—the whole mess. **Finally,** we'll go out for ice cream and a movie."

Take a look at page 70 for a bulletin board of transitional words and phrases that provide the ideal logic links between your key points and the mind of the reader. In most cases, favor the short, spoken ones over the long, bookish ones. For example, use *but* more than *however*, *so* more than *therefore*, and *also* more than *in addition*. (Note that different transitions require different punctuation. If you're uncertain about the rules, check out guidelines for comma and semicolon usage on pages 282 and 307.)

> "The movie was too long; **therefore**, we left after three hours."

> "The movie was too long, **so** we left after three hours."

There are many ways to bridge gaps in thought and move the reader from one idea to another. One classic transitional approach involves repetition of key words at the beginning of individual sentences. This is especially popular in formal or ceremonial writing or speaking. Notice how the writer of the following paragraph repeated *simplicity*, *incisiveness* and *focus* to make points clear:

> The effective presentation of concepts depends on simplicity, incisiveness and focus. Simplicity is necessary under time constraints when there's insufficient time for complicated relationships. Incisiveness fixes an idea in the listener's mind, appeals to common sense and facilitates understanding. Focus limits the subject to essentials, promoting the presenter's objectives.

Internal transitions, in the form of one or more related words, are key to a well-written paragraph because they guide the reader between related ideas. But how do we move from paragraph to paragraph? We need *external transitions* to knit together their main points.

EXTERNAL TRANSITIONS

External transitions are typically sentences or paragraphs that guide the reader **between separate paragraphs** and **major sections** of your communication.

Transitional paragraphs are usually reserved for long papers, books, and reports that contain major sections or chapters. They are used to summarize one section and lead the reader to the next section, or they introduce the next section and tie it to the preceding section. Transitional paragraphs are not commonly used in staff writing, but are often seen in books and academic essays.

The short paragraph immediately above this section ("Internal transitions, in the form of one or more related words…") is an example of a transitional paragraph. As you can see, it sums up the previous section on internal transitions and then introduces the new section on external transitions.

Let's look closer at transitional sentences, which you'll probably use more frequently than transitional paragraphs. A transitional sentence is often used to bridge man points in two separate paragraphs (though not every new paragraph requires an external transition). There are three options of a transitional sentence bridging paragraph 1 and paragraph 2:

1) It can be a stand-alone sentence at the end of paragraph 1.

2) It can be a stand-alone sentence at the beginning of paragraph 2 (In this case, paragraph 2's topic sentence is the second sentence in the paragraph).

3) It can be merged with the topic sentence of paragraph 2 (In this case, the "transitional" part of the sentence is a separate clause at the beginning of the sentence).

Let's look at a situation where a transitional sentence is appropriate. Suppose we have two paragraphs:

> Paragraph 1 describes parking problems.

> Paragraph 2 describes potential solutions to the parking problems.

Here's an example of a stand-alone transitional sentence for these paragraphs:

> **Fortunately, we can solve these parking problems if we offer our people some incentives to use car pools.** *(transitional sentence)*

If this sentence were at the end of paragraph 1 (option 1), paragraph 2 would start with a topic sentence written something like this:

> **We can offer our personnel three incentives to participate in car pools: preferred parking spaces, guaranteed duty hours and distant parking for nonparticipants.** *(topic sentence)*

If our transitional sentence were at the beginning of paragraph 2 (option 2), then our topic sentence would be the second sentence in paragraph 2, like this:

> **Fortunately, we can solve these parking problems if we offer our people some incentives to use carpools.** *(transitional sentence)* **We can offer them three incentives: preferred parking spaces, guaranteed duty hours, and distant parking for nonparticipants.** *(topic sentence)*

Now let's look at our third option where we merge the transition with the topic sentence of paragraph 2. In this case, we have one sentence instead of two, like this:

> **Fortunately, we can solve these parking problems** *(transitional clause)* **by offering our people three incentives to participate in car pools: preferred parking spaces, guaranteed duty hours and distant parking for nonparticipants** *(topic of paragraph 2)*.

Whether used at the end or beginning of a paragraph, transitional sentences can make your writing smoother and make your reader happier!

HEADINGS

Another effective way to transition from one major area to another, especially in a longer report, is to use *headings*. They allow your reader to follow along easily, even at a glance. Headings are also helpful when topics vary widely. Be informative and avoid relying on headings that use one or two vague words. You'll note that headings are used effectively in this publication! Here are a couple examples.

> For: Procedures
>
> Try: *How to Complete AF Form XXXX*
>
> For: *Contractors*
>
> Try: *How Much Contractors May Charge*

Now that you have a good idea of how to draft "the big picture" part of your communication—your introduction, conclusion and paragraphs in the body—it's time to dig a little deeper. It's time to look at building effective sentences within your paragraphs.

DRAFTING EFFECTIVE SENTENCES

To draft clear and concise sentences, choose clear and concise words and phrases to make up your sentences. In this section, we'll cover some of the most important considerations when writing effective sentences: active voice, smothered verbs, parallel construction, misplaced modifiers, using the right word for the job and avoiding wordy words and phrases. Let's get started with probably the most common pitfall to clear and concise sentences—not writing actively.

WRITE ACTIVELY: DOERS BEFORE VERBS

Is your active voice all bottled up? Active voice shows the subject as the actor. For example: *The girl sang a song.* By using mostly active voice, your writing is clear, concise, and alive—it reaches out to the reader and gets to the point quickly with fewer words. Unfortunately, many writers overuse passive voice. Passive voice shows the subject as receiver of the action. For example: *A song was sung by her.* Besides lengthening and twisting sentences, passive verbs often muddy them. Whereas active sentences must have doers, passive ones are complete without them. When you overuse passive voice and reverse the natural subject-verb-object pattern, your writing becomes lifeless.

> Your support is appreciated ...
> Requisitions should be submitted ...
> The IG team will be appointed ...
> It is requested that you submit ...

Yawn. The actor (or doer) in the sentence is either obscure, absent altogether or just lying there. Who appreciates? Who should requisition? Who appoints? Why not write …

> I appreciate your support …
> Submit your requisitions …
> Colonel Hall will appoint the IG team …
> Please submit …

THE SYMPTOMS OF PASSIVE VOICE AND THREE CURES

How can you diagnose passive voice? You don't have to be a grammarian to recognize passive voice. First, find the verb by asking yourself, "What's happening in this sentence?" Then find the actor by asking, "Who's doing it?" If the actor comes after the verb, its passive voice. Also, watch for these forms of the verb *to be* (*am, is, are, was, were, be, being, been*) *and* a main verb usually ending in *-ed* or *-en*. Let's look at a few examples:

> **Passive:** The mouse *was* eat*en* by the cat.
> **Active:** The *cat ate* the mouse.
>
> **Passive:** Livelier sentences will be written by you.
> **Active:** You will write livelier sentences.
>
> **Passive:** Water is drunk by everybody.
> **Active:** Everybody drinks water.

To correct a passive sentence, try one of these cures:

1. Put the actor (*doer*) before the verb.

> **This:** The *handlers* must have broken the part.
> **Not:** The part must have been broken by the *handlers*.

2. Drop part of the verb.

> **This:** The results *are* in the attachment.
> **Not:** The results *are listed* in the attachment.

3. Change the verb.

> **This:** The replacement has not *arrived* yet.
> **Not:** The replacement has not *been received* yet.

Though most writers overuse passive voice sometimes it's appropriate. Clear and forceful language may be inappropriate in diplomacy or in political negotiations. Passive voice is also used to soften bad news, or when the doer or actor of the action is unknown, unimportant, obvious or better left unnamed. Here are a few examples:

> The part was shipped on 1 June. (The *doer* is unimportant.)
> Presidents are elected every four years. (The *doer* is obvious.)
> Christmas has been scheduled as a workday. (The *doer* is better left unnamed.)

The bottom line: Passive voice is wordy, indirect, unclear and reverses the natural order of English. Active voice is clear and concise. So, activate your writing!

As you can see, using verbs correctly—actively—is key to writing clear, concise and interesting sentences. For that reason, let's look at another way to keep your verbs active—don't smother them!

WATCH OUT FOR SMOTHERED VERBS

Make your verbs do the work for you. Weak writing relies on general verbs that take extra words to complete their meaning. Don't use a general verb (make) plus extra words (a choice) when you can use one specific verb (choose). For example:

Wordy:	The IG team *held a meeting* to *give consideration to* the printing issue.
Better:	The IG team *met to consider* the printing issue.
Wordy:	They *made the decision* to *give their approval*.
Better:	They *decided* to *approve it*.

Here's another tip on verbs—watch out for words ending in *-ion* and *-ment*. These are verbs turned into nouns. Whenever possible, change these nouns to verb forms, and your sentences will be shorter and livelier. For example:

Wordy:	Use that format *for the preparation of* your command history.
Better:	Use that format *to prepare* your command history.
Wordy:	*The settlement of* travel claims involves *the examination of* orders.
Better:	*Settling* travel claims involves *examining* orders.

We've spent a lot of time looking at verbs because they're the most important words in your sentences. **The bottom line:** keep verbs active, lively, specific, concise, and out in front, not hidden. Another potential stumbling block for readers is "unparallelism."

USE PARALLEL CONSTRUCTION (PARALLELISM)

Use a consistent pattern when making a list. If your sentence contains a series of items separated by commas, keep the grammatical construction similar—if two of three items start with a verb, make the third item start with a verb. Violations occur when writers mix things and actions, statements and questions, and active and passive instructions. The trick is to be consistent. Make ideas of equal importance look equal.

Needs work:	The functions of a military staff are to *advise* the commander, *transmit* instructions and *implementation* of decisions. [Advise and transmit are verbs, while implementation is a noun.]
Acceptable:	The functions of a military staff are to *advise* the commander, *transmit* instructions and *implement* decisions. [Parallel ideas are now written in the same grammatical form.]
Needs work:	The security policeman told us *to observe the speed limit* and *we should dim our lights*. [Parallel ideas are not written in the same grammatical form.]
Acceptable:	The security policeman told us *to observe the speed limit* and *to dim our lights*.
Needs work:	Universal military values include that we should act with integrity, dedication to duty, the belief that freedom is worth dying for and service before self.

Acceptable:	Universal military values include commitment to integrity, dedication to duty, service before self, and the belief that freedom is worth dying for.

If one of the items in a list can't be written in the same grammatical structure, place it at the end of the sentence. In the previous example, "the belief that freedom is worth dying for" does not match the three-word construction of the other items, but its placement helps the sentence's readability.

Active voice, strong verbs and parallelism can help make your sentences clear and concise. Now, let's look at some more things you can do to write effective sentences—using the right word for the job.

USE THE RIGHT WORD FOR THE JOB

BE CONCRETE. Without generalizations and abstractions, lots of them, we would drown in detail. We sum up vast amounts of experience when we speak of dedication, programs, hardware and lines of authority. But such abstract language isn't likely to evoke the same experiences in each reader's mind. Lazy writing overuses vague terms such as *immense dedication, enhanced programs, viable hardware* and *responsive lines of authority*. It especially weakens job descriptions and performance evaluations, etc.

Do not write "The commander will give guidance," or "The equipment must meet specs." Your reader might wonder what kind of guidance and what kind of specs? Neither you nor your readers can tackle the problem until you are specific. Be as definite as the situation permits. Include only the ideas your reader needs and then give those ideas no more words than they deserve.

For	Try	For	Try
commanders	MAJCOM commanders	Ford	Ranger
headache	migraine	emotion	love
car, vehicle	Ford	plane	F-117
computer	Pentium	socialize	mingle, meet

KNOW VARIOUS SHADES OF MEANING. Use different words to express various shades of meaning. The writer with an adequate vocabulary writes about the *aroma* of a cigar, the *fragrance* of a flower, the *scent* of perfume or the *odor* of gas instead of the *smell* of all these things.

JUDGE THE JARGON. The aim of all communication is to make a personal contact in the simplest possible way, and the simplest way is to use familiar, everyday words. Above all, it must be adapted to specific circumstances with a minimum of jargon. Jargon consists of "shorthand" words, phrases or abbreviations that are peculiar to a relatively small group of people. *DEROS* (Date Eligible to Return from Overseas) and *AWOL* (Absent Without Leave) are examples of military jargon. Every profession has it. *NPO* which means Nil Peros (nothing by mouth) and contusion (bruise) are examples of medical jargon. Writers often use jargon in their sentences to fill space and impress the naive. Unfortunately, overuse of jargon can backfire on you by actually confusing your reader. **CAUTION!** Before you use jargon, make sure you have carefully assessed the audience! Keep it simple with everyday words and phrases, or at least explain any jargon you must use. If you use an abbreviation, spell it out the first time it appears. If it appears only twice or infrequently, spell out the term every time and avoid the abbreviation entirely. For more on abbreviations see Appendix 1.

CLICHÉS. Clichés are expressions that have lost their impact because they have been overused. Strive for originality in your choice of words and phrases. The list below is not exhaustive. You just may not find your favorite here.

acid test
add insult to injury
armed to the teeth
as a matter of fact
at a loss for words
banker's hours
battle royal
beat a hasty retreat
beauty and the beast
benefit of the doubt
better late than never
bewildering variety
beyond the shadow of a doubt
bite the dust
blazing inferno
blessed event
blessing in disguise
blissful ignorance
brave as a lion
break of day
bright and early
bull in a china shop
burn one's bridges
burn the midnight oil
burning issue
bury the hatchet
busy as a bee
by the same token
calm before the storm
cherished belief
clear the decks
club-welding police
colorful scene
conspicuous by its absence
cool as a cucumber
coveted award
crack of dawn
crack troops
cutting edge
dramatic new move
dread disease
dream come true
drop in the bucket
easier said than done
fame and fortune
feast or famine
fickle fortune
food for thought
from the face of the earth
gentle hint
glaring omission
glutton for punishment

gory details
grief stricken
grim reaper
hammer out (an agreement)
hand in glove
happy couple
hard as a rock
head over heels in love
heart of gold
heavily armed troops
honest as the day is long
hook, line and sinker
hungry as wolves
in short supply
in this day and age
intensive investigation
iron out (problems)
irony of fate
it goes without saying
Lady Luck
lash out
last but not least
last-ditch stand
leaps and bounds
leave no stone unturned
lend a helping hand
light at the end of the tunnel
lightening speed
limp into port
lock, stock and barrel
long arm of coincidence (the law)
man in the street
marvels of science
matrimonial bliss (knot)
meager pension
miraculous escape
moment of truth
more than meets the eye
Mother Nature
move into high gear
never a dull moment
Old Man Winter
on more than one occasion
paint a grim picture
pay the supreme penalty
picture of health
pillar of (the church, society)
pinpoint the cause
police dragnet
pool of blood
posh resort
powder keg

predawn darkness
prestigious law firm
proud heritage
proud parents
pursuit of excellence
quick as a flash
radiant bride
red faces, red-faced
reign supreme
reins of government
round of applause
rushed to the scene
sadder but wiser
scantily clad
scintilla of evidence
scurried to shelter
selling like hotcakes
sharp as a razor
sings like a bird
spearheading the campaign
spirited debate
spotlessly clean
sprawling base, facility
spreading like wildfire
steaming jungle
stick out like a sore thumb
storm of protest
stranger than fiction
supreme sacrifice
surprise move
sweep under the rug
sweet harmony
sweetness and light
tempest in a teapot
tender mercies
terror stricken
tip of the iceberg
to no avail
too numerous to mention
tower of strength
tragic death
trail of death and destruction
true colors
vanish in thin air
walking encyclopedia
wealth of information
wave of the future
whirlwind campaign
wouldn't touch with a 10-
    foot pole

EASILY CONFUSED WORDS. Many writers and speakers frequently confuse the meaning of some words. Even the dictionary isn't clear-cut and can add to your confusion. Here's a small list of some easily confused words. Be on the lookout for others.

accept	verb, receive
except	verb or preposition, omitting or leaving out
advice	noun, counsel given, an opinion
advise	verb, to give counsel or advice
affect	verb, to influence or feign
effect	noun, result; verb, to bring about
aggravate	make worse or intensify
annoy	disturb or irritate
all ready	everyone is prepared
already	adverb, by specific time
all together	collectively or in a group
altogether	wholly or entirely
alright	not acceptable spelling
all right	satisfactory
allusion	indirect reference
delusion	false belief
illusion	a false impression
alumni	men graduates or group of men and women graduates
alumnae	women graduates
among	used when more than two alternatives
between	used when only two alternatives
amount	qty that can't be counted/measured in units
number	quantity counted and measured in units
anxious	worry or fearfulness
eager	keen desire
apt	suitable, quick to learn, natural tendency
liable	legally responsible
likely	refers to the probable, probability
as	a subordinate conjunction
like	a preposition
avocation	hobby
vocation	customary employment
beside	preposition, next to or near
besides	adverb, in addition; preposition, addition to
bi-	occurring every two (units of time)
semi-	occurring twice (during the time period)
bring	action toward the speaker
take	action away from the speaker
can	ability
may	permission
capital	city or money
capitol	a building

censor	examine in order to forbid if objectionable
censure	condemn or to reprimand
compliment	praise
complement	supplies a lack; it completes
compose	to constitute
comprise	to include or consist of
consul	foreign representative
council	a group
counsel	advice, to give advice
contemptible	base, worthless, despicable
contemptuous	expressing contempt or disdain
continually	closely recurrent intervals
continuously	without pause or break
credible	believable
creditable	deserving credit or honor
credulous	ready to believe anything
disinterested	impartial or objective
uninterested	indifferent
eligible	qualified to be chosen
illegible	unable to read
emigrate	to leave a country to settle in another
immigrate	to enter a country to settle there
eminent	noted or renowned
imminent	impending
enervating	weakening
invigorating	stimulating
ensure	guarantee
insure	obtain insurance for
exceptional	out of the ordinary
exceptionable	objectionable
farther	expresses distance
further	expresses degree
fewer	refers to numbers; countable items
less	refers to mass; items can't be counted
formally	in a formal manner
formerly	in the past
hanged	to execute; criminals are hanged
hung	suspended or nailed up; pictures are hung
healthy	possessing health
healthful	conducive to health
wholesome	healthful as applied to food or climate
imply	to hint at or suggest
infer	to draw a conclusion based on evidence

incredible	unbelievable, improbable	practical	useful, sensible
incredulous	skeptical, doubting	practicable	feasible; a person cannot be practicable
ingenious	clever or resourceful	principal	adjective, foremost; noun, main person
ingenuous	innocently frank or candid	principle	noun, precept or idea
instance	example	raise	to lift or cause to be lifted
instant	moment of time	rise	to move to a higher position
incident	event or an occurrence	respectively	in the order given
later	after the usual time	respectfully	full of respect
latter	to designate the second of two things mentioned	set	to put or to place
		sit	to occupy a seat
lay	to place	shape	condition of being
lie	to recline; to stretch out	condition	state, situation
likely	a favorable probability	sometime	at some unspecified time
liable	legally responsible	some time	a period of time
apt	a natural fitness or tendency	sometimes	now and then
lose	a verb	specie	coin
loose	primarily an adjective	species	a kind or variety
luxuriant	abundant growth	stationary	in a fixed place
luxurious	pertains to luxury	stationery	writing paper, envelopes
may be	a modal verb	than	conjunction of comparison
maybe	perhaps	then	adverb, at that time
moneys	currency	their	third person plural pronoun, possessive
monies	amount of money	there	adverb or interjection
morale	refers to a spirit or a mood	they're	contraction of they are
moral	refers to right conduct	verbal	applies to that which is communicated in words, spoken or written
persecute	to afflict or harass	oral	applies only to that which is spoken
prosecute	to pursue until finished or to bring legal action against a defendant	who	refers to people
		which	refers to things

WORDY WORDS AND PHRASES

Many people use certain words and phrases because they think it makes them appear learned or they think padding emphasizes or rounds out a passage. Don't force your reader to trudge through a dictionary. Also, many needless phrases are introduced by prepositions like *at, on, for, in, to* and *by*. They don't give sentences impressive bulk; they weaken them by cluttering the words that carry the meaning. So prune such deadwood as *to the purpose* (to), etc. The longer it takes to say something, the weaker you come across. Pages 81-87 list big words or phrases and simpler ones to try.

DOUBLEHEADERS. *The Word* by Rene J. Cappon details how to avoid writing a project's *importance and significance* when importance will do. Even a person's *success and achievement* is okay with just success. Pairs of words with similar meanings add needless bulk. Whatever the differences are between *test and evaluate*, for example, they aren't worth calling attention to if you just want to give a general idea. When you're tempted to use two words, try one to say it all. Thomas Jefferson said: "The most valuable of all talents is that of never using *two* words when *one* will do."

aid and abet	each and every	ready and willing
beck and call	fair and just	right and proper
betwixt and between	few and far between	safe and sound
bits and pieces	irrelevant and immaterial	shy and withdrawn
blunt and brutal	nervous and distraught	smooth and silky
bound and determined	nook and cranny	success and achievement
clear and simple	null and void	sum and substance
confused and bewildered	part and parcel	test and evaluate
disgraced and dishonored	pick and choose	various and sundry

REPETITIVE REDUNDANCY. Not every noun needs an adjective. Not every adjective needs an adverb. Not every writer has gotten the message. Keep your pencil from adding modifiers to those nouns that need no additional voltage. *Serious* danger, *stern* warning, *deadly* poison, *grave* crisis are examples; the nouns operate better without the modifiers.

absolutely conclusive	entirely absent	old antique
advance planning	erupt violently	opening gambit
agricultural crops	exact counterpart	organic life
anthracite coal	fellow colleague	original founder
ascend upward	few in number	original prototype
assemble together	first beginning	passing fad
awkward dilemma	founder and sink	past history
basic fundamental	free gift	patently obvious
big in size	from whence	personal friend
bisect in two	fuse together	personal opinion
blend together	future plan	pointed barb
both alike	gather together	present incumbent
capitol building	general public	protrude out
chief or leading or main	grateful thanks	real fact
protagonist	habitual custom	recall back
close proximity	hired mercenary	recoil back
coalesce together	hoist up	recur again or repeatedly
collaborate together or jointly	individual person	short in length or height
complete monopoly	invited guest	shuttle back and forth
completely full	irreducible minimum	single unit
completely unanimous	join together	skirt around
congregate together	knots per hour	small in size
connect together	large in size	tall in height
consensus of opinion	lonely hermit	two twins
continue to persist	meaningless gibberish	temporary reprieve
courthouse building	merge together	true facts
current or present incumbent	mutual cooperation	ultimate outcome
descend downward	necessary need	universal panacea
divisive quarrel	new innovation	violent explosion
doctorate degree	new record	visible to the eye
end result	new recruit	vitally necessary
endorse (a check) on the back	old adage	

SIMPLER WORDS AND PHRASES

Instead of	Try
a great deal of	much
a minimum of	at least
a number of	some, many, few
a period of (2 days)	for
abandon	give up
abet	help, assist, aid
abeyance (hold in)	delay, postpone, wait
abridge	shorten, condense
abrogate	do away with, abolish, cancel, revoke
accelerate	speed up, hasten
accept	take, receive
accommodate	make fit, make room for, allow for
accompany	go with
accomplish	carry out, do, complete
accomplish (a form)	fill out, complete, produce, fill in, make out, prepare
according to (an instruction)	per
accordingly	so, then, therefore
accrue	add, gain
accumulate	gather, amass, collect
accurate	correct, exact, right
achieve	do, make
achieve the maximum	get the most from, excel
acquire	get, gain, earn, win
activate	start, drive, put into action, turn on
active consideration (to give)	consider
activities	actions
actual	real, true
actual emergency	emergency
actual facts	facts
actuate	induce, move, drive, impel
additional	added, more, other, further
address	speech, speak of, speak to, deal with (a problem)
adequate	enough, plenty
adjacent to	next to
advanced plans	plans
advantageous	helpful, useful, favorable, beneficial, good
adverse to	against, opposed to
advise	recommend, tell, inform
advised (keep me)	informed (or "inform me")
affirmative (answer in the)	agree, assent to, say "yes"
affix	put, stick, attach, place, add
affix a signature	sign
afford an opportunity	allow, let, permit
after the conclusion of	after
agency	office
aggregate	all, total, sum, combined, whole, entire
aircraft	military plane
all of	all
allegation	charge, claim, assertion
alleviate	ease, relieve, lessen
allotment	share, portion
along the lines of	like, similar to
alter, alteration	change
alternative	choice, option, substitute
amalgamate	merge, combine, unite, mix
ambient	surrounding
ameliorate	better, improve
and/or	and, or (use whichever fits; if both fit, use both)
annually	yearly
antedate	precede
anticipate	expect, foresee
antipathy	dislike, distaste

Instead of	Try
antithesis	opposite, contrast
anxiety	fear
any or	any
apparent	clear, plain, visible
apparently	seemingly, clearly
appear	seem
appellation	name
append	add, attach
applicable	which applies, proper, correct, suitable
application	use (noun)
appreciable	many
appreciate	value
apprise	tell, inform
appropriate	proper, right, apt, suitable, pertinent, relevant (or delete it), fit
approximately	about, nearly, almost
are desirous of	want to
are in receipt of	received
as a matter of fact	in fact
as a means of	to
as a result of	because
as against	against
as and when	as, when (not both)
as at present advised	as advised
as of (this date)	by (today), today
as prescribed by	under, per
as to whether	whether, if
ascertain	find out, learn, make sure
assert	claim, declare
assimilate	absorb, digest, join, include
assist, assistance	aid, help
at a later date	later
at a much greater rate	faster, more quickly
at all times	always
at an early date	soon
at present	now, currently, presently
at such time	when
at the present time	currently, at present, now
at the time of	when
at this juncture (time)	now
at this time	now
at your earliest convenience	as soon as you can
attached herewith is	here's
attached please find	here's, attached is, enclosed is
attain	reach, gain, achieve
attempt	try
attempts to	tries
attention is invited to	note, see
attired	dressed
augment	add, increase, extend, enlarge, expand, raise
authored	wrote
authoritative	valid, official
authority	sanction, control, guidance
authorize	allow, let, permit, empower, prescribe
autonomous	independent
avail yourself of	use
availability	presence, use
based on the fact that	because
be acquainted with	know
be cognizant of	know
be of assistance to	assist, help, aid
befall	happen, occur
behest	request, order
behoove	(avoid this pompous term)
benefit	help

Instead of	Try
bestow	give
betterment	improvement
biannual	twice a year
biennial	once in 2 years
bilateral	two sided
bona fide	real, genuine, sincere
brief (in duration)	short, quick, brief
brook (interference)	allow
burgeoning	increasing, growing
by means of	by, with
by virtue of	because, by, under
came to an end	ended
cannot	can't
capability	ability
capable	able
care should be taken	be careful, take care
category	class, group
characteristic	trait (n), typical of (adj)
characterize	describe, portray
circuitous	roundabout
classify	arrange
close proximity	close, near
cognizant of	aware of, know, understand, comprehend
coincidentally	at the same time
collaboration	(see "cooperation")
colloquy	discussion, talk
combine	join
combined	joint
comes into conflict	conflicts
commence	begin, start
commensurate	equal to, to agree with
commensurate with	corresponding to, equal to, to agree with, according to
communicate verbally	talk, discuss, say, tell
compensate (compensation)	pay
comply (with)	follow, carry out, meet, satisfy
component	part
comprehend	grasp, take in, understand
comprehensive	all-inclusive, thorough
comprise	form, include, make up, contain
comprised of	made up of, consists of
concerning	about, on
conclude	close, end, think, figure, decide
conclusion	end
concur	agree, approve
condition	state, event, facts
conduct (verb)	carry out, manage, direct, lead
confront	face, meet, oppose
conjecture	guess
connection	link, tie
connotation	meaning
consensus of opinion	agreement, verdict, general, view
consequently	so, therefore
consider	look at, think about, regard
considerable (amount)	large, great
consolidate	combine, join, merge
constitutes	is, forms, makes up
construct	build, make
consult	ask
consummate	finish, complete
contained in	in
containing	has, that have, etc.
contains	has
contemporaneously	at the same time
contiguous	next to, near, touching
continue	keep on
contractual agreement	agreement, contract
contribute	give
cooperate	help

Instead of	Try
cooperate together	cooperate
cooperation (in)	jointly, with
coordination	staff action, relate, agree, conform
couched	phrased, worded
course of time	time
criteria	standards, rules, yardsticks
criterion	standard, norm
currently	now (or leave out)
de-emphasize	play down
decelerate	slow down, reduce speed
deem	think, judge, hold, believe
deficiency	defect, shortage, lack
definitely	final
definitize	make definite
delegate authority	empower, assign
delete	cut, drop
delineate	draw, describe, portray, outline
delinquent	late
demeanor	manner, conduct
demise	death
demonstrate	prove, show, explain
depart	leave
depict	describe, show
deprivation	loss
deprive	take away, remove, withhold
derive	receive, take
derogatory	damaging, slighting
descend	go down
designate	appoint, choose, name, pick, assign, select
desire	wish, want
detailed	more, full
deteriorate	run down, grow worse
determination	ruling
determine	decide, figure, find
detrimental	harmful
develop	grow, make, take place
dialogue, dialog	talk, discussion
dichotomy	split, separation
difficult	hard
dimension	size
diminish	drop, lessen, reduce, decrease
disadvantage	drawback, handicap
disallow	reject, deny, refuse
disclose	show, reveal, make known
discontinue	drop, stop, end
disseminate	issue, send out, pass out, spread, announce, get out
distribute	spread, share, allot
divulge	make known, reveal
do not	don't
donate	give
downward adjustment	decrease
due in large measure	because, due to
due to the fact that	because of, hence, since, due to
duplicate	copy
duration	time, period
during such time	while
during the periods when	when
echelon	level, grade, rank
edifice	building
educator	teacher, trainer
effect (verb)	make, cause, bring about
effect an improvement	improve
effectuate	carry out, put into effect
elaborate (on)	expand on, develop
elapsed (time has)	passed
elect	choose, pick
elementary	simple, basic

Instead of	Try
elevated	height, altitude
elicit	draw out, bring out, prompt, cause
eliminate	cut, drop, end, remove, omit, delete
elimination	removal, discarding, omission
elucidate	explain, clarify
emanates	emits, comes from, gives out
emphasize	stress, point out
employ	use
enable	let
encompass in	include, enclose
encounter	meet, find, meeting
encourage	urge, promote, favor, persuade
end product	result, product, outcome
end result	end, result, outcome
endeavor	try, effort, action
enhance	increase, raise, heighten, improve
ensue	follow, result
ensure	make sure, see that
enumerate	count, list
envisage	picture, view, have in mind, regard
equally as	as
equanimity	poise, balance
equitable	fair, just
equivalent	equal
eradicate	wipe out, remove, destroy, erase
erroneous	wrong, mistaken
especially	chiefly
essential	basic, necessary, vital, important
establish	set up, prove, show, make, set, fix
estimate	conclude, appraise, judge
evaluate	check, rate, test, fix the value of, measure, analyze, think about, price
evaluation	rating
eventuate	result
every effort will be made	(I/you/we/they) will try
everybody, everyone	each, all
evidence	fact
evidenced	showed
evident	clear, plain, obvious
evince	show, display, express
evolution	change, growth
exacerbate	make more severe, violent or bitter; to make worse; to aggravate
examination	checkup, test, check, search, questioning
examine	check, look at, test, study, inspect, look into
exceed	go beyond, surpass
exceedingly	notable, extremely, very
excessive	too much, too many
execute	sign, perform, do act
exercise (authority)	use
exhaustive	thorough, complete
exhibit	show, display
exigency	urgent demand, urgent need, emergency
exorbitant	too much, abnormal
expedite	hurry, rush, speed up, fast, quick, hasten
expeditious	fast, quick, prompt, speedy, exercise care, watch out, take care, use care
expend	pay out, spend, use
expendable	normally used up or consumed, replaceable
expenditure	(see "expense")

Instead of	Try
expense	cost, fee, price, loss, charges
experience has indicated	experience shows, learned
experiment	test, try, trial
expertise	expert opinion, skill, knowledge
explain	show, tell
expostulate	demand, discuss, object
extant	existing, current
extend	spread, stretch
extensive	large, wide
extenuating	qualifying, justifying
external	outer
extinguish	quench, put out
fabricate	construct, make, build, invent
facilitate	ease, help along, make easy, further, aid
factor	reason, cause
failed to	didn't
familiarity	knowledge
familiarize	inform, learn, teach
fatuous numskull	jerk
feasible	possible, can be done, workable, practical
females	women
final	last
finalize	complete, finish, conclude, end
firstly	first
foe	enemy
for example	such as
for the purpose of	for, to
for the reason that	because, since
for this reason	so
for your information	(usually not needed)
forfeit	give up, lose
formulate	make, devise, repair
forthcoming	coming, future, approaching
forthwith	at once, right away
fortuitous	by chance, lucky, fortunate
forward	send
fragment	piece, part
frequently	often
fullest possible extent	as much as possible, fully
function	act, role, work
fundamental	basic, main, primary
furnish	give, send, provide, supply
furthermore	besides, also
future date	sometime, later
gained from the following	obtained, learned, source
gainsay	deny, dispute, contradict
generate	produce
germane	relevant, fitting, related
give consideration to	consider
give encouragement to	encourage, urge (see "encourage")
give feedback	respond
give instructions to	instruct, direct
give rise to	raise, cause, bring about
goes without saying	(unnecessary)
govern	rule
habituate	accustom, make use to, adapt, adjust
has the ability	can
has the capability	can
has the capability of	is capable of, can, is able to
have the need for	need
have to	must, need to
held a meeting	met
henceforth	until now
hereby	by this
herein	here (often unnecessary)
heretofore	until now

Instead of	Try
hiatus	gap, lapse
higher degree of	more
hitherto	up to now, until now
hold in abeyance	suspend, delay, wait
homogeneity	unity, agreement
hopefully	I hope
however	but
identical	same
identification	name, designation
identify	find, name, show, point out, recognize
if and when	if, when (not both)
ilk	sort, kind
illustrate	show, make clear
immediately	at once, now, promptly, quickly
imminent	near
impact	affect (verb), effect (noun)
impacted	affected, changed, hit
impediment	block, barrier
imperative	urgent
impetus	drive, power, force
implement	carry out, do, follow, complete, fulfill
implication	impact meaning, effect
important	major, greater, main
impugn	assail, attack, criticize
impulse	drive, push, thrust
in a manner similar to	like, in the same way, as
in a number of cases	some, often, at times
in a position to	can
in a satisfactory manner	satisfactorily
in a situation in which	when
in accordance with	by, under, per, according to
in accordance with the	AFI 37-XXX requires, authority contained in AFI 37-XXX
in addition to	also, besides, too, plus
in an effort to	to, so that, so
in case of	if
in close proximity	near, nearby, close
in compliance with the	as directed, as requested, request
in conjunction with	with, together
in connection with	in, with, on, about
in favor of	for
in its entirety	all of it
in lieu of	instead of, in place of
in order that	for, so, so that
in order to	to
in process of preparation	being prepared
in recent past	lately
in reference to	regarding, about, on, concerning
in regard to	about, concerning, on
in relation to	about, concerning, on
in respect to	regarding, about, concerning, on
in sufficient time	early enough, soon enough, far enough ahead
in the amount of	for, of
in the course of	during, in, when
in the event of	if
in the event that	if, in case
in the immediate future	soon
in the majority of instances	the time (case)
in the matter of	in, on
in the nature of	like
in the near future	soon
in the negative	no, denied, disapproved
in the neighborhood of	about, around
in the time of	during
in the vicinity of	near, around
in view of	since
in view of the above	so, since, therefore

Instead of	Try
in view of the fact that	because, as
in this day and age	today, nowadays
in this instance	here (often necessary)
in-depth	(avoid if possible) thorough, complete
inaccurate	wrong, incorrect
inadvertently	accidentally, mistakenly
inasmuch as	since, because
inaugurate	start, begin, open
inception	start, beginning
incident to	pertaining, connected with
incidental	related, by chance
incombustible	fireproof, (it) will not burn
incorporate	blend, join, merge, include, combine, add
increase	rise, grow, enlarge, add to
increment	increase, gain, amount
incumbent upon	must
indebtedness	debt
indefinite	vague, uncertain
indeterminate	vague, uncertain
indicate	show, write down, call for, point out
indication	sign, evidence
individual (noun)	person, member
individually	each, one at a time, singly
ineffectual	futile, useless, ineffective
inexpensive	cheap, low-priced
infinite	endless
inflammable	(it) burns, flammable, burnable
inherent	basic, natural
inimical	hostile, unfriendly, opposed
initial (adjective)	first
initially	first, at first
initiate	start, begin, act
innate	basic, native, inborn
innuendo	hint
input (provide)	data, thoughts. comment on, advise, respond
insignificant	slight, trivial, unimportant
insofar as	since, for, because
insomuch as	since
instance	case, example
instantaneously	instantly, at once, suddenly
institute (verb)	set up, start
integrate	combine
interface	connect, talk, coordinate, join, work together, merge, joint, point of contact, frontier, junction, common boundary
interpose no objection	don't object
interpose objections to	disapprove, disagree with, do not concur with, object to
interpret	grasp, explain, understand
interrogate	question
investigate	examine, study
irrespective (of the fact that)	regardless
is dependent upon	depends on
is in receipt of	receives, got
is responsible for obtaining	obtains
is responsible for selection	selects
(is) symptomatic of	shows
it is	(leave out)
it is essential	must
it is important to note that	note
it is obvious that	clearly, obviously
it is possible that	may, possibly
it is recommended	I, we recommend
it is requested	please, request
jeopardize	endanger

Instead of	Try
jurisdictional authority	control
justification	grounds, reasons
justify	prove
juxtaposition (in)	alongside, next to
knowledge	experienced, well-trained
legislation	law
limitations	limits
limited number	few
locate	find
location	place, scene, site
magnitude	size, extent
maintain (maintenance)	keep, support (upkeep)
majority	greatest, longest, most
make a decision	decide
make a reply	reply
make a request	request, ask for
make a statement	state
make an adjustment	adjust, resolve
make every effort	try
make provisions for	provide, do
mandatory	must, required
manifest (to be)	clear, plain
manufacture	make, build
materialize	appear, take form
materially	greatly
maximal	highest, greatest
maximize	increase
maximum	most, greatest
meets with approval	is approved
mention	refer to
metamorphosis	change
minimal	least, lowest, smallest
minimize	decrease, lessen, reduce
minimum	least. lowest, small
mitigate	lessen, ease
mode	way, style
modify	change, moderate, qualify
monitor	check, watch, oversee, regulate
multitudinous	populous, large (crowd)
more specifically	for example
most unique	unique
negligible	small, trifling
neophyte	new, novice
nevertheless	however, even so, but
nebulous	vague
necessitate	cause, need, make, require, cause to be
not infrequently	often
not later than	by, before
not often	seldom
not withstanding the fact that	although, nonetheless, nevertheless
notification	announcement, report, warning
notify	let know, tell
numerous	many, most
objective	aim, goal
obligate, obligatory	bind, compel
observe	see
obtain	get
obviate	prevent, remove, rule out
obvious	plain, clear
of great importance	important
of large dimensions	large, big, enormous
of late	lately
of no avail	useless, no use
of the opinion (to be)	to believe, think
often times	often
on account of	because
on behalf of	by, for, representing
on the basis of	based on

Instead of	Try
on the grounds that	because
on the part of	for
operate	run, work
operation	action, performance
operational	working
optimize	improve, strengthen
optimum	greatest, most favorable, best
option	choice, way
opus	work
organization	makeup, work site
orifice	hole, vent, mouth
originate	start, create, begin
outlook	view
outstanding (debt)	unpaid, unresolved
over the signature of	signed by
overlook	view, sight
parameters	limits, factors, boundaries
paramount	superior, supreme, principal, chief, outstanding
partake	share, take part in
participate	take part
particularize	(avoid using) state in detail, specify, itemize
patently	evidently
peculiar to	unusual
penitentiary	prison
per annum	each year, a year
perform	do, act, produce, complete, finish
period of time	period, time
periodic	cyclic, recurring
periphery	confines, limits, perimeter
permit	let
pernicious	deadly, harmful
personnel	people, staff
pertaining to	about, of, on
pertinent	to the point
peruse	read, study
phenomenon	fact, event
pictured	shown, imagined
place	put
plaudits	praise, applause, approval
plethora	(pompous) excess, too much
point in time	time, now, then
point of view	(usually unnecessary)
portend	predict, mean
portent	sign, omen
portion	part, share, lot
position	place
positively	(often unnecessary)
possess	have, own
posterior	end, rear
postpone	put off, delay
postulate (verb)	claim, assert, suggest
posture (on an issue)	view, position, attitude
potential (adjective)	possible
practicable	possible, workable
practically (done)	almost, nearly
precept	order, command, principle, rule of action
precipitate (adjective)	rash, sudden, hasty, abrupt
preclude	prevent, shut out
predicament	fix, dilemma
predicated on	based on
predominant	dominant, main, chief
predominantly	mainly, chiefly, mostly
preeminent	chief, outstanding, foremost, first
preliminary to	before
premier	first, leading
preparatory to	before
prepared	ready

Instead of	Try
preponderantly	mainly, chiefly
presently	now, soon
preserve	keep
prevail upon	persuade
prevalent	widespread
primary	first, chief
prime	best
prior to	before
previous to	before
previously	before
probability	chance, likelihood
problematical	doubtful
procedures	rules, ways
prioritize	(no such word) rank, rank in order
preventative	preventive
previous	earlier, past
proceed	do, go on, try
procure	get, gain
proficiency	skill, ability
profound	deep
programmed	planned
prohibit	prevent, forbid
project (ed) (verb)	planned
promulgate	announce, issue, set forth
proportion	share, part, size, amount
proposal	plan, offer
prototype	first or original, model, pattern
provide	give, say, supply, furnish
provide for	care for
provided that	if
provides guidance for	guides
provisions (of a law)	terms
proximity	nearness, distance
purchase	buy
purport	claim, mean
pursuant to	to comply with, in, under, per, according to
purvey	supply, provide, sell
purview	range, scope
quantify	count, measure, state the amount
rationale	reason
reach a decision	decide
reason for	why
reason is because	because
recapitulate	sum up, summarize, report
recipient	receiver
recommend	propose, suggest, advance
recommendation	advice, thought, counsel, opinion
reduce	cut
referred to as	called, named
reflect	show, say
regarding	about, of, on
regardless	in spite of, no matter
reimbursement	payment, repayment
reiterate	repeat
related with	on, about
relating to	about, on
relative to	on, about, for
relocation	move
remain	stay
remainder	the rest, what remains
remedy	cure
remittance	payment
remove	take away, take off, move
remuneration	pay, payment
render	give, make, report
repeat again	repeat, do again
replete	full, filled
represent	stand for, depict
reproduce, reproduction	copy
request	ask, please

Instead of	Try
require	must, need, call for
requirement	need
requisite	needed
reside	live
retain	keep
return	go back
review	check, go over
rudiments	first steps, basics
salient	main, important
salutary	good, healthy
sans	without
satisfactory	fine, good, good enough
saturate	soak, fill
scant	little, only
scrupulous	careful
scrutinize	study carefully, look into
segment	part
seldom ever	seldom
selection	choice
serves to	acts, helps, works
significance	meaning, point, importance
significant	main, great, major, marked,
signify	mean, show (verb)
similar to	like
sine qua non	essential
situated	placed, located, situation, work assignment, state
small in size	small
so as to	to
solicit	ask for
solitary	lone, single
somewhat	(usually padding)
specifications	terms, details, conditions
specify	list
square in shape	square
state (verb)	say
statutory	legal
still remains	remains
stimulate	stir, arouse
stipend	salary, payment, fee
strict accuracy	accuracy
subordinate (verb)	to lower, subdue
subordinate commands	their commands
subsequent to	after, later, next
submit	offer, give, send
substantial	large, real, strong, much, solid
substantiate	prove, support
substitute (verb)	replace
succor	help, aid
succumb	die, yield
such	similar, like
such as	like, that is
sufficient	enough, ample
subsequently	after, later, then, next
stringent	tight, strict
subject	the, this, your
subject to examination	check, examine, verify
sufficiently in advance	early enough
sum total	sum, total
superfluous	extra, too much, useless
supervise	manage
supposition	belief, thought, idea
surmise	think, guess, suppose
susceptible to	open to, subject to
symptom	sign
synthesis	merging, combining
synthesize	put together, group, assemble
tabulation	table
take action	act
take appropriate measures	please
take necessary action	act

Instead of	Try
take necessary steps	do
technicality	detail, fine point
technique	way, method
tender (verb)	offer, give
tentative	uncertain
terminate	end, stop
terrible disaster	disaster
that	(leave out)
that aforesaid	(usually unnecessary) given or said above
the fact that	(usually unnecessary) that
the following	this, these
the foregoing	these, those, (something) above
the fullest degree possible	fully, as much as possible
(the) provisions of	(leave out)
the question as to whether	whether
the undersigned is desirous of	I want
(the) use of	(leave out)
thence	from there
therapy	treatment
there are	(leave out)
there is	(leave out)
thereafter	after that, afterwards, then
thereby	by that, by it
therefore	so
therein	in (usually unnecessary)
thereof	of, its, their
thereon	on (usually unnecessary)
thereto	to that, to it
thereupon	at once
thirdly	third
this office	us, we
this point in time	now
thither	there
through the use of	by, with
thus	so
thwart	frustrate, block, stop, hinder
time period/frame	time, period, span
timely basis	promptly, fast, quickly
to be aware of	know
to effectively direct	to direct
to the extent that	as far as, so much that
transcend	go beyond
transformation	change
transmit	send
transparent	clear
transpire	happen, occur
transport	carry, move
transverse	crosswise
trauma	shock
true facts	facts
type	(leave out)
ultimate	final, end
ultimately	in the end, finally
under advisement	(avoid) being considered
under separate cover	(usually necessary)
underprivileged	poor, deprived
understand	know
unintentionally	by mistake, mistakenly, accidentally
until such time as	until, when
upgrade	improve
upon	on

Instead of	Try
upward adjustment	raise, increase
usage	use
utilize, utilization	use, employ
validate	confirm
value	cost, worth
variation	change
velocity	speed
vend	sell
verbatim	word for word, exact
veritable	(padding—usually unnecessary)
very	(usually unnecessary)
very far	distant, remote
very hot	torrid, scorching, fiery
very large	enormous, immense, huge, spacious, vast
very last	last
very least	least
very near	adjacent, close
very pretty	gorgeous, beautiful
very quiet	still, silent
very small	tiny, puny
very strong	powerful, potent, forceful
very stupid	dense, moronic, idiotic, stupid
very weak	exhausted, frail, flimsy, inadequate
via	in, on, through, by way of
viable	workable, capable of growing or developing (does not mean: feasible, advisable, workable, achievable, effective or practical)
vicinity of	close, near
vicissitudes	ups and downs, changes, difficulties
vie	compete
virtually	almost
visualize	see, imagine, picture
vitiate	weaken, spoil, impair, debase
voluminous	bulky, large
warrant	call for, permit
whence	from where
whenever	when, each time
whereas	since, while
whereby	by which
wherein	in which, where
wherever	where
wherewithal	means
whether or not	whether, if
will be effected	will be done
will make use of	will use
with a view to	to, for
with due regard for (or to)	for
with reference to	on, about
with regard to	about, on, regarding, concerning
with the exception of	except, except for, but
with the purpose of	to
with the result that	so
within the purview of	under
withstand	stand, resist
witnessed	saw
/	and, or

We've been looking a lot at how to write clear and concise sentences—what to do and what to avoid. Before we leave this section on effective sentence writing, there are two more areas we need to cover that have an impact on readability—sentence length and using questions.

SENTENCE LENGTH

The purpose of words on paper is to transfer thoughts in the simplest manner with the greatest clarity. You should avoid long, complicated sentences over 20 words (average is 17 words). Break up long, stuffy sentences by making short sentences of dependent clauses or by using lists. Short sentences increase the pace; long ones usually retard it. The key is to vary your pattern since constant use of either form can be monotonous.

ASK MORE QUESTIONS

Use questions now and then to call attention to what you want. You're actually reaching out to your reader when a sentence ends with a question mark. In a longer communication, a question can definitely be a welcome change. Can you hear how spoken a question is?

Well, that's it—the general guidance for writing your first draft. You now have some great guidance on using the three part structure (introduction, body and conclusion) and writing effective paragraphs and sentences. As promised earlier in this chapter, under Drafting Basics, we briefly mentioned "writer's block." We didn't give you any solutions, but we're going to give you some now. When you sit down to start on your first draft, don't waste time staring at a blank screen or paper—try some of the tips on page 89.

ADVICE ON OVERCOMING WRITER'S BLOCK

If you occasionally suffer from writer's block, you're not alone—even experienced writers have a hard time getting started. Before we get to some cures, exactly what is writer's block? It's a temporary inability to get words on paper (or on the computer). Like many other problems, it has a life cycle—denial, despair, acceptance and recovery. What leads to writer's block, anyway? There are five fears that can cause it: fear of failure, fear of rejection, fear of success, fear of offending and fear of running dry (out of ideas). So now that we know what writer's block is and what causes it, what can we do about it?

In most cases we just need a gentle nudge to get us back on track. In her book ***The Complete Idiot's Guide to Creative Writing***, Laurie E. Rozakis, PhD, provides several suggestions on how to overcome writer's block. Here are some of her ideas, as well as some of our own:

◆ Brainstorm or "free write" to get your creative juices flowing. Just get the words down as fast as they come, preferably on the computer so they will be easy to edit. Spill your brains, don't worry about punctuation—just get it down. Stick pretty close to your outline. Don't revise. Don't polish. If your outline is comprehensive, you may only need to string the ideas together with brief transitions. If your outline is a series of key words in a logical pattern, you'll have to fill in the larger blanks.

◆ Start wherever you want. Don't feel you need to start with the introduction; some writers do that section last. The key here is to just start writing. Try starting with the part that's easiest to write.

◆ On a similar note, try writing just the topic sentences for each paragraph. Once you do this, the other, support sentences will start really flowing.

◆ Avoid procrastination. Waiting until the last minute just increases your "blockage"!

◆ If page length, word count, or some other constraint is holding you back, forget about it on the first draft. You can reshape later, once you have something to revise.

◆ Tell your ideas to a friend.

◆ Briefly do some mindless activity—but only briefly!

◆ Try changing your writing mode—if using the computer, try writing longhand and visa versa.

◆ Use visuals, like pictures or diagrams, to show what you mean. This can help ignite your ideas and thoughts. Then, you can write them.

◆ Develop rituals or routines to get in the mood for writing—a cup of coffee, an early arrival at the office, etc.

◆ If you work in a crowded or noisy office, try using earplugs to cut down on noise and distractions. It may sound strange, but it really works!

SUMMARY

Writer's block is very common and usually very temporary and curable. There are lots of ways to overcome it. Hopefully, the tips here will help you. Always remember—writing should be fun, not frightening!

Congratulations! The most difficult task is over—you've successfully written the dreaded first draft. Take a break and step back from your draft. When you come back, you'll be ready to revise and edit it.

CHAPTER 8

STEP 6:
EDITING
YOUR DRAFT

This chapter covers:

- Editing fundamentals: our goals and some advice on how to get there.

- Edit in multiple steps—always start with the big picture.

- Common grammatical problems to check for when editing.

Spotting problems in our own writing is not easy. Many of us take great pride in what we write. Once our words are on paper, we resent the suggestion that something could be wrong. We don't like to check and change the words, the organization, the limits of the subject, the spelling, the punctuation or anything else, and we often have trouble making time to edit properly.

Yet editing is critical. Take the time to make sure you have a cohesive, clear, error-free product that the audience can relate to. Here's the good news: if you completed the steps described in Chapters 3-7, the editing will be a lot easier... and at this point, you're almost home free.

"... When you revise from the top down, from global structure to paragraphs to sentences to words, you are more likely to discover useful revisions than if you start at the bottom with words and sentences and work up."

— The Craft of Research

A NOTE ON EDITING (STEP 6) VERSUS FEEDBACK (STEP 7)

In the "Seven Steps for Effective Communication," we recommend that you edit your own writing before asking for feedback from someone else. There are many reasons to do this. For one, it develops your own editing skills—you'll be better prepared for those times when you don't have access to a second opinion. Second, it shows respect for the people you're seeking feedback from. Why should someone else invest time and effort to improve your writing if you aren't willing to do so yourself? Finally, you'll catch the worst mistakes and avoid embarrassing yourself in front of your coworkers. It never hurts to put the most professional product you can out for review, even if the review is an informal one. In this chapter, we'll assume it's just you and your draft. In Chapter 9, we'll talk more about seeking feedback from others.

> **FOCUS Principles**
> **Strong Writing and Speaking:**
>
> **Focused**
> Address the issue, the whole issue, and nothing but the issue.
>
> **Organized**
> Systematically present your information and ideas.
>
> **Clear**
> Communicate with clarity and make each word count.
>
> **Understanding**
> Understand your audience and its expectations.
>
> **Supported**
> Use logic and support to make your point.

WHAT'S OUR EDITING GOAL?

This is the easy part. Remember the FOCUS principles from Chapter 1? Good editing all relates to those principles and will tell you how well you followed the steps for effective writing. As you read through this chapter on editing, keep FOCUS in mind. Here are the principles again for your review:

EDITING FUNDAMENTALS

When you edit, there are a few key rules to remember:

1. EDIT WITH FRESH EYES. Give yourself some time between drafting and editing. By that we mean put the draft on a shelf, in a desk drawer or under a paperweight and let it sit a spell, preferably for several hours for shorter projects and at least a day for longer ones. After this down time you'll come back fresh and will be more likely to catch errors.

2. REVIEW THE BASICS. Take the time to review earlier sections on writing tone (pages 23-24), drafting clear and concise sentences (pages 73-88), common grammatical errors (pages 97-101), and any other material that represents a problem area for you. Editing is your last chance to apply the guidelines you've read about in earlier chapters. If the concepts are fresh in your mind when you start editing, you'll be better able to spot problems in your draft.

3. SLOW DOWN AND TAKE YOUR TIME. You aren't in a race. If you read at your normal pace you're more likely to miss errors. Try different approaches to slow yourself down, including reading aloud and reading one line at a time using a "cover" to hide the rest of the page. If you're checking for misspelled words, move backwards through a sentence.

4. REMEMBER YOUR READERS. Try to put yourself in the role of your audience as you edit. You may catch some areas that may need revision if you read it from their perspective and knowledge base. Also consider your secondary audience—even if you've got your primary audience targeted correctly, are you unnecessarily insulting others that may end up reading this?

5. START WITH THE BIG PICTURE, *then* work down to the details. When you begin to edit, don't focus in on the nitty-gritty—*look at the big picture first*. Misuse of "there" and "their" is really not that important if your paper lacks cohesion, is poorly organized or fails to include a clear purpose statement. Again, anyone can use spellchecker, but a well-edited paper requires much more and begins with *The Big Picture.*

EDITING EFFICIENTLY ... A THREE STEP APPROACH

One way to make sure you edit efficiently is to read your document *at least three times* to allow yourself to really look hard at the problem areas that could botch your product. In the first pass, look at the big picture; in the second pass, look at paragraph construction; and the third pass, look at sentences, phrases and words.

FIRST PASS: THE BIG PICTURE

"If it needs major surgery, it's best to know early!"

In this first "go around" you should be paying attention to the arrangement and flow of ideas. Here are some areas to think about:

Check your tasking and purpose.

- What was my original tasker? Check the wording one more time.

- What is my purpose statement? For short assignments, underline it in your draft. For longer assignments, write it down on a separate sheet of paper and refer to it throughout the editing process.

- Does the purpose statement "answer the tasker," or does it miss the point?

Check your introduction.

- Does it exist and does it contain my purpose statement?

- Is it an appropriate length? (typically one paragraph long for assignments)

- Does my purpose statement and introduction give the readers a good idea of what they are about to read?

> **Review:**
> **Elements of an Introduction**
>
> In Chapter 7 (pages 66-67) we described how an introduction often begins with optional **stage-setting remarks** that grab the reader's attention. The introduction should include your **purpose statement**, which informs the reader where you are going and why you are going there (readers love this!). The introduction often contains an **overview** of the main point(s) covered in the body.
> These are just guidelines: the composition of an introduction should be tailored to the assignment.

Compare your introduction and conclusion.

- First read your introduction and then read your conclusion.

- Do they sound like they go together without being identical? Does the introduction declare your purpose and does your conclusion show your readers you've accomplished your purpose?

- Do you let your readers down gradually? Or do you stop with a jerk?

- Does the conclusion sum up your point? Don't introduce any new ideas here—you'll leave your readers hanging in limbo!

Check overall page count and length.

- What are my audience's expectations regarding page count? Am I on target? Will I have to make this draft significantly longer or shorter?

- Check the scope and flow of paragraphs in your body.

Check for relevance and completeness.

- Do the paragraphs clearly relate to the thesis statement?

- Are some paragraphs irrelevant or unnecessary?

- Am I missing any main points in this assignment?

- Are paragraphs arranged in a consistent order?

- How does your draft compare with your original outline?

> **Organizational Editing Check**
>
> Some writers can write powerful and clear sentences but have trouble keeping "on target" in their communication. Their main editing challenge isn't grammar; it's the big picture. If this sounds like you or someone you know, try this simple editing check.
>
> *Note: This editing check assumes you followed the paragraph construction guidelines from the previous chapter and placed the topic sentence at the beginning of each paragraph.*
>
> Read the following sections out loud:
>
> - Your complete introduction.
>
> - The first sentence of each paragraph in the body, in order of appearance.
>
> - Your complete conclusion.
>
> Does it answer the question? Does it stay on message? Does it flow well? If so, congratulations! It looks like you've got the big picture in place ... now you need to check your paragraph construction.

SECOND PASS: PARAGRAPH STRUCTURE AND CLARITY

After your first pass, you know the paper contains what it needs to do the job. In the second pass, you will check whether the main points and supporting ideas are appropriately organized in paragraphs.

Let's take a close look at individual paragraphs in the body of your writing. For each paragraph, ask the following questions:

Unity of Focus

- Is there one, and only one, main point of the paragraph?

- Is all the information in the paragraph related enough to be in the same paragraph?

- Can you identify the central idea of each paragraph?

Topic Sentence

- Does the paragraph have a topic sentence—one sentence that captures the central idea of the paragraph?

- Is the topic sentence the first sentence of the paragraph? (Or, if you're starting with a transitional sentence, the second sentence?)

Supporting Ideas

- Do sentences expand, clarify, illustrate and explain points mentioned or suggested in each main idea? Your goal is to lead the reader in a smooth, step-by-step process to each main idea.

- Are there enough details in the paragraph to support the central idea?

- Are there any "extra sentences" that seem to be irrelevant to the main point?

- Do all transitional words, phrases, and clauses improve the flow and show proper relationships?

- Do most paragraphs contain three to seven sentences?

If you did a lot of rearranging of paragraphs in this step, try the organizational editing check on the previous page—just to make sure you're on track.

THIRD PASS: SENTENCES, PHRASES AND WORDS

Now you're ready to look at the details. Though you've probably corrected some minor errors in the first two passes, now is the time to really concentrate on the "small stuff" that can sabotage your communication: passive voice, unclear language, excessive wordiness, grammatical errors and spelling mistakes. Some of these concepts were covered in the chapter on drafting; while others will be introduced in this section.

Let's start with some general advice. **Read the paper *out loud*.** Reading the paper out loud will increase your chances of catching errors because it requires you to slow down and use two senses—seeing and hearing. What one sense misses, the other may pick up!

Listen to the sound of words, phrases and sentences. Remember, the quicker your audience can read and understand it, the better. If you have to read a sentence two or three times, chances are they will too. Not good!

DRAFTING BASICS: DID YOU APPLY THEM?

As part of your editing step, you need to check some of the same concepts discussed in the previous chapter on drafting. Remember these guidelines, and refer back to the referenced pages if you need more details or additional examples.

1. Write in active voice (pages 73-74). In active voice, the subject comes first in the sentence. In passive voice, the "doer" or subject of the sentence shows up late in the sentence or is missing entirely. Avoid overusing passive voice; it often creates lengthy and confusing sentences.

> **Passive:** Captain Smith was given a choice assignment by his career field monitor.

> **Active:** The career field monitor gave Captain Smith a choice assignment.

2. Avoid smothered verbs (page 75). Use one specific verb instead of a general verb and several extra words.

> **Smothered:** This directive *is applicable* to everyone who *makes use of* the system.

> **Better:** This directive *applies* to everyone who *uses* the system.

3. Check for misspelled or commonly misused words (pages 78-79). In today's computer age, your software's spell checker is your first line of defense against misspelled words. Still, you can get into trouble by misusing synonyms or easily confused words like "there" and "their" and "accept" and "except." Spell check will not flag these words because they are spelled correctly, but used in the wrong context. When in doubt, check the dictionary—it still exists!

4. Use parallel construction (parallelism) in lists and series (pages 75-76). Use a similar grammatical construction within a list or series. Make items of equal importance look equal. If one starts with a verb, start the other with a verb. If three items in a list are commands, make the fourth a command. Parallelism helps make sentences clear.

> **Needs work:** Remember the following when editing: *write in active voice, parallelism, smothered verbs should be avoided*, and *spelling.*

> **Better:** Remember the following when editing: *write in active voice, use parallel construction in lists, avoid smothered verbs*, and *check for misspelled words.*

5. Avoid unnecessary redundancy and word doublings (page 80).

Don't use one word to modify another unless both add value.

> **Needs work:** *Repetitive **redundancy** hurts readability.*

> **Better:** *Redundancy hurts readability.*

Don't use two nearly identical words unless both add value.

> **Needs work:** We must comply with the *standards and criteria* for *controlling and reducing* environmental pollution."

> **Better:** We must comply with the *standards* for *reducing* environmental pollution."

COMMON GRAMMAR TRAPS...

Grammatical errors can confuse your readers and undermine the credibility of your communication. We've listed some of the most common mistakes below—look out for them when editing your work.

1. MISPLACED MODIFIERS

A modifier is a group of words that describes another group of words within the sentence. Modifiers should be placed near the words they describe. Improperly placed modifiers can create ambiguity or imply an illogical relationship. There are two kinds of misplaced modifiers: dangling and ambiguous.

a. Dangling modifiers literally hang illogically on sentences, usually at the beginning. They are placed so they seem to modify the wrong word and, thus, show an illogical relationship. To correct a dangling modifier, either place the modifier next to the word it modifies or change the subject of the sentence to clarify your intent.

Confusing:	Approaching the flight line from the east side, the operations building can be easily spotted by a pilot. [The operations building doesn't approach the flight line—the pilot does!]
Better:	A pilot approaching the flight line from the east side can easily spot the operations building.
Confusing:	To make a climbing turn, the throttle is opened wider. [The throttle doesn't make a climbing turn.]
Better:	To make a climbing turn, open the throttle wider. [The subject you is understood.]

b. Ambiguous modifiers seem to modify two different parts of a sentence. Readers can't tell whether they modify words that come before or after them. To correct an ambiguous modifier, place the word so its relationship can't be misinterpreted.

Confusing:	People who drive cars to work *occasionally* can expect to find a parking space.
Better:	People who *occasionally* drive cars to work can expect to find a parking space.
Confusing:	Although working conditions improved *slowly* employees grew dissatisfied.
Better:	Although working conditions *slowly* improved, employees grew dissatisfied. [Case #1: the conditions improved slowly]
Better:	Although working conditions improved, employees *slowly* grew dissatisfied. [Case #2: employee morale dropped slowly]

2. ERRORS IN SUBJECT-VERB AGREEMENT

Plural subjects take plural verbs and singular subjects take singular verbs. Another way to state this rule using grammatical terms is "Subjects and verbs must agree in number."

The key to avoiding most problems in subject-verb agreement is to identify the subject of a sentence, determine whether it's singular or plural, and then choose a verb in the same number and keep it near its subject.

Generally subjects that end in *s* are plural, while verbs that end in *s* are singular. (There are exceptions to this rule—for example, the word *ballistics* is singular.)

a. **Phrases between the subject and verb** do not change the requirement that the verb must agree in number with its subject.

> An inspection *team* consisting of 36 people *is* investigating that problem.

> A *general*, accompanied by 3 colonels and 15 majors, *is* attending the conference.

b. A **linking verb** agrees with its subject, not with its complement.

> The commander's main *problem is* untrained Airmen.

> Untrained *airmen are* the commander's main problem.

c. A **compound subject** consists of two or more nouns or pronouns joined by one of these conjunctions: *and, but, or, for* or *nor*. Some compound subjects are plural; others are not. Here are guidelines for subject-verb agreement when dealing with compound subjects:

- **If two nouns are joined by *and*, they typically take a plural verb.**

  > The *Air Force* and the *Army are* two components of the nation's defense forces.

- **If two nouns are joined by *or*, *nor*, or *but*, the verb should agree in number with the subject nearest it.**

  > Either the *President* or his *cabinet members are* planning to attend.

  > Either the *cabinet members* or the *President* is planning to attend.

- **Use a singular verb for a compound subject that is preceded by *each* or *every*.**

  > *Every* fighter pilot and his aircraft *is* ready for the mission.

  > *Each* boy and girl *brings* a snack to school.

- **Use a singular verb for a compound subject whose parts are considered a single unit.**

  > The *Stars and Stripes* was flown at half-mast at the Headquarters building.

  > *Ham* and *eggs is* a delicious breakfast.

d. Use a singular verb with **collective nouns** (and noun phrases showing quantity) **treated as a unit**, but a plural verb when treated as individuals.

> The *thousand wounded is* expected to arrive soon. [A quantity or unit]

> A *thousand wounded were* evacuated by air. [Individuals]

e. Use singular verbs with **most indefinite pronouns**: *another, anybody, anything, each, everyone, everybody, everything, neither, nobody, nothing, one, no one, someone, somebody* and *something.*

> *Everyone* eats at the cafeteria.
>
> The *President* said *everybody was* welcome to join.
>
> *Everyone* in the squadron *takes* a turn leading a service project.

f. With **all, any, none** and *some*, use a singular or plural verb, depending on the content.

> *All* of the money *is* reserved for emergencies. [singular-equivalent to "The money is reserved for emergencies."]
>
> When the men arrive, all *go* straight to work. [plural—equivalent to "The men go straight to work"]
>
> *All are* expected to have a tour of duty overseas.

3. ERRORS IN PRONOUN REFERENCE ("Pronoun-Antecedent Agreement")

A common error in pronoun use involves agreement in number. If the noun is singular, the pronoun is singular. If the noun is plural, the pronoun should be plural, too.

Let's look at an example of an incorrect pronoun reference:

> **Incorrect:** *Everyone* should bring *their* books to class. [*Everyone* is singular, while *their* is plural.]

When correcting such a sentence, **try for gender inclusive language**. Often the best approach is to make the subject plural and keep the pronoun unchanged:

> **Correct:** *All students* should bring *their* books to class.

Grammar Review: Pronouns

Pronouns are words that replace nouns and refer to a specific noun. The noun being referred to or replaced by the pronoun is called the **antecedent**. Some examples:

SSgt Smith is our nominee for the award and he has a good chance of winning.

[*SSgt Smith* is the antecedent; *he* is a pronoun replacing the noun later in the sentence.]

Three lieutenants arrived late to the meeting. Their boss was not happy with them.

[*Three lieutenants* is the antecedent; *Their* and *them* are pronouns replacing the antecedent in the next sentence.]

Of course, using *his or her* is also acceptable, but it gets cumbersome when overused:

> **Also correct:** *Everyone* should bring *his or her* books to class.

With a compound subject joined by *and*, use a plural pronoun:

> My *advisor and I* can't coordinate *our* schedules. [*our* is a plural pronoun]

When parts of an antecedent are joined by "or" or "nor," make the pronoun agree with the nearest part:

> John *or* Steve should have raised *his* hand.
>
> Neither the student *nor* his roommates will have *their* deposit returned.

Avoid awkward phrasing by placing the plural part second if one part of the antecedent is singular and one part is plural.

> **Awkward:** Neither my parents *nor* my sister has stayed on *her* diet.

> **Better:** Neither my sister *nor* my parents have stayed on *their* diet.

Remember that embedded descriptive phrases can be tricky:

> **Incorrect:** He is one of those ambitious ***people*** who ***values*** promotion over personal ethics. [***Values*** should be ***value*** because the pronoun ***who*** refers to ***people,*** not ***one.*** Clarification: he is one, but not the only one, of many ambitious people.]

Here are some other examples of incorrect pronoun reference:

- The Air Force maintains different *types* of numbered forces, but the organization of *its* headquarters is similar. [*Its* should be *their* to refer correctly to *types*.]

- The *committee* plans to submit *their* report by the end of the month. [*Their* should be *its* because *committee* functions as a single unit in this sentence.]

4. COMMA PLACEMENT AROUND PARENTHETICAL EXPRESSIONS

There are many rules about using commas to punctuate sentences, and we recommend you check out Appendix 1 for the complete list. One class of common mistakes is nearly universal and worth covering in this chapter—placement of commas around groups of words that interrupt the sentence's flow. Here's the basic rule:

Enclose nonrestrictive or parenthetical expressions with commas.

What does this mean? If an expression (a group of words) can be removed from the sentence without changing its meaning, then enclose the expression with commas.

Though the rule is simple, applying the rule requires some judgment. Advocates of open punctuation would argue that if the group of words does not "significantly" interrupt the sentence, you don't need commas. (See pages 274-275 for more about the open and closed punctuation debate). Another judgment area is deciding which expressions are restrictive and which are nonrestrictive. A *restrictive expression* limits or restricts the meaning of the words it applies to, so it can't be removed from the sentence without changing the meaning. A *nonrestrictive* expression merely adds additional information. **Here's the bottom line:** If you can remove the expression from the sentence without changing the meaning, it is a nonrestrictive expression that should be enclosed by commas.

Which punctuation is correct?

1. People who live in glass houses shouldn't throw stones.
2. People, who live in glass houses, shouldn't throw stones.

Answer #1 is correct. The expression *who live in glass houses* is restrictive. If you eliminate it, the sentence changes meaning: *People shouldn't throw stones.*

Which punctuation is correct?

1. My friend the architect who lives in a glass house has a party every year.

2. My friend the architect, who lives in a glass house, has a party every year.

The correct answer depends on your situation. If you are in the Air Force and have one friend who is an architect, then answer #2 is correct. The expression *who lives in a glass house* is nonrestrictive—it provides information that is not essential to the sentence's meaning, and it can be removed without impact. On the other hand, if you work in an architecture firm and all your friends are architects, then answer #1 is correct. In this case the expression identifies which one of your architect friends has a party every year—it's the one who lives in a glass house.

Though there's some judgment involved in deciding if something is nonrestrictive, **once you decide to enclose an expression, don't forget one of the commas.**

Incorrect:	The new faculty, including the civilians must show up at 0600 tomorrow for physical training.
Correct:	The new faculty, including the civilians, must show up at 0600 tomorrow for physical training.
Incorrect:	Grammar errors including misplaced commas, inhibit writing clarity.
Correct:	Grammar errors, including misplaced commas, inhibit writing clarity.

PROOFREADING MARKS AND ABBREVIATIONS

Is there a simpler way to make a point? Did I use acceptable grammar? Are all ideas clearly stated? Have I applied standard practices in sentence construction and mechanics? You have your work cut out for you! Here are some proofreading marks and abbreviations that will help you to edit your own or someone else's written work.

Mark	Description	Mark	Description
℘	Delete letter	└	Move to left
℘	Delete or delete/change	┘	Move to right
℘	Delete/close up	eq#	Equalize space (margin notation)
┼┼┼┼	Delete underscore	⌐	Move up
. . . .	Retain deleted material (text symbol)	└┘	Move down
stet	Retain deleted material (margin symbol)	◯	Move as indicated
#	New paragraph	⑤	Ident 5 spaces
no#	No paragraph	∧ or ∨	Insert
⌒	Bring together	∨	apostrophe
#	Separate	✶	asterisk
∿	Transpose	[ ]	brackets
≡	Capital (text symbol)	⊙	colon
caps	Capitals (margin symbol)	$\frac{!}{N}$ or $\frac{!}{M}$	dash
lc or /	Lowercase	!	exclamation mark
ital or ___	Italic type	-	hyphen
rom	Roman type	( )	parentheses
‿	Boldface (text symbol)	⊙	period
bf	Boldface (margin symbol)	∨	quotation mark
⊔	Center vertically	⌃	semicolon
⊐⊏	Center horizontally	#	space
‖	Align vertically	/	virgule
⹀	Align horizontally	∨	Superscript (raise above line)
◯	Spell out; abbreviate; change word to number, change number to word	∧	Subscript (drop below line)

ABBREVIATIONS:

agree	Pronoun and antecedent or subject and verb do not agree
amb	Ambiguous meaning
awk	Awkward construction
Cap	Faulty capitalization
clear?	Meaning unclear
CS	Comma splice
dead	Deadhead word; eliminate it
dng	Dangling modifier
EX	Examples: the writer needs to provide examples
frag	Sentence fragment
gr	Faulty grammar
imp	Too impersonal; needs personal pronouns
jarg	Jargon
lc	Use lower case (not capital) letter
mod	The writer has a misplaced or dangling modifier
PL	Construction is not parallel (symbol // can also be used)
pass	Passive voice; should be active
point	Doesn't get to the point
P	Punctuation faulty or needed
ref?	Indefinite reference; What does this pronoun refer to?
Rep	Repetitious
RO	Run-on sentence
sv	Smothered verb
source	Source of this data unclear or needed
sp	Incorrect spelling
tense	Change tense of verb
TS	Problem with topic sentence or one does not exist
trans	Transition needed for coherence
trite	Word or expression overworked, monotonous
wc	Poor word choice; use simpler word
wordy	Should be shortened

SUMMARY

Always edit! Editing is crucial to producing professional communication. Without solid editing your writing can be disjointed, your reader becomes confused, and your message may be lost. Does it take time? Absolutely! Budget time for editing—especially for time-critical assignments—and with practice the whole process will seem second nature.

Editing isn't the final step, however. Yes, someone else needs to look at your work of art. Get ready to put on your thick skin, as this is not for the meek and timid. Read on to the final step to better communication ... how to *fight for feedback.*

CHAPTER 9

FIGHTING FOR FEEDBACK AND GETTING APPROVAL

This chapter covers:

- Informal feedback—how to get it, how to give it.
- Staff coordination—the players.
- Staff coordination—the process.

Fighting for feedback and getting approval for your communication are both activities that are part of life in the Air Force. When you fight for feedback, you voluntarily seek out someone else's views on your speaking or writing product. Feedback can be very informal and it doesn't have to be from people with impressive job titles. When you get staff approval, you're going through a more formal process that allows individuals to review and comment on your communication product. Feedback and coordination are closely linked: if you do a good job at fighting for feedback, you'll find that the coordination process becomes much smoother.

FIGHTING FOR FEEDBACK

WHY FIGHT FOR FEEDBACK?

So, why should you fight for feedback? Perhaps the biggest benefit is getting a second set of eyes to review your work. Even the best writers and speakers can become so close to their projects that they can't see where they can be made stronger. They may omit vital information, fail to see a weakness in their argument or just overlook the need for a transition between two main points. Their closeness to the material and pride of authorship can distort or obscure their viewpoint. Smart communicators realize this tendency and seek objective feedback from a fresh set of eyes. If you seek out and listen to feedback you are much more likely to produce accurate, understandable communication that "answers the mail" and resonates with your audience.

Another reason to fight for feedback is that it often saves time during the coordination process. Whether it's the staff package you've been working hard on for 3 weeks or the briefing you have to present Friday, getting feedback from someone else's point of view will help smooth things out later as your package makes its way up the chain of command. In some cases having someone review and provide feedback can also be a smart political move—if individuals "buy in" early in the process, they may be a source of support later if it becomes necessary to defend the material.

WHERE CAN I GET FEEDBACK?

Meaningful feedback can come from many different sources. Coworkers or fellow staffers may be a good choice because of their familiarity with the issue and its jargon. They may have also had to brief or write for the same people on similar issues; if so, they can give you some tips or lessons learned. You may also choose to go to different people for different aspects of your work. For example, you may find it helpful to solicit suggestions for improving grammar, organization and content from three different "trusted agents" who are strong in those particular areas. You may even want to use a trusted agent who's totally outside your organization to see if your message makes sense to someone with no clue about the material.

WHAT KIND OF FEEDBACK SHOULD I ASK FOR?

Once you've picked out your feedback sources, you should let them know what kind of feedback you're looking for. (We're not suggesting you say, "Tell me how wonderful this is—I've worked so hard on it!") Unless you give them clear guidance, reviewers may focus on details like spelling, grammar and margins. Though these are important, make it clear that you want feedback on the big picture, too. Here are some examples of items you should ask them to address:

Is my purpose clear and am I properly targeting my audience? For starters, you want to give your reviewers a sense of your audience and your purpose. Will the audience positively receive the message you intend to convey? Ask them to tell you the bottom line they walked away with after reviewing your material. Was it what you intended?

Did I address the issue at the right level of detail? Too many details can obscure your message while too few details can lead to confusion, questions and delays. Ask your reviewers if you've addressed the issue(s) without going too far into the weeds. You could also have them ask you questions on the material. Have you anticipated possible questions? If you don't feel comfortable answering their questions, you may need to go back and do more research. Along the same lines, reviewers may help you pinpoint inconsistencies or unclear material that your final audience might find as well.

Are there other viewpoints I need to consider? Finally, your reviewer may offer differing viewpoints on the material. If that's the case, ask for clarification on their viewpoint if necessary, but don't argue with them. Instead, ask yourself if their ideas may come up again later. If so, you probably need to address them in your material.

The bottom line to getting feedback is having an open mind and being able to accept criticism. Don't take comments personally, even though they are attacking your baby. Accept feedback willingly and use it constructively—it's part of the process of developing a quality product.

HOW TO GIVE FEEDBACK

There are certain things to keep in mind when giving feedback. First, effective feedback is consistent, objective and sensitive to the stated purpose. If someone asks you to review a package, make sure you understand what the person wants from your review and stick to it. Second, distinguish between necessary, desirable, and unnecessary changes. A page full of red marks is hard to interpret. Instead, give the author a sense of what really needs to be changed versus the "happy to glad" kinds of suggestions. Next, helpful suggestions pinpoint specific problems, such as awkward sentences, grammar, etc. A general statement like "you need to work on your sentence structure" isn't as helpful as underlining specific sentences that need help. Finally, you should concentrate on improving the message's content, not the style or personal preferences of the author (unless the author has asked you specifically to comment on writing style).

FEEDBACK PHILOSOPHY

Feedback should describe rather than judge. Authors are more likely to listen and incorporate feedback if it's phrased constructively. Avoid judgmental language—it places people on the defensive. Remember, feedback should be directed at a person's work or behavior, not at the person.

Feedback is both positive and negative. A balanced description of other people's work takes both strong and weak points into account. Both types of feedback give useful information to those people who want to change and improve their work.

Feedback should be specific rather than general. General statements about other people's work do not indicate the performance elements they may need to change and the elements that may serve as models. Highlight or underline specific items you want to bring to the author's attention, and make annotations or comments in the margins. See page 102 for commonly used proofreading marks.

Feedback should consider the needs of both the receiver and the giver of the feedback. Feedback often reflects the state of mind of the reviewer, not just the quality of the work. If you're seeking feedback from someone else, try to pick an appropriate time to make the request and be realistic about the time required for the review. Similarly, reviewers should make sure they are in the right frame of mind before analyzing the material and offering feedback.

Feedback should be directed at behavior the receiver can control. Only frustration results when people are reminded of shortcomings they cannot control. A suggestion to improve the briefing room's temperature, for example, is probably beyond the individual's control. However, briefing skills and mannerisms *are* within the person's ability to control.

Feedback should be analyzed to ensure clear communication. Discuss or clarify any feedback you're not sure of to clear up any misinterpretations. The sender's intended message is not always what the receiver hears.

Feedback should be solicited rather than imposed (except for the supervisor-subordinate situation). Feedback is most useful when the receiver asks for it. The receiver is more likely to be receptive to your inputs in that case, as opposed to an attitude of "Who asked *you*?"

A WORD ON SUPERVISOR-SUBORDINATE FEEDBACK

As a supervisor, you need to be tactful and patient, especially when approving and disapproving the communications of subordinates. As a supervisor, you are obligated to help your people improve their work. This obligation may mean helping them to revise or rewrite their communication, especially if they are inexperienced. Whatever your role, tact and patience come more easily to people once they really understand feedback in its broadest context.

GETTING APPROVAL (a.k.a. COORDINATION)

A formal coordination process gives interested individuals an opportunity to contribute to and comment on a communication product. Though most staff coordination involves written products (i.e., "the staff package"), important briefings can also be subjected to a formal review process. Formal coordination gives affected individuals a chance to comment, and helps ensure the best course of action is presented to the decision maker. Coordination also lets the decision maker know who supported or disagreed with the position stated in the paper and who agreed to take subsequent actions within their areas of responsibility.

Getting a staff package fully coordinated takes a lot of time, diligence, and hard work. The coordination process has to be closely monitored by the package's OPR to make sure the package gets seen by everyone who needs to see it and doesn't get forgotten in someone's in-box. Here are several considerations that may reduce headaches as you work your package up to the decision maker.

THE "WHO" OF COORDINATION

One of the first things you need to decide as you get ready to put your package through the approval process is who needs to see the package.

Check your organization's policies on coordination. In many organizations, policies exist on coordination requirements for routine packages: find out if your product falls in this category. Your boss may have a list of people who need to see your package. Also check to see what guidance is available electronically—many units have an Action Officer's Guide posted on their website.

Check with key contacts in the organization. Contacts throughout the organization can be very helpful during the staffing process. Fellow staffers, executive officers and secretaries can provide advice on who needs to coordinate on your package, including individuals or agencies that you hadn't thought of. Also, they can give you perspectives on what the bosses will and won't accept. As you build and get to know this "underground of expertise," use it to your advantage.

Realize that the coordination list may grow with time. Don't be surprised if other offices are added as the coordination process occurs. Depending on the material and which level of staff you're dealing with (wing staff versus Air Staff), you may be unaware of all the offices that need to see your package.

Is there a person on your coordination list who carries a lot of clout? When planning your coordination strategy, you should probably determine who the "Heavy" is. Who is the one person that can make or break your package during coordination? Who is the one person who could kill your project with a nonconcur? These people often have strong feelings about how and when they are approached. Some of them may hate a "surprise package" and always want to be the first to be consulted on an issue. In this case you may want to get early buy-in—which can help ensure that others fall into line. Others may want to see what others have to say about the issue before it ever reaches their desk. Use your contacts to find out who these influential people are, what their preferences are and how they like to be approached.

THE "HOW" OF COORDINATION

Aside from deciding who needs to coordinate on your package, you have other things to consider prior to releasing your package for coordination.

Do you want to send out a preview copy? You may want to send a draft package out early to potential coordinators, especially for issues that are complex or for offices whose inputs are crucial to your package's success. Doing so allows them time to study the issue and also saves time later during formal coordination. Also, you may want to coordinate by telephone for small packages, for people who are extremely familiar with issue or with off-base agencies.

How are you going to route your copy/copies to the various coordinating offices? You need to consider how you're going to send your package around. Will you have only one copy that is routed to all affected offices? This may work for high-level packages that don't have a lot of offices to go through, but the more offices you add to the coordination list, the longer this process will take. Instead, you may want to "shotgun" out a number of copies of the package so a number of offices can coordinate on the package at the same time. This will speed up the process, but you will have more copies of the package to keep track of and the various offices won't get to see what each other are saying about the material. You will have to figure out which way will work better for your particular case. Also, don't forget to determine when the "Heavy" will coordinate on your package.

Consider the boss and the schedule. You're not done yet. There are some more things you want to do while preparing to go for coordination. For one, get your boss's blessing before going out-of-office. You want to make sure the boss agrees with what you are saying. You probably want to establish a tentative schedule, based on any deadlines you are up against. If you have a deadline for completion, build a schedule backwards from that point to allow for reviews, changes and recoordination. As part of your schedule planning, check to see if any key personnel coordinating on your package are going to be TDY or on leave for extended periods.

If you're using e-mail to send your package around, consider the following. Specify who is coordinating, who is getting an informational copy, and who will be approving the package. Consider using "COORD," "INFO," "APPROVE," or other keywords in the Subject line. Use clear instructions (i.e., how do you want comments documented and what do you want the offices to do with the package when they're finished with it?) and finally, attach all attachments.

WAIT! One final check. *Before* you hit the SEND button or *before* you go to make all of those copies to send out, get someone else to review the package to make sure you haven't forgotten something obvious. Do one last check for spelling and grammar errors—don't make others proofread your work.

FOLLOWING YOUR PACKAGE

Know where your package is at all times. Use secretaries, contacts or automated tracking systems to track your package(s) and follow up, follow up, follow up. Keep the package moving according to your schedule. You want to avoid suspensing higher offices, but at the same time you want to let them know what the situation is so they can help push your package along. As you get the packages back, keep all correspondence and comments and make sure you retrieve all coordination copies before going for signature. Make sure you've incorporated any appropriate suggestions into the final product. You may want to summarize the comments and inputs for the approving authority.

NONCONCURRENCES

How do you handle a nonconcur? Generally, you only want to send up packages that have received concurrence from all offices. So, do you change the package or do you just include the nonconcurrence in the final package? That may depend on where the nonconcur is coming from. You probably would concede the point to someone with a lot of clout; otherwise your package is as good as dead. Short of that, what can you do? You may be able to persuade the other party to see your point of view. If that doesn't work, do you give in by making small concessions or make a stand? Before you decide to make a stand, you probably want to think about a couple of things. For one, is the issue THAT important, or can you make a concession? Remember also that with the give and take of staff work, a compromise now may help you later. Choose your battles carefully; however, there comes a time to stand firm when you know you're right. Finally, do you have the full support of your boss? Any unresolved disputes at your level may need to be highlighted for resolution at a higher level.

STARTING OVER

Too many substantial changes may require you to start the coordination process all over. You'll have to decide if you've crossed this threshold once you see what kind of inputs you're getting on the package. If you decide to start over, recirculate both the original and changed packages to illustrate the changes you've made and why you've made them.

FINISHING UP

Don't give up. You'll eventually work your way up the chain until you reach the final audience. Remember what we said at the beginning: getting that staff package fully coordinated takes a lot of time, diligence and hard work. Don't get frustrated. Along the way, you will receive lots of suggested improvements to your package, but remember why you are coordinating this material in the first place—to present the boss with the best course of action and to tell him or her who agreed with it and had inputs to it. So don't be surprised by inputs that keep your message consistent with previous decisions by the boss and the other supervisors in your chain of command.

COORDINATING ON OTHERS' PACKAGES

When asked to coordinate on someone else's package, don't put it off. Review it, make your inputs to it and keep it moving. It helps keep your desk clear and the other person may remember your efficiency when he or she gets *your* package. This will also keep you from busting the suspense. If you need more time to review the material, ask for an extension. But don't wait until the suspense to ask for an extension—be proactive. Finally, ask if there are any nonconcurrences on the package so you can take that into consideration.

SUMMARY

As a staffer, you have to remember the big picture. Your job is to get the corporate stamp on the package. You do this by first getting feedback from a few key individuals to make sure your message is loud and clear. Then, you get other offices to approve what you are proposing through the coordination process. Only after the package is fully coordinated can you provide the boss with the best course of action and tell him who is supporting that action.

Proper coordination is the oil that lubricates complex organizations and enables efficient operations. How you view the process and your critical role is crucial. Is coordination just a bureaucratic hassle that you have to endure, or will you meet the challenge head-on by doing a professional, proper job? Your attitude is key to success.

COORDINATOR'S CHECKLIST

Before routing to others, prepare the package and all of its attachments. As necessary...

- ☑ Prepare a "camera-ready" copy of the material.
- ☑ Regarding the material you are presenting, ask yourself:
    - ☑ Are all facts—pro and con—given, and are they accurate?
    - ☑ Are some of your "facts" actually assumptions or opinions?
    - ☑ Is the material concise and clearly presented?
    - ☑ Does the final recommendation represent a sound—the best—position?
    - ☑ Can I justify my proposal if asked?
    - ☑ Are all administrative procedures completed and accurate?
- ☑ Get feedback on your package from a trusted agent.
- ☑ Would you sign the paper yourself if you were the approving official?
- ☑ Modify your paper according to the above guidance and questions.

After taking the necessary actions above, you should:

- ☑ Decide who will coordinate (and in what order), ensuring each office has a continuing interest in the substance of the paper. Seek your chief's guidance, if necessary.
- ☑ Decide how you will route the package through the above offices.
- ☑ Map out a schedule for coordination, based on your routing plan.
- ☑ Coordinate the paper within your own directorate first, and get the chief's signature on the paper so that coordinators will know they're coordinating your office's position, not yours.
- ☑ Hand carry coordination whenever time permits or when the subject is complex.
- ☑ Coordinate by telephone and e-mail when feasible and practical.
- ☑ Attach any nonconcurrence to the package if it cannot be resolved with the appropriate official and send a summary of the disputed issues to your chief to resolve. Show future coordinators the nonconcurrence.
- ☑ Provide copies only if required or upon request. Save paper!
- ☑ Keep all correspondence and comments.
- ☑ Retrieve all coordination copies and incorporate/address suggestions before going for final signature/approval.

PART IV:

FACE-TO-FACE: SPEAKING AND LISTENING

CHAPTER 10

AIR FORCE SPEAKING

This chapter covers:

- The basics of verbal communication: It's all in the delivery!

- Conquering those crazy quirks.

- Delivery formats.

- Using visual aids to your advantage.

Sooner or later, you will have to speak in public. It comes with being in the military, there's little you can do to avoid it, and the requirements will increase as you climb the ranks. If the thought makes you nervous, you're not alone! Research shows that most people place fear of public speaking second only to fear of dying. If you are inexperienced, the fundamentals and tips for polished speaking in this chapter will help you solve these problems. If you are an accomplished speaker, use this as a review … or skip it.

One goal should be to improve your self-concept as a speaker. Think positively, and focus on improvement, not perfection. Like writing and listening, speaking is a skill; once you grasp the basics, the rest is practice, polish and style. You may be embarrassed by your initial mistakes, but you'll survive. Few of us will become guest speakers, but all of us can become more effective if we practice the basics. Learn all you can from your contemporaries; some of them are accomplished speakers. If you are already a speaker extraordinaire, share your views, tips and personal hang-ups about speaking with others. Everyone improves when they receive timely and objective feedback.

"Practice doesn't make perfect; perfect practice makes perfect."
– Joan Ballard and Steve Sifers

Before you consider the fundamentals unique to speaking, you may want to review the Chapter 2 summary of the Seven Steps to Effective Communication.

Remember, the fundamentals above are just as necessary for good speaking as they are for good writing. Although there are subtle differences in the drafting and editing sections, the general concepts apply. Indeed, these basic principles are a good place to start when preparing an oral presentation, but we all know there's more to it than that. Let's talk about delivery techniques.

```
SEVEN STEPS FOR EFFECTIVE
COMMUNICATION

1.  Analyze Purpose and Audience
2.  Research Your Topic
3.  Support Your Ideas
4.  Organize and Outline
5.  Draft
6.  Edit
7.  Fight for Feedback and Get Approval
```

IT'S ALL IN THE DELIVERY:
PART ONE—VERBAL COMMUNICATION

How effectively do you use your voice to drive home your ideas or information? You have control over rate, volume, pitch, pause and other aspects of your speech. So, use your voice to create interest in your presentation. Read on to find out what we mean.

RATE

There is no correct rate of speed for every speech. However, you might consider this: People can listen 4 to 5 times faster than the normal spoken rate of 120 words a minute. So, if you speak too slowly, you will lose the interest of an audience who is processing information much faster than you are delivering it! On the other hand, you don't want to use the same rate of speech all the time. Use the rate of speech that you need to add emphasis to what you want during your presentation. Consider speaking at a faster rate to indicate excitement or sudden action or a slower rate to hint at calm or a more serious tone.

VOLUME

Volume is another verbal technique that can give emphasis to your speech. If possible, check out the room to know how loudly you must talk, remembering you will need to talk louder with a crowd since the sound is absorbed. Ask someone in the back of the room if you can be heard. Remember your voice will carry further when the room is empty versus full. If the audience must strain to hear you, they will eventually tune you out from utter exhaustion. A portable microphone may be a good idea if you know you tend to speak quietly, and one is often required in large auditoriums. Speak louder or softer to emphasize a point—a softer level or lower volume is often the more effective way to achieve emphasis.

PITCH

To use pitch effectively, you need to practice the talents of a singer. Pitch is really the use of notes (higher or lower) in voice range. Start by speaking in a voice range that is comfortable for you and then move up or down your scale for emphasis, using pitch changes in vowels, words or entire sentences. You can use a downward (high to low) inflection in a sentence for an air of certainty and an upward (low to high) inflection for an air of uncertainty. Variety in speech pitch helps to avoid monotone and capture the listener's attention.

PAUSE

The pause gives you time to catch your breath and the audience time to collect your ideas. Never hurry a speech; pause occasionally so your audience can digest your comments. The important question is this: Where? Pauses serve the same function as punctuation in writing. Short pauses usually divide points within a sentence, and longer pauses note the ends of sentences. You can also use longer pauses for breaks from one main point to another, from the body to the conclusion of your speech, or to set off an important point worthy of short reflection. A pause may seem long to you, but it's usually much shorter than you think … and your audience will appreciate it. However, don't get pause-happy and make your speech sound choppy.

ARTICULATION AND PRONUNCIATION

Your articulation and pronunciation reflect your mastery of the spoken English language. Articulation is the art of expressing words *distinctly*. Pronunciation is the ability to say words *correctly*. Of course, you may be able to articulate your thoughts and still mispronounce words while doing so. Unfortunately (and unfairly), many people consider word pronunciation or mispronunciation a direct reflection on your intelligence. Listen to yourself and make your words distinct, understandable and appropriate to your audience. If you are not sure of your pronunciation, consult a current dictionary—before you get up and do your thing. You can even look up dictionaries on-line with audio links that will pronounce the word for you!

LENGTH

"Be clear, be quick, and be gone."
— from Presenter's University

Have you ever taken a PME course in-residence and wondered why the instructors were so big on the timing of your briefings? It is because the length of your presentation is crucial. In our military environment, you must be able to relay your thoughts and ideas succinctly. A key rule in verbal communication is to *keep it short and sweet*. There are few people who will tolerate a briefer or speaker who wastes the audience's time. Have your stuff together before you speak. Know what you want to say and then say it with your purpose and the audience in mind.

OK, we've just hit the highlights of controlling and managing your voice to optimize the delivery of your briefing. Your mission, should you choose to accept it, is to actually practice these tips so that you will soon have the "radio voice" that public speakers dream about. But wait, you're not done yet! There's more to the fine art of public speaking than your managing your voice. You need to manage your gestures, movement, and that pesky thing called *nervousness*.

IT'S ALL IN THE DELIVERY:
PART TWO—NONVERBAL COMMUNICATION

"You never get a second chance to make a first impression."
– Anonymous

Numerous studies have shown that people remember less than 10 percent of what is verbally presented, but first impressions are largely based on nonverbal communication such as how you dress, carry yourself, and use gestures and other body language. Your biggest nonverbal challenge to conquer will be your nervousness, so you must be prepared to overcome (or at least diminish) stage fright. Stage fright is often nothing more than misdirected energy, meaning the excitement and/or possible anxiety you feel is displayed in some form or fashion for all to see. You probably have witnessed a great presentation "gone bad" solely due to nerves that have gone unchecked. Here's a checklist on how to overcome stage fright and put your best foot forward ... or at least fool your audience. Remember, it may be impossible to be entirely free from nervousness—that's OK! But you don't want the nervousness to disable you from sending your message.

OVERCOMING "SWEATY PALMS SYNDROME"

☑ Analyze your audience: listening traits, needs, desires, behaviors, and educational background. This will reduce your fear of the unknown and the resulting nervousness.

☑ Check out the place where you're speaking. Is it large enough to accommodate the number of people? Does it have a blackboard, microphone, arrangement for visual aids, tables, chairs, ventilation, lighting, pencils, paper, telephones, extra projector bulbs, etc? Does the equipment operate properly?

☑ Practice, practice, practice. Using a tape recorder, a video camera, a full-length mirror or even your peers can be really helpful. Try doing a "dry run" at the office or where you'll be.

☑ Memorize your introduction and transition into the main point. It'll help you through the first and most difficult minute.

☑ Smile and be positive! Your audience wants you to succeed! Keep your nervousness to yourself ... chances are your audience won't even notice if you don't mention it.

☑ Take a short walk right before you "go on stage" to help release some energy.

☑ Deliver your message. Focus your attention where it belongs ... not on yourself.

☑ Make eye contact and look for feedback. Play your audience. Let them know you are looking at and talking to them. It holds their attention. If you look only at your notes, you may lose your listeners—and you can't wake them up if you don't know they're asleep!

☑ Use simple, everyday language appropriate for your audience. Use contractions and keep sentences short. Use personal pronouns, if appropriate. Repeat key words and follow with specific examples if you get into abstract or complicated reasoning

☑ Involve members of your audience by soliciting their answers and information.

☑ Enhance your presentation through creative use of newspaper clippings, cartoons, music, appropriate quotes and relevant, self-deprecating experiences to get a point across.

☑ Use your excess energy naturally: facial expressions, pertinent gestures, walking, or pressing fingertips or thumbs against lectern or chair. Use your facial expressions, hands and arms to reinforce your speech and your points of emphasis—just don't overdo them. Leaning on the lectern, rocking back and forth or side to side or slouching on one leg and then the other is never a positive way to use your excess energy. Read on for more tips on those dreaded nervous habits.

☑ Looking good builds confidence and builds your credibility with the audience. Do you need a haircut? Is your uniform pressed? Are your ribbons, nametag and insignia attached correctly? Your buttons buttoned? Your shoes shined? Are you standing erect and feeling alert, but relaxed? Remember, in your audience's mind, a frumpy uniform and sloppy bearing equals an equally frumpy presenter. Fair or not, that's the way your audience's mind works. We're all critics!

COMMON NONVERBAL QUIRKS

Here are some final thoughts on nervousness. Most of us have quirks when we are put in the limelight. The key is to be cognizant of yours and don't overdo them. Keep yourself in check, and as always, seek feedback. In time you will have it down to an art. We've named a few of the more common quirks below. Do any hit home for you?

- The **life raft.** This is the speaker who clings for dear life to podium or lectern. Their ultimate fear is leaving the comfort and security of that piece of wood in front of them, so they hold onto it with both hands in desperation. Walking about the room is incomprehensible to these speakers.

- The **fig leaf.** This is the speaker who is recovering from the above habit, is now venturing out in front of the audience, but is still not quite sure what to do with those pesky things called hands. He or she wants to run back to the safety of their life raft, but instead may lay one hand over another at the end of stiff arms, with hands resting … where a fig leaf would be. Do you get the picture?

- The **hand washer.** This is a speaker who stores all nervousness in their hands. While they speak, they wash and they wash. You may think with all that friction they would suffer thermal injury to their palms, but they're indestructible. Too bad you missed their message while you focused on the quirk.

- The **caged tiger.** Listening to these speakers is like watching a tennis match. These speakers continually pace from one side of the room or stage to the other, never even stopping to check their pulse. They expend so much energy that their briefing qualifies as aerobic exercise. Their calorie expenditure goes up even higher when they combine this technique with hand washing.

- The **rocker.** Rockers are caged tigers on the road to recovery. They have conquered the worst phases of stage fright and no longer fear visible perspiration or dry mouth, but they have some residual fear of standing still and simply talking. Speaking experience has tamed this beast, but not to the point of total comfort with the art of public speaking. There are two style variations: 1) the fore-and-aft rocker and 2) the side-to-side rocker.

- **Pocket maniacs.** OK, it's against regulation, but these folks should consider sewing their pants pockets shut because whenever they speak they start counting the change from their last trip to the soda machine. Many of these speakers are trying desperately not to be hand washers or fig leafers and end up jamming their hands in their pockets. They are often oblivious to the new distraction and annoyance they inflict on their audience. These speakers may find that holding something—anything—will help them refrain from this habit and prevent them from "hand washing."

- **Pen clickers.** These speakers are related to the pocket maniacs. They have to be doing something with their hands. All pens, pencils and other similar objects should be removed from their possession before they get up to give a presentation because the temptation is just too great. They are compelled to manipulate and click any pen in their possession, which does not make for high marks on audience critiques.

By themselves, these quirks won't make you fail as a speaker, but they can create problems if they are severe—the audience begins to focus on the quirk instead of your message. Again, most of us have done some of these things at one time or another. Try to be aware of your own mannerisms, keep them in check and make sure they are not becoming habit and detracting from your message.

I hope this podium doubles as a flotation device...

Life Raft Guy

Please don't hurt me... I'll never get up here again.

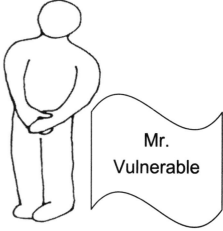

Mr. Vulnerable

With my new laser pointer, I can highlight main points AND blind anyone who dares to ask me a question.

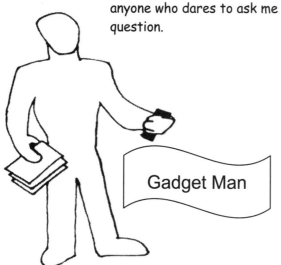

Gadget Man

If I'm going to bomb up here, I might as well be comfortable.

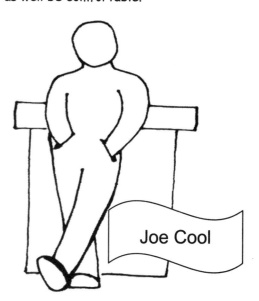

Joe Cool

Memorable Speakers

DELIVERY FORMATS:
IMPROMPTU, PREPARED AND MANUSCRIPT

Your approach to delivery of the spoken message may be affected by several factors, including the time you have to prepare and the nature of the message. Three common delivery formats are listed below.

Impromptu speaking is when we respond during a meeting or "take the floor" at a conference. It's what we do when we speak publicly without warning or on a few moments' notice. To do it well requires a great amount of self-confidence, mastery of the subject and the ability to "think on your feet." A superb impromptu speaker has achieved the highest level in verbal communications.

> Both students and instructors have had trouble differentiating impromptu and extemporaneous speaking in the PME classroom. Why? Because dictionaries vary in how they define these terms. Some, including The Cambridge Dictionary of American English, define extemporaneous as "done or said without preparation" (synonymous with impromptu). The Merriam-Webster dictionary, on the other hand, states that extemporaneous speech "*is* prepared in advance but delivered without notes." To confuse matters further, extemporaneous briefers in the Air Force frequently use notes or an outline (but not a script) when speaking. So, for ease of discussion in this edition of *Tongue and Quill*, the term *prepared speaking* will replace extemporaneous speaking. 'Nuff said.

Prepared (formerly *extemporaneous* speaking or briefing) refers to those times when we have ample opportunity to prepare. Most military briefings are done in this format. This doesn't mean we write a script and memorize it, but it does require a thorough outline with careful planning and practicing. The specific words and phrases used at the time of delivery, however, are basically spontaneous and sound very natural.

A **manuscript** briefing is used in situations that require every word to be absolutely perfect. To do this, you prepare a manuscript, a word-for-word script of what you are going to say. Such a script ensures you get it right every time. Manuscripts are often used at higher management levels for complex or controversial issues (policy briefings, press conferences, source selection briefings to unsuccessful officers, etc.). They're also used for routine briefings that must be repeated several times a year (base orientation, etc.), or at formal ceremonies (retirements, medal presentations, etc.) that must adhere to established customs and courtesies. Manuscript-style briefings provide several advantages:

- Ensures key information won't be omitted.

- Avoids repercussions caused by a briefer's inadvertent ad-libbing.

- Imparts exact definitions and precise phrasing, if these are important.

- Allows anyone to present a "canned briefing" without extensive preparation and rehearsal time—including less knowledgeable personnel.

CAUTION! Manuscripts make a briefing a piece of cake, right? Wrong. Unless you are a talented speaker, reading words aloud sounds dull. People frequently tend to lack spontaneity, lack eye contact and stand behind the lectern with their script. You may also lose credibility. The audience may think you're hiding behind the script and reading something you really don't know anything about. Finally, know your audience! They may feel insulted if you read something they could have just as well read themselves. If you can deliver a manuscript briefing without error and still maintain a natural and direct contact with your audience, you have a masterful command of "speak-reading."

Here are some key points in preparing and presenting a manuscript briefing.

- Prepare the briefing:
  - Use a large, easy-to-read font (at least size 12) in all capital letters
  - Write as if you were speaking
  - Fill only the top two-thirds of page so that your eyes won't drop and you won't lose eye contact with your audience
  - Double or triple space; never break a word at the end of a line or a sentence at the end of a page
  - Number pages with bold figures
  - Underscore or highlight words you wish to emphasize … insert double slashes (//) where you plan a major pause
  - Use a loose-leaf binder (e.g. in case of windy days or nervous fingers) or stack pages loosely to turn pages
  - Mark script with red dots to show visual aid changes

- Practice, practice, practice:
  - Read and reread until you've practically memorized it
  - Add the ingredients of volume, inflection and eye contact
  - Avoid combinations of words that are difficult to say
  - Look at your audience when uttering emphatic words and during the closing words of a sentence
  - Practice using gestures—strive for enthusiasm
  - Dry run your visual aids

- Close with confidence:
  - Never explain why you choose to read (it won't be apparent if you've prepared!)
  - Be flexible; if necessary (and appropriate), know where you can shorten the speech
  - Avoid being lengthy after saying "In conclusion."
  - Don't add new information at the end

Individuals who can strongly present briefings in all three types of delivery formats are the envy of everyone. They appear knowledgeable and comfortable in their roles as speakers because they have done their homework. They may be experts on their subject and know how to present their views with clarity on a moment's notice (impromptu speaking). They have researched and rehearsed their presentations (prepared and manuscript briefings). They think carefully before they speak, outline their main ideas, say what has to be said, conclude and shut up. Remember, there is no substitution for preparation. If you have the time to prepare—do it!

FYI ... BRIEFINGS IN THE JOINT ENVIRONMENT

In the Air Force, the term *briefing* is loosely applied to almost any oral presentation given in the context of military activities. The following list was adapted from Army FM 101-5, and serves as a reference for those who may find themselves in a joint assignment. This terminology is not commonly used in the Air Force; the intent is to familiarize you with what else is "out there." If you are in a joint assignment, you may hear briefings defined and named according to their purpose. Here are four types of briefings and their characteristics.

Informative briefing: Purpose is to keep listener abreast of the current situation and supply specific information.

- Designed to inform the listener and gain his understanding
- Deals with
  - High priority facts and information requiring immediate action
  - Complex information on complicated plans, systems, statistics, or visuals
  - Controversial information requiring explanation
- May have conclusions or recommendations

Decision briefing: Purpose is to produce an answer to a question or obtain a decision on a specific problem.

- Briefer must be prepared to present
  - Assumptions
  - Facts
  - Alternative solutions
  - Reasons/rationale for recommended solutions(s)
  - Coordination involved
  - Visual information
- Briefer states he/she is looking for a decision; asks for decision if one is not forthcoming at conclusion.
- Advises appropriate staff elements of commander's decision after the briefing.

Staff briefing: Main purpose is to secure a coordinated effort and
  - Rapidly disseminate information orally
  - Aid group decision-making
  - Secure a united effort

- Most widely used and most flexible briefing—used at all levels of command
- Visuals will make complex issues clearer
- Keeps commanders/staffs abreast of situation
- May involve an exchange of information, issuance of directives, or presentation of guidance

Mission briefing: Purpose is to impart information that is used to elaborate on an order, give specific instructions, or instill an appreciation for the mission.

- Briefer must exercise care to avoid confusion or conflict with orders
- Use maps and graphic representations of the situation
- Mission briefing format varies from command to command

A PICTURE IS WORTH A THOUSAND WORDS...

"Enough with the bells and whistles—just get to the point."
– General Hugh Shelton

We've spent the majority of the chapter on how to speak and how to get over your nervousness. But there's more you can do to make an oral presentation more professional and useful for your audience. Visual aids can enhance your oral presentation by helping the audience remember and understand the content of your message. Remember, comprehension rises dramatically when we *see* something rather than when we just hear about it. The average person retains 5 percent of what is heard and 65 percent of what is seen. More dramatically, the human brain processes visuals 400,000 times faster than text! In other words, "show and tell" is better than just "tell" alone.

> As always, check with your organization for local policies on formats. These are only general guidelines.

Slides are the most common visual aid used for briefings in today's Air Force. They help the briefer to remember key points and help to keep the presentation brief. The presenter makes the slide simple and fills in the "white space" with concise spoken words.

Before we launch into Slides 101, let's take a second to stress one point—*you* are also a visual aid. If you are well groomed, professional and well prepared, you'll be the most effective visual aid in your presentation. It doesn't matter how awesome your slides are. If you look like a slob and appear insecure and awkward, the audience most likely won't take you or your message seriously. Take the time to put on a freshly pressed uniform and look sharp!

PREPARING YOUR SLIDES

The capabilities of computers today make it easy to produce a blizzard of flashy visual aids. Instead of drawing up a dozen slides on a legal pad and running them over to the printing department, we can create hundreds of slides in a few hours without ever leaving our desks.

> **Attention PowerPoint Rangers!**

Each organization is different when it comes down to how your slides should look. However, *most* organizations in today's Air Force agree that the information we present shouldn't be too complex. Don't pack your slides with every detail and custom animation feature you can dig up. You don't need Venetian blind effects and fancy backdrops; all you need is the information!

COLOR. Color is a very important communication tool. Good designers limit their color palettes even if 256 different colors may be available.

- Use colors in a standard manner throughout your presentation.

- Limit your choices to 4 or 5 colors.

- Use light colors on a dark background and vice versa.

- Use colors to emphasize key elements, but try to avoid red lettering.

- Use the same background color on all images.

- Avoid red/green and blue/red color combinations.

- Use bright colors (yellow, orange, etc.) sparingly.

- Maintain good contrast between important information and background.

TEXT. The first rule here is "less is more." Less experienced briefers are often tempted to pack presentations with every detail they can think of for fear they might leave something out during the real thing. Slides should have minimal content and lots of "white space." Slides aim at the visual portion of the brain and will only confuse the audience if they are jam-packed with data. The slides should *not* be self-explanatory. If they are, you probably have too much stuff on them. Remember, you should add value to the presentation and should supplement the slides with your eloquent speaking abilities.

- Keep it simple. Use the "7 x 7 rule:"

    - No more than 7 words per line

    - No more than 7 lines per slide

- Spelling is important!

- Don't read slides—it's insulting.

- If you have more than one slide per main point, add "Continued" on subsequent slides to keep flow.

- Have only one thought per slide—it gives the audience time to refocus on you.

- Avoid hyphenation at end of lines.

- Use upper and lower case for three reasons:

    - Helps identify acronyms, which we love in the military!

    - More comfortable for audience, because this is how we read.

    - Makes your presentation look more professional.

- Emphasize key words with boldface, italicized, underlined or colored text.

- Left-justify your text.

- Font size: Use the same type font throughout the presentation.

    - Title: 40 point

    - Subtitle: 30 point

    - Text: 22-26 point

- Most importantly, you fill in the information. So, you should always know and be ready to present one level of detail below a piece of information. If you don't, you will end up reading slides or the audience will have many questions left unanswered.

GRAPHICS. Whether designed for a briefing or written report, no graphic should be so elaborate it becomes an end in itself and obscures your intended message. However, when used wisely graphics can certainly add to your presentation. Although text is important, audiences remember more when content is graphically presented. A 60-minute briefing can be pretty boring if it is all done in text. But, on the flip side, cool graphics do not guarantee an effective briefing. Read below for some tips to keep you in check when using graphics.

- Use only artwork suitable for your presentation. Know your audience!

- Use graphs (bar charts, pie charts, etc.) to convey statistics.

- Be careful with graphs: too many can be confusing.

- Line graphs show trends over time.

- Bar graphs compare values.

- Pie charts compare values against a whole.

- Tables: Don't use if you can convey information verbally or in a graph—they usually appear overly "busy" on a slide.

  - Limit to 4 rows and 7 columns.

  - Use footnotes to remove distracting data from tables.

  - Round off numbers if possible.

  - Don't put decimal points in numbers like 10 or 100. The audience may interpret "100.00" as "10,000."

- Place your graphics off-center—use them to lead your audience to important text information.

- Be careful with animation:

  - Sound: Use sparingly and make sure they add impact.

  - Slide transitions: Most briefers overuse slide transitions. If you decide to use transitions, use the same type throughout your entire briefing, and make sure they add, not detract from your presentation.

- Video can be quite effective, but again, use sparingly.

Make it **Big**, keep it **Simple**, make it **Clear**, and be **Consistent**!

OK! We've just highlighted how to develop effective slides for your briefing or presentation with proper use of color, text and graphics. Remember, local policy will always supersede the guidance we have provided. Let's close this part of the chapter with some general tips on slide presentations.

DO's	DON'Ts
Stand beside your visual aid. Better yet, get away from slides and walk around. This may depend on your purpose and audience, and location of your briefing.	Stand between the audience and visual aid and block the audience's view.
Start out with well-thought out opening statement and try to elicit audience involvement by asking a relevant question or two. Use a personal story or experience (if appropriate) to bridge to your topic.	Jump right into slides.
If referring to screen, stand aside, use a pointer, and put it down when done.	Talk at the screen with back to your audience.
Give the audience time to read slides.	Change slides too quickly.
Read the slide silently or watch to see if audience has finished reading. If slide contains a long quote, paraphrase or underline important parts.	Read the slide to the audience.
Speak naturally and use gestures.	Give a memorized briefing.
Make the slide simple and fill it out with concise spoken words.	Show a complicated slide *and* give a complicated explanation of it.
Show only necessary slides.	Use slides as gimmicks or crutches.
Turn off projector or use a cover slide.	Leave projector on with blank screen.
Check for spelling and punctuation...more than once!	Forget to check spelling and punctuation.
Practice handling slides and gauging time needed to read them. Use an assistant to advance slides if available.	Disrupt presentation to handle slides.
Anticipate likely questions and rehearse possible answers. Keep answers short and simple. Listen carefully to questioner and clarify question if needed.	Be caught off-guard by questions from the audience. Don't give quick replies. Don't direct questions to specific members of audience.
Know your purpose, audience, and any time constraints.	If time is limited (5-15 minutes), you may want to consider NOT using slides. In these cases, it's more important to establish connection with your audience than it is to show a few slides.
Practice, practice, practice! And test visual aids prior to your briefing.	**Wing it,** and never apologize because you didn't prepare. This makes you look irresponsible and ruins your credibility before you even start!

SUMMARY

Use vocal characteristics such as rate, volume, pitch and pause to enhance the impact of their message.

Be aware of any nonverbal quirks; reduce nervousness through solid preparation.

Select an appropriate delivery format: impromptu, prepared, or manuscript.

Visual aids can enhance oral presentations, but don't lose sight of the big picture.

Remember the basics of delivery, keep focused, and do your homework.

The more often you speak in front of or with a group, the more self-confident you become. High confidence and thorough knowledge of your subject are important prerequisites for speaking. Remember, if you have the time to prepare—do it!

CHAPTER 11

EFFECTIVE LISTENING STRATEGIES

This chapter covers:

- Why listening is important to Air Force personnel.

- Three active listening strategies: Informative, Critical and Empathic approaches.

- Cultivating the motivation to listen well.

Listening is a valuable communication skill, but one that gets little respect and attention. Some of us spend hours preparing and practicing our briefings; how much time and effort do we spend on preparing and practicing to listen? Because listening is so undervalued, we'll first make the case for why it's so important, and then describe ways to improve your listening skills.

> "Conversation in the United States is a competitive exercise in which the first person to draw a breath is declared the listener."
>
> – Nathan Miller

UNDERSTANDING HEARING AND LISTENING

To better understand the listening process, let's begin by distinguishing between hearing and listening. Hearing occurs when your ears pick up sound waves being transmitted by a speaker or some other source. Hearing requires a source of sound and an ear capable of perceiving it. It does not require the conscious decoding of information. Each day, you hear numerous sounds—background music in an elevator, the hum of the computer, cars passing by outside—sounds you may not even be aware of unless someone draws your attention to them.

Listening, on the other hand, involves making sense out of what is being transmitted. Listening involves not only hearing; it involves attending to and considering what is heard. As you listen, you receive sounds, and you consciously and actively decode them. Effective listening is an active process, and active listening involves exerting energy and responding appropriately in order to hear, comprehend, evaluate, and remember the message.

WHY IS LISTENING SO IMPORTANT?

Though listening is usually ignored in formal education, it is the most common form of communication. According to Dr. John Kline, an authority on Air Force communication and the author of *Listening Effectively*, studies have shown that we spend 70 percent of our waking hours in some form of communication. Of the time we spend communicating, 10 percent is spent writing, 15 percent is spent reading, 30 percent is spent speaking and 45 percent is spent listening.

When speakers and listeners fail to communicate, the outcomes can be disastrous. Planes can crash, unit morale can deteriorate, "routine" surgeries can have horrendous outcomes and families can self-destruct. Though we often focus on the speaker's role in communication, good listening skills can often make the difference between success and failure.

Both Air Force and civilian managers put a premium on good listening. In a recent study, 282 members of the Academy of Certified Administrative Managers identified listening as the most crucial management skill. In another study published in *Communication Quarterly*, business people were polled on the most important communication skills they used at work and the one they wished they had studied in college: Listening was the #1 answer to both questions.

Listening is especially important in the Air Force, and actually in any military unit. Success is literally a matter of life and death, and we routinely maintain and operate equipment worth millions of dollars. Receiving, comprehending, and remembering spoken information is critical, and any miscommunication is potentially catastrophic. Effective listening helps to build the trust and mutual respect needed to do our job. Military personnel must understand their team members and the situation, and leaders with good listening skills often make better decisions and have a stronger bond with their troops.

To summarize, listening allows us to learn the facts, evaluate situations, understand our team members, and build trust with others. These are all really important, so why do so few people listen well? Why is it so hard? And how do we get better at it?

MOTIVATIONAL BARRIERS TO LISTENING

Listening takes a lot of effort. Our willingness to exert that effort is linked to our motivation to listen. Techniques to improve listening skills are worthless without cultivating the motivation to use them.

To help yourself evaluate your own listening motivation, scan the questions below.

? Do I pretend to listen?

? Do I seek distractions?

? Do I criticize the speakers?

? Do I stereotype topics as uninteresting?

? Do I prejudge the meaning and intent of speaker's messages?

? Do I avoid difficult and complex topics?

? Do I get emotionally charged up about minor points a speaker has made?

? In a conversation, do I tune out the speaker because I'm preparing my response?

? Do I interrupt others? Has anyone told me I sometimes interrupt others?

? Do I spend a disproportionate amount of time talking in meetings or social events?

? Does everything a speaker describes remind me of something that happened to me?

Though we've all been guilty of these once in a while, if you answered "yes" to three or more questions, then you might benefit from some of the suggestions to improve your self-motivation later in the chapter.

Sometimes we have blind spots and overestimate our success as listeners. In his book, *Listening Effectively*, Dr. Kline describes how most individuals in groups will rate their own listening skills as a 7.5 out of 10, while rating the rest of the group as a 4.1 out of 10. These data indicate that most people see listening as a problem in others, but not in themselves.

Here's another set of "workplace climate" questions to consider. If you have "yes" answers to several of these questions, you might want to take a closer look at how you listen and interact with others.

? Do employees go around you to talk to others about work issues?

? Do you often learn about important events after the fact?

? Do you frequently find yourself putting out organizational fires?

? Are you rarely tasked with complicated responsibilities?

? Do you receive lots of information in writing … including some items that would normally be handled face to face?

Sometimes it takes a deliberate effort to cultivate our motivation to listen. In today's culture, a "good communicator" is often defined in terms of one's speaking ability, yet many would argue that listening is just as important, if not more so. Today's effective listeners command respect and admiration because listening is crucial to solving problems and making decisions.

PICK THE RIGHT TOOL FOR THE JOB: INFORMATIVE, CRITICAL, AND EMPATHIC LISTENING

There are different situations where listening is important and different reasons to listen. It's important to acknowledge and identify these differences because appropriate listening behaviors in one situation might be inappropriate in another situation.

Part of the challenge of listening is sorting out what situation you're in and what response is appropriate. One way to approach this sorting process is to look at three major reasons why we listen in the workplace. We'll talk more about each of these three categories in later sections, but here's a summary:

We listen to collect information from others (Informative Listening):

- Receiving radio instructions in an aircraft cockpit.
- Listening to an instructor lecture on good writing technique.
- Receiving a briefing on a change to the assignment process.
- Obtaining medical instructions from a doctor.
- Learning your boss's expectations during an initial feedback session.

We listen to judge—to evaluate a situation and make decisions (Critical Listening):

- Investigating causes of a fatal mishap.
- Determining which Airman to nominate for a quarterly award.
- Helping the user formally specify requirements for a new weapon system.
- Deciding which disciplinary action to administer.

We listen to understand and help others in situations where emotions are involved and the speaker, not just the message, is important (Empathic Listening):

- A subordinate is seeking advice on whether to reenlist.
- Your spouse is worried about your next deployment.
- A coworker is unsure how to deal with subtle discrimination.
- Your teenager doesn't want to move during senior year; you've got orders.
- You're trying to mend fences with a coworker after a major conflict.

Informative, Critical, and Empathic listening categories can be used in two different ways. First, they can be used to characterize listening situations—that is, WHY we should be listening in this situation. Second, they can be used to describe listening approaches—the behaviors we use in a given situation and HOW we are listening.

The real fun begins when the approach and situation are mismatched—for example, using an informative listening approach (taking notes, asking focused questions) in an empathic listening situation (an angry spouse who feels you're neglecting the family).

INFORMATIVE LISTENING

In informative listening, the listener's primary concern is to understand information exactly as transmitted. A successful listening outcome occurs when the listener understands the message exactly as the sender intended. Informative listening is important when receiving information from an established authority, or receiving information that is not open for debate.

For example, if you're receiving a briefing on changes to the assignment system, your goal should be to understand exactly what the rules are, not to sort out the reasons why you think the rules are stupid. If you're receiving formal training that will be followed by a test, your primary goal should be to understand the material, not to evaluate whether the material should be taught at this level.

SUGGESTIONS FOR IMPROVING INFORMATIVE LISTENING:

1. Keep an open mind. If your primary goal is to understand the message, set aside your preconceptions about the topic and just listen.

2. Listen as if you had to teach it. Many education and training specialists suggest this technique. Typically we expend more effort to understand a subject when we know that we have to teach it to someone else. By taking this approach, we have the mental fortitude to focus longer, ask questions when we don't understand, and think more deeply on a topic.

3. Take notes. Focus on main points, and don't attempt to capture everything. This classic technique is used in situations where you are trying to capture objective information, such as in classes, staff meetings, etc. Note that if the listener and speaker are in a less formal, emotionally charged situation, note taking might be misconstrued as hostile behavior (i.e., being put "on the record," or being documented for future adverse action) so use your judgment on whether it is appropriate to the situation.

4. Exploit time gap between thinking and speaking speeds: The average speaking rate is: 180 words per minute, while most listeners can process 500 words per minute. Use this extra time to mentally repeat, forecast, summarize and paraphrase the speaker's remarks.

5. Respond and ask appropriate questions: Good informative listening questions help you clarify and confirm you understand the message. Remember that you are trying to absorb as much information as possible, and you are less focused on making value judgments on the material. Here are some examples of appropriate questions or responses:

- Repeat exact content back to the speaker—(this is common in radio communication).

  "We need toner, copy paper, and a three-hole punch from supply. I'll get them now."

- Paraphrase in your own words.

  "So if I understand you, then…."

- Ask for more specifics or details.

"When you say a draft is due on 15 April, is that before or after the internal coordination process?"

"How does the situation change if you're married to another military member?"

- Request an example for clarity.

"If asking good questions can help listening performance, can you provide some examples?"

CRITICAL LISTENING

Critical listening can be thought of as the sum of informative listening and *critical thinking*. In this case, the listener actively analyzes and evaluates the message, and listening success requires understanding the message and assessing its merit. The listener is evaluating the support offered by the speaker—as well as the speaker's credentials and logic—and may either agree or disagree with the message.

Critical listening may be appropriate when seeking input to a decision, evaluating the quality of staff work or a subordinate's capabilities, or conducting research.

SUGGESTIONS FOR IMPROVING CRITICAL LISTENING:

Several of the suggestions made for improving informative listening are equally important for critical listening: after all, you need to understand the message before you can critically analyze it.

1. Take notes. See "Improving Informational Listening" on the previous page.

2. Listen as if you had to grade it. Teaching a topic is tough, but grading another's presentation of a topic is even tougher. Is the message clear and precise? Is the supporting material relevant and convincing? Does this make sense? Your attempt to mentally answer questions like these may help you stay focused on the speaker.

> **Critical Thinking**
>
> *Critical thinking* is the act of exercising careful judgment in forming opinions or conclusions. In Chapter 4, page 46-47, we described how intellectual standards like accuracy, precision, relevance and clarity can be used to evaluate the quality of information sources. Critical thinking attempts to improve the quality of thought by using these same standards. It is self-directed, self disciplined, and self-corrective thinking applied to an important problem.

3. Exploit time gap between thinking and speaking speeds: Part 2: Use the time gap described on the message. Remember to try to understand first and then evaluate second. Even when you are listening critically, don't mentally argue with the speaker until the message is finished.

4. Ask appropriate questions: Good critical listening questions will be probing in nature so you can better evaluate the intellectual content of the speaker's message. Some examples of good critical listening questions can be found in *The Miniature Guide to Critical Thinking* by Dr. Richard Paul and Dr. Linda Elder. The questions help the listener evaluate how the message stacks up with universal intellectual standards like accuracy, relevance, fairness, etc.

    Accuracy: How could we verify or test that? Are others reporting the same results?
    Relevance: How does that fact relate to the problem?
    Breadth: Do we need to consider another point of view?
    Logic: Do our conclusions flow from our evidence?
    Significance: Is this the central idea? The most important problem?
    Fairness: Do we have any vested interests here? How would our opponents view this issue?

EMPATHIC LISTENING

In empathic listening, we listen with the primary intent to understand the speaker and his or her frame of reference. Empathic listening is often useful when communication is emotional, or when the relationship between speaker and listener is just as important as the message. It is often used as a first step in the listening process, a prerequisite to informational or critical listening. Empathic listening is often appropriate during mentoring and nonpunitive counseling sessions, and can be very helpful when communicating with family members. Depending on the situation, it may also be useful in negotiation and teambuilding activities.

In his best-selling book, *The 7 Habits of Highly Effective People,* Dr. Stephen Covey suggests that listening be used to "diagnose before you prescribe" when dealing with others, a concept captured by the phrase: "Seek First to Understand, then to be Understood." Empathic listening is described as a powerful tool to understand others and to build relationships. If you allow yourself to fully listen for both content and feeling, and then reflect that back, there are several productive outcomes:

> **Empathic Listening:**
> **Not just for sissies**
>
> Though most people quickly see that empathic listening skills are useful in dealing with a spouse or a child, some might think this listening approach is too "sensitive" for the military environment.
>
> Realize that the same reasons these skills are relevant within a family make them relevant within the workplace. Empathic listening builds trust and encourages cooperation, and may help small group cohesion—a critical factor in team performance in combat and crisis situations. Though you may not see many empathic listening behaviors when the bullets are flying, it lays the groundwork for success in that environment.

    The listener truly understands how the speaker feels.

    The speaker feels understood.

    The listener can give better advice.

    The speaker will be more open to it.

"Everybody wants to talk, few want to think, and nobody wants to listen."
        – Anonymous

SUGGESTIONS FOR IMPROVING EMPATHIC LISTENING SKILLS

For empathic listening to be successful, the listener must understand the content of the message and the speaker must feel understood. It's this second half—making the speaker feel understood—which requires some specialized skills.

Don't you hate it when others want to "one-up" your story with something else? Or jump in and give advice before they understand? If you'd like to improve your emphatic listening skills, try to avoid the following invasive responses—they get in the way when you're "seeking to understand:"

AUTOBIOGRAPHICAL STORIES: "Your situation is just like something that happened to me…" "When I was your age…"

ADVISING: Immediately providing counsel based on our own experience—whether or not it was requested.

PROBING: Asking questions solely from our own frame of reference.

EVALUATING: Immediately agreeing or disagreeing.

These types of responses are discouraged because they distract the speaker from the critical part of the message, and they allow the listener to derail the conversation. Extroverts have more difficulty avoiding autobiographical responses than introverts do. Parents often use these responses with their children—the trick is to decide how much is too much. These responses are appropriate *after* the speaker feels understood and is looking for advice or help … it's just important to wait until that point. The listener should return to empathic listening if emotions rise again.

The skills involved in empathic listening are easy to describe, but difficult to practice. To help you remember them, think about the acronym "HEAR"

> **H = HEART:** Commit to listening sincerely, avoid manipulation; remember the importance of the person, as well as the issue.
>
> **E = EMOTIONS:** Look and listen for speaker's emotions as much as their words.
>
> **A = AVOID:** Avoid advice, autobiographies, evaluation; don't interrupt or change the subject.
>
> **R = REFLECT:** Reflect back meaning and feeling until the speaker feels understood.

SOME FINAL THOUGHTS ON LISTENING: MOTIVATION MATTERS!

Think about the times you did a good job of listening…. Why did you do well? One of the reasons was probably your level of motivation. Here are four suggestions on how to enhance your motivation to listen in future situations.

1. Recall why this listening situation is important. Listeners should try to remember why this particular situation is important. (Speakers should try to remind their audience why it's important to listen, too).

Informative and listening situations:

- Is it critical to mission success? Will this help me do my job better?
- Will this information help me make a better decision?
- Is it on the test? Is it a prerequisite for other material that is on the test?
- Can someone get hurt if we mess this up?
- Will our outfit look bad if I don't "get" it? Will I look bad?
- Will we be discussing this later? Will I have to teach this to someone else?
- Is understanding this important to my personal or family goals?

Critical listening situations:

- Do I have to decide which position to take?
- Is the evidence strong and the logic sound?
- What unanswered questions surround this issue? What things are being left unsaid?

Empathic situations:

- Is this speaker having trouble communicating the facts because of strong emotions?
- Do I need to mend fences with this coworker?
- In this negotiation, am I really sure I know what is most important to the other party?
- Does this family member count on me for emotional support?
- Do I have a personal commitment to this person, as well as this issue?

2. Identify and correct barriers to listening motivation. Both listeners and speaker can benefit from a hard look at factors that may inhibit listening. What factors can you control? What factors do you just have to live with? Fix what you can, and acknowledge what you can't.

Physical barriers may block listening—noisy equipment, visual distractions, etc. Avoid distractions when possible: If you can, sit up front. This puts less noise and visual distraction between you and the speaker, and allows you to more clearly see any visual aids. Sitting next to quiet (or boring!) people also helps you to focus on the speaker.

Personal barriers such as physical fatigue, illness and discomfort, as well as psychological distractions like work, family or financial problems, can also affect listening.

Semantic barriers may create obstacles: words or phrases with more than one meaning; ideas, objects or actions with more than one word image; slang, jargon or organizational acronyms. For example, the word "crusade" may have one connotation with an audience in Ohio, but have a different connotation when used in a press conference in the Middle East.

3. Look for common ground. Our listening motivation is crippled when we adversely stereotype speakers or topics. Are you carrying around biases that might be triggered by a speaker's age, race, religion, gender, ethnicity or personal appearance? Have you already decided that this topic is uninteresting, irrelevant or taught at a level that is below you? This kind of thinking can be lethal to listening motivation.

To break the pattern, ask yourself some more empowering questions. What are the underlying commonalities and interesting interrelationships between your interests and those of the speaker? Relate the topic at hand to your own interests. Sometimes this small, simple exercise creates an interesting challenge and bridges the gaps that exist.

4. Treat listening as a learning opportunity and an intellectual challenge. So, how do you become motivated to listen to a topic that may, at first glance, seem boring, unrelated or irrelevant?

If you're listening to an informational briefing on a topic you're already an expert on, listen to improve your ability to teach the topic to someone else. How did the speaker organize the talk? What terms were defined? What terms were not defined? Did the speaker use interesting examples that you could "borrow" when you try to explain to someone else?

If you're listening to a speaker who's failing miserably, ask yourself how could he or she do better? Make some notes, and offer tactful but constructive feedback later.

Translate the problem to a personal, intellectual challenge. Effective and active listening is an exercise in critical thinking and can serve to sharpen your concentration skills. Develop a "remember game" and make listening a learning activity.

SUMMARY

In our eagerness to make our point or make our mark, we often forget there's more to the story than what we know. Listening is a critical communication skill that doesn't get the attention it deserves. Misunderstandings and mistakes in communication can be lethal in the military environment, and listening helps build trust between members of the Air Force team.

In this chapter we have talked about three approaches to listening—informative, critical and empathic listening. Different situations call for different approaches and the listening skills used should be tailored to the situation.

PART V:

WORKPLACE CHALLENGES

CHAPTER 12

ELECTRONIC COMMUNICATION

This chapter covers:

- E-mail basics.

- Tips for effective e-mail messages.

- Voice mail and telephone systems.

- A glossary of computer terms.

Electronic communication technology and software are constantly evolving. We would only muddy the waters if we tried to provide definitive guidance for everything you need to know, and most large organizations have their own written guidance. For that reason, we'll hit the basics here, then refer you to local experts and policies for the rest. (If you see any unfamiliar words in this chapter, check the small glossary of electronic communication terms located on pages 152-154.)

A *computer* was something on TV
From a science fiction show of note.
A *window* was something you hated to clean...
And *ram* was the cousin of a goat ...
Meg was the name of my girlfriend
And *gig* was a job for the nights.
Now they all mean different things
And that really *mega bytes*.

E-MAIL AT WORK: KNOW WHEN IT'S AUTHORIZED

Air Force Instruction (AFI) 33-119, *Official Messaging*, provides guidance on what constitutes official and authorized use of e-mail. Official use includes communications (including emergency communications) the Air Force determines necessary in the interest of the Federal Government. This category includes communications by deployed Air Force personnel, if approved by commanders in interest of morale and welfare.

Authorized e-mail includes personal e-mail approved by the "agency designee," who is the first commissioned officer or civilian above GS/GM-11 in the chain of command or supervision of the employee concerned. AFI 33-119 identifies several common-sense guidelines that should be considered when approving personal e-mail—it should be of reasonable frequency and duration, and it should not add significant cost or overburden the computer system and so on.

Commanders and supervisors may authorize any use that is directed by the Uniform Code of Military Justice (UCMJ) and Joint Ethics Regulation, and does not interfere with mission performance.

AFI 33-119 specifically prohibits the following on government communication systems:

- Distributing copyrighted materials without consent from the copyright owner.

- Using e-mail for financial gain.

- Intentionally misrepresenting your identity or affiliation.

- Sending harassing, intimidating, abusive, or offensive material.

- Using someone else's UserID and password without proper authority.

- Causing congestion on the network by sending inappropriate e-mail messages (propagation of chain letters, etc.) or excessive data storage on the e-mail server.

Remember that the network, like the phone, is subject to monitoring. Your e-mail is saved on backup tapes and servers, and it can be used against you!

"Do unto others in cyberspace as you would do unto them face to face."
– Anonymous

ELECTRONIC MAIL

E-MAIL BASICS

Though some communication guidelines are universal (FOCUS principles, the Seven Steps to Effective Communication, etc.), e-mail is a truly unique medium.

Remember the three e-mail *advantages:*

> It's fast.
>> It can get to more people.
>>> It's aperless.

Remember the three e-mail *disadvantages:*

> It's fast … a quickly written e-mail can fan as many fires as it extinguishes.

> It can get to more people … too many copies can clog the network and can be forwarded into the wrong hands.

> It's paperless … but it does leave an electronic trail and power fluctuations can make things disappear.

E-mail started off as a very informal communication technique, but today it covers a spectrum from personal to professional. **Personal e-mail** often includes language that is a mixture of shorthand and slang. (Check out page 149 for some examples ☺). It doesn't necessarily affect the Air Force (unless you make a very big mistake) and you probably wouldn't include a signature block at the end of the message.

Professional or **official e-mail** is different—it's more like a business memo. It *does* affect the Air Force and the rules you follow should conform to military courtesy. Though professional e-mail is often less formal (and less carefully reviewed) than most "paper copy" correspondence, there are some basic guidelines to help keep you on track.

From: Forbes Lance Lt Col ACSC/DEOT
To: *@*
Cc:
Subject: Max1 Gateway Outage

Good morning,

The Max1 mail gateway is down and has been since yesterday. The 42 Comm Squadron is working the problem and will notify me when it is back in operation.

// SIGNED //
Lt Col Lance Forbes
ACSC/DEOT
Director of Technology Operations
3-6937

PROFESSIONAL E-MAIL

AIR FORCE E-MAIL: SPECIAL CONSIDERATIONS

Use appropriate greetings and closings. Address people with their titles when appropriate. AFI 33-119 states that all official e-mail will include //SIGNED// in upper case before the signature block to signify it contains official Air Force information (e.g., instructions, directions or policies). At a minimum, official signature blocks should include name, rank, position and organization, but often include telephone numbers (both commercial and DSN) and addresses (both commercial and e-mail).

- **Follow the chain of command.** Comply with standard procedures to correspond with superiors. Be professional and watch what you say since e-mail is easily forwarded.

- **Think of the e-mail address as the recipient's personal phone number.** If the topic is important enough that you'd call the general without talking with the colonel, then send the message to the general. (Don't quibble—sending the colonel an info copy of the message doesn't count as following the chain of command.)

- **Get approval before sending to large groups or the public.** Check local policies for the wickets you must jump through before using large mail groups (all base personnel, all users, etc.) or the general public. Excessive e-mail sent to large mail groups can waste a great deal of time and overload the server. Also, just because the material is attached to an e-mail doesn't mean you can avoid the standard clearance procedures.

> **Signature Blocks**
>
> If you haven't already done so, you may want to set up your e-mail formatting to automatically add a standard signature block to your e-mail. Most programs allow you to select and save different versions for different purposes (formal, informal, etc.).
>
> Don't overdo the cute logos and snappy fonts though—they can look frivolous and possibly overload the system.
>
> **NOTE:** An e-mail signature block does not verify an e-mail came from the person indicated. However, the new Common Access Card (CAC) supports an encrypted electronic signature that verifies the e-mail truly came from the sender and has not been altered—reference AFI 33-119 for more detail.

ELECTRONIC STAFFING

If properly managed, using e-mail for coordination and staffing can increase efficiency. Use organizational accounts when sending correspondence to offices for coordination or action. Each MAJCOM typically issues their own guidance on the details of how electronic staffing should be implemented, and local commanders may provide additional guidelines that take into account the local conditions and unit operating procedures. Check out pages 222-223 for the ABCs of staff packages, then see your local guidance for implementation details.

E-MAIL PROTOCOL

"When thou enter a city, abide by its customs."
– The Talmud

E-mail protocol or "netiquette" provides guidelines for proper behavior while on-line. There are many ways to make social blunders and offend people when you are posting. To make matters worse, there is something about cyberspace that causes a "brain burp" and erases the reality that we are dealing with live human characters and not some ASCII characters on a screen. Respect the social culture, and remember that the net is multicultural. Nuances get lost in transmission. For the newbies floating around out there, cruise through the Netiquette suggestions on the next few pages. As always, remember some practices are dictated by aspects of the e-mail system itself (software, gateways, hardware).

RULE #1: BE CLEAR AND CONCISE

⌨ **Make the "Subject" line communicate your purpose.** Be specific: avoid titles like "Status"—call it "OI 36-109 Update Status." Some audience members keep a lot of messages in their inboxes—a clear title will help them find your message later.

⌨ **Lead with your most important info.** If your goal is to answer a question, then paste the question on top for clearer understanding.

⌨ **Use topic sentences** if the e-mail has multiple paragraphs (see *Drafting Effective Paragraphs* [pages 68-69]).

⌨ **Be brief and stick to the point.** Follow all the basic rules for drafting clear and concise messages (pages 73-88). They're even more important in e-mail because it's tougher to read from a computer screen. Address the issue, the whole issue, and nothing but the issue … and you get extra points if you can get your message in 24 lines (size of one screen on most computers).

⌨ **Use bold, italics or color to emphasize key sentences.** If your e-mail doesn't allow these, a common convention is to use a star on either side of key passages for emphasis … this highlights the *key points.*

⌨ **Choose readable fonts.** Use 12 point or larger font size. Use easy-to-read fonts and save the script fonts for your signature.

RULE #2: WATCH YOUR TONE

"Life is not so short but that there is not enough time for courtesy."

– Ralph W. Emerson

⌨ **Be polite.** Treat others as you want to be treated. Use tact. Then use more tact. Then, for good measure, use more tact. Think of the message as a personal conversation. If you were face to face, would you say the same words and be as abrupt? If not—rewrite the message with a more positive tone (see pages 23-24).

⌨ **Be careful with humor, irony and sarcasm.** Electronic postings are perceived much more harshly than they are intended, mainly because you cannot see body language, tone of voice and other nonverbals that make up 90 percent of interpersonal communications. Positive enthusiasm can be easily mistaken for angry defiance when you use capital letters, exclamation points and strong adjectives and adverbs.

⌨ **DON'T SHOUT.** Do not write in CAPITALS—it's the e-mail version of shouting and it's considered very rude.

⌨ **Keep it clean and professional: E-mail is easily forwarded.** Harassing, intimidating, abusive, or offensive material is obviously unacceptable, but aim for a higher standard. If you wouldn't want it overheard or posted on the office bulletin board, it probably doesn't belong in an e-mail.

⌨ **Don't send in haste, repent at leisure.** E-mail can get you into trouble—its informality encourages impulsive responses (like a conversation), but your words can be printed out and forwarded. If you're really mad about an issue, go ahead and draft an e-mail, but don't send it until you calm down and read it over. *Never* flame! If you do, be prepared to apologize.

RULE #3 BE SELECTIVE ABOUT WHAT MESSAGES YOU SEND

⌨ Don't discuss controversial, sensitive, official use only, classified, personal, privacy act or unclassified info requiring special handling of documents. You just may one day see yourself on CNN or *America's Most Wanted*.

⌨ Remember OPSEC (Operational Security). OPSEC is a continuous analytical process which involves identifying sensitive information, recognizing that information could be valuable to an adversary, and making changes in the way we do things to reduce our risk that the information will be compromised. Even unclassified information, when brought together with other information, can create problems in the wrong hands.

⌨ Don't create junk mail, forward it, or put it on a bulletin board.

⌨ Don't create or send chain letters. They waste time and tie up system.

⌨ Don't use e-mail for personal ads; put these comments on an appropriate BBS or web page.

⌨ Don't fire/promote someone by e-mail. Some messages aren't appropriate.

RULE #4 BE SELECTIVE ABOUT WHO GETS THE MESSAGE

⌨ Reply to *specific* addressees to give those not interested a break.

⌨ Use "reply all" sparingly.

⌨ Get permission before using large mail groups.

⌨ Double-check the address before mailing, especially when selecting from a global list where many people have similar last names.

RULE #5 CHECK YOUR ATTACHMENTS AND SUPPORT MATERIAL

⌨ Ensure ALL info is provided the first time to keep from repeating e-mail to "add just another fact!" *Support Your Ideas* (pages 41-53).

⌨ Before sending, check your attachments. (It's the most common mistake.)

⌨ Cite all quotes, references, and sources. Respect copyright and license agreements.

RULE #6 KEEP YOUR E-MAIL UNDER CONTROL

⌨ Sign off the computer when you leave your workstation—defamers may read your mail or send hostile messages under your e-mail address!

⌨ Create mailing lists to save time. REMEMBER: Time is a dime!

⌨ Read and trash files daily. Create an organized directory on your hard drive to keep mailbox files at a minimum. Ensure record copies are properly identified and stored in an approved filing system.

⌨ Acknowledge important or sensitive messages with a reply to sender: "Thanks," "done," "I'll start working it immediately," etc.

⌨ If you will be away from your e-mail for an extended period, consider setting up an Auto Reply message that gives people an alternate point of contact.

PERSONAL E-MAIL

Personal e-mail sometimes contains shorthand and slang that would be unacceptable in a professional communication. Emoticons (facial expressions) or abbreviations are sometimes used with humor or satire in an attempt to make sure the audience doesn't "take things the wrong way." They are a resource, but use them sparingly. Some of them are more clever than clear, and much of your audience may only know the smiling face. (In fact, some software automatically substitutes the icon "☺" for the emoticon below.)

⌨ :-)—I'm smiling

⌨ :-(—I'm unhappy

⌨ ;^)—satire or sarcasm

⌨ :'-)—I'm crying

⌨ :-<—very sad

⌨ >:-(—angry

⌨ :-@—screaming

⌨ :-&—tongue-tied

⌨ :-x—my lips are sealed

⌨ :-_|—I'm being tongue-in-cheek

⌨ %-)—I've been staring at the screen too long

⌨ <g>—grin

⌨ <bg>—big grin

⌨ <vbg>—very big grin

⌨ BRB—be right back

⌨ BTW—by the way

⌨ CU—see you (as in see you later)

⌨ FWIW—for what it's worth

⌨ FYI—for your information

⌨ GMTA—great minds think alike

⌨ IMHO—in my humble or honest opinion

⌨ IMNSHO—in my not-so-humble opinion

⌨ JK—just kidding

⌨ LOL—laugh out loud (it's a way to tell folks you laughed at their last comment)

⌨ RE—regarding

⌨ REQ—for requests

⌨ ROTFL—rolling on the floor laughing

⌨ URGENT—time-critical messages

⌨ YMMV—your mileage may vary

"A good listerner is not only popular everywhere but, after a while, he gets to know something."

– Wilson Mizner

VOICE MAIL AND TELEPHONE SYSTEMS

Well, if this is the wrong number … why did you answer it?"
– James Thurber

Voice mail, fax, and telephone systems are key to staff communication, so we'll review some of the common courtesies associated with using this equipment.

VOICE MAIL & ANSWERING MACHINES

Different systems have different features, so check out your manual for all the bells and whistles. Here are some basic guidelines.

Do ... record the message in your own voice.

Do ... identify yourself and your organization.

Do ... check your system regularly.

Do ... return all messages as quickly as possible.

Don't ... leave amusing messages on an official system.

If you will be unavailable for an extended period, consider these suggestions:

Identify whom callers can contact in your absence.

Set up your voice mail to answer after the first ring.

A sample message: "This is SSgt_____. If you need some immediate information on _____, call 555-7084."

TELEPHONE PROTOCOL

Telephone protocol is something most of us take for granted, but here are a few tips for people new to the workplace.

If you're answering the phone...

Do ... answer the phone on the first ring and in the way you would like to be called after identifying the organization.

Do ... be pleasant and professional—you are representing your organization, as well as yourself, when you answer the phone.

Do ... introduce everyone in the room if you are on a speakerphone—callers may object to the lack of privacy.

Do ... put the radio and TV on hold until you're off.

Do ... speak clearly, keeping your lips about 1 inch from the mouthpiece. Good posture (or standing while speaking) will also improve your vocal quality.

Do ... have a pencil, a memo pad, and your directories within easy reach.

Do ... adjust your speaking tempo to match the other person's to establish instant rapport.

Do ... ask if someone else can help if the person isn't there.

Do ... take a number and call back instead of putting them on hold if you are finding something.

Do ... give the caller the phone number before you transfer the call.

Do ... allow the person initiating the call to bring it to a close.

Do ... record important conversations, especially those that result in a decision, in a memo for record and place it in a file.

Don't ... transfer an angry caller. Listen carefully, never interrupt and ask questions that require more than a "yes" or "no." Also, make notes and let the caller know since this shows you're interested and are willing to help.

Don't ... put the phone over your chest to put someone on hold—your voice goes over the wires loud and clear—use the "hold" button.

If you're making the call...

Do ... have your act together. Organize your thoughts and make notes before you place a call— especially if you're representing your organization, seeking help or information, calling long distance or talking to someone more senior in rank.

Do ... call during core hours (0900-1100 and 1300-1500) to reduce phone tag. What's their time zone? When will they return?

Do ... identify yourself and your organization before asking to speak to _____.

Do ... be pleasant and professional.

Do ... ask if the person has time to talk if you plan a lengthy conversation. But you need to keep it as brief as possible!

Do ... record important conversations in a memo for record and place it in a file.

Don't ... put the phone over your chest to put someone on hold—your voice goes over the wires loud and clear—use the "hold" button.

REACH OUT & FAX SOMEONE

As e-mail capabilities grow, faxes are becoming less common. If you fax documents regularly, remember the following:

Do ... make it readable by using pica type (10 characters per inch): Times New Roman, Courier New, Century Schoolbook, etc. and number your pages.

Do ... protect your document by ensuring correct receiver information is entered.

Do ... use black and white.

Do ... use a fax when you cannot get someone to return your call, including a short explanation, deadline to return call and a "Thanks for your time."

Do ... use Post-It® stickies to save those trees!

Do ... send a return cover sheet with complete return address to encourage a quick reply.

Don't ... send a legal-sized document unless you know it can be received or personal, confidential, or financial info unless you know it will be protected.

Don't ... use italic type. It looks ragged and makes it difficult to read.

... A SMALL GLOSSARY OF ELECTRONIC AND COMPUTER TALK

American Standard Code for Information Interchange (ASCII)—Pronounced "asky." The most common international standard for representing alphanumeric text on a computer.

bandwidth—The number of bits that can be passed along a communications channel in a given period of time. Usually expressed as bits per second (bps). Each military installation has a limited amount of bandwidth—don't waste it with frivolous e-mail.

bulletin-board system (BBS)—An electronic system allowing individuals with similar interests to post and view messages in a public electronic form.

binary file—A digital file format used to store non-text data. The information stored includes executable programs, sounds, images and videos.

binary digit (bit)—The smallest unit of storage in a digital computer. All programs and data in a digital computer are composed of bits.

browser—A software program that allows users to interact with World Wide Web sites. Example includes Microsoft Internet Explorer, Mozilla and Opera.

client—A computer or program that can download, run or request services from a server.

compact disc (CD, CD-ROM, CD-RW): A disk that stores digital information using a pattern of microscopic pits and lands to represent ones and zeros. One CD-ROM holds from 650 to 700 megabytes of data or the equivalent of approximately 250,000 pages of text.

data compression—A procedure used to reduce the size of a file to reduce the disk space required to store the file or the bandwidth required to transmit the file. Many different compression formats are available and each requires a program to compress and expand the file—zip format is one of the most common.

digital video disk (DVD, DVD-R, DVD-RW)—A disk that stores digital information using a pattern of pits and lands to represent ones and zeros. A specially formatted DVD is used to store movies. One DVD-ROM holds approximately 4.7 gigabytes of information or the equivalent of approximately 1.8 million pages of text.

Electronic mail (E-mail)—A message sent electronically over a computer network, such as a LAN or the Internet.

emoticon—Facial expressions originally drawn using ASCII characters and more recently drawn using an extended character set. (See page 149 for examples).

encryption—Changing the contents of a message in a manner to obscure the contents while still allowing the intended audience to read the message.

executable—A file containing a set of instructions to perform some process on a computer. A word processor and Internet browser are examples of an executable.

facsimile (fax)—A method of transmitting images of printed matter that predates digital computer networks. This method traditionally used phone lines but can now be implemented using computer networks.

flame mail—An e-mail message critical of some person or position taken by a person. Usually more derogatory than constructive.

frequently asked questions (FAQ)—A list of questions and corresponding answers focusing on a specific topic. The FAQ is typically provided to members of a community to avoid the repetitious answering of questions asked by new users.

home page—The web page providing the entry point for a web site (see web page and web site).

hyperlink—A way to link access to information of various sources together within a web document. A way to connect two Internet resources via a simple word or phrase on which a user can click to start a connection.

hypertext—A method for storing, retrieving, and presenting information based on the processing power of computers. Allows computerized linking and almost instantaneous retrieval of information based on a dynamic index.

Hypertext Markup Language (HTML)—The native language of the WWW.

Instant Messaging (IM)—A type of communications service that enables you to create a kind of private chat room with another individual in order to communicate in real time over the Internet, analogous to a telephone conversation but using text-based, not voice-based, communication. Typically, the instant messaging system alerts you whenever somebody on your private list is online. You can then initiate a chat session with that particular individual.

Internet—The overarching global computer network connecting computers, servers and local area networks across the globe.

intranet—A network with restricted availability. An intranet may provide web pages, printing and e-mail services similar to those available using the Internet, but only for a restricted set of users. Most military bases run an intranet that is only available to personnel on that base.

Internet Relay Chat (IRC)—A communications program that allows real-time text-based conversations along multiple users.

list server—A computer running an electronic mailing list subscribed to by individuals with some common area of interest. Individuals typically subscribe by sending an e-mail asking to be placed on the list. Once added to the list, subscribers automatically receive messages sent to the list by other subscribers and may send their own messages to the list which are then relayed to all other subscribers.

local area network (LAN)—A system occupying a relatively small geographic area providing digital communications between automated data processing equipment, such as computers and printers.

modem—A device allowing a computer to send and receive data over telephone lines.

netiquette—Commonly accepted etiquette used when communicating over a computer network—network etiquette. Specific forms of communications might include e-mail, list server or IRC.

newbie—An individual new to using computers or new to a specific group.

newsgroup—A network service allowing individuals to post, read messages, and respond to messages posted by other users. Newsgroups may be moderated or unmoderated. If the newsgroup is moderated, messages may be removed by the moderator and user posting privileges controlled by the moderator.

organizational e-mail account—An e-mail account used to receive and send messages on behalf of an organization. This type of account allows an organization to maintain a single address for correspondence despite changing responsibilities within that organization.

server—A computer that responds to requests for information from client computers—see client.

web browser—See browser.

web page—An electronic document available on the Internet or an intranet that is viewable using a web browser (see browser, Internet and intranet).

web site—A collection of related web pages.

World Wide Web (WWW)—The entire web pages on all of the web sites available through the Internet.

The Ten Commandments For Computer Ethics

1. Thou shalt not use a computer to harm other people.

2. Thou shalt not interfere with other people's computer work.

3. Thou shalt not snoop around in other people's files.

4. Thou shalt not use a computer to steal.

5. Thou shalt not use a computer to bear false witness.

6. Thou shalt not use or copy software for which you have not paid.

7. Thou shalt not use other people's computer resources without authorization.

8. Thou shalt not appropriate other people's intellectual output.

9. Thou shalt think about the social consequences of the program you write.

10. Thou shalt use a computer in ways that show consideration and respect.

– from the Computer Ethics Institute

CHAPTER 13
MEETINGS

This chapter covers:

- Meeting planning—how to prepare for success.

- Meeting agendas and execution.

- Preparing the meeting minutes.

- Tips on group dynamics and making meetings fun.

All of us attend meetings. Some of us feel we attend too many of them. Others may be conducting more business electronically, and attending fewer meetings than in the past. But, in today's world of trying to do things "faster and smarter," technology is not always all it's cracked up to be. How many times have you e-mailed someone who works 20 feet or one office away? Have you ever spent 20 minutes to write an e-mail to four people to discuss a topic that would take 2 minutes of a meeting? Despite the surge in electronic staffing there is no substitute for the human element found in meetings; thus it is unlikely they will ever be replaced.

So why does the mere word "meeting" strike a nerve in so many of us? Probably we've all attended so many that were a huge waste of time. But meetings don't have to be that way; there are ways to make them better! If they are done right, they can go a long way in helping your organization run more efficiently. Simply put, they're used to share information, solve problems, plan, brainstorm or motivate. Whatever the purpose, it's good to know some basics about conducting an effective meeting. That's what this chapter is all about.

RUNNING AN EFFECTIVE MEETING

At some time or another, you may be the one calling the meeting. A meeting doesn't always have to be a major production, but there are some key points to consider during planning, execution and follow-up. Here are some tips on making the most of everyone's time—including your own!

PLANNING YOUR MEETING

Success or failure in a meeting can usually be traced to the planning phase. Do your homework, and you're well on your way to success. If you don't do your homework, you'll pay a heavy price during the execution and follow-up phases.

Listed below are the key issues associated with planning a meeting. As you step through these items, remember to check on what is standard operating procedure in your organization. Meetings come in all flavors—from totally spontaneous to highly structured and ceremonial. Most are in the middle. If this group has been meeting regularly for a while, try to find out how they've done business in the past.

1. DECIDE IF A MEETING IS APPROPRIATE

In the book *How to Make Meetings Work,* Michael Doyle and David Strauss identify seven situations when having meeting might be a good idea:

- You want information or advice from the group.
- You want to involve a group in solving a problem or making a decision.
- An issue needs to be clarified.
- You want to address concerns with the entire group.
- The group itself wants a meeting.
- There is a problem that involves people from different groups.
- There is a problem, but it's not clear what it is or who is responsible for dealing with it.

In these situations, face-to-face discussion can help speed up the process. If your goal is just to pass on information, ask yourself if e-mail is a viable and *appropriate* substitute for the meeting. The purpose of many meetings is simply to share information and keep people up to date on a project. In these cases, try to substitute the meeting with an e-mail. This saves everyone's time and still keeps everyone in the loop, and you're still meeting the goal for the meeting you just avoided!

Local policy may dictate that some groups meet weekly, monthly, or quarterly, but if you're not directed to meet and don't need to … don't!

2. DEFINE YOUR PURPOSE

Every meeting should have a purpose. If it doesn't, you shouldn't meet. When you think about your purpose, try to define it in terms of a product that you want at the end of the meeting, and what it will be used for. "Talking about Issue X" is not an ideal purpose statement for the meeting, because it describes a process, not a product—try these alternatives:

- To identify why Issue X is a problem. [The product is a clearer understanding of the problem.]

- To brainstorm ways to resolve Issue X. [The product is a list of ideas about potential solutions.]

- To discuss different options for resolving Issue X. [The product is list of pros and cons ... the decision will happen later.]

- To decide how the unit will handle Issue X. [The product is discussion of pros and cons and a decision. Make sure attendees know who makes the decision. Is it the team, or is it the boss?]

Most Air Force professionals want to feel like they've accomplished something in a meeting. A clear purpose for the meeting is the first step towards success.

3. DECIDE WHO SHOULD BE INVITED

Have you gone to a meeting and after 5 or 10 minutes asked yourself, "Why am I here?" Remember that when you are holding a meeting! Invite only those directly involved in the issues being discussed. Meetings can be a time waster if too many or too few participants attend. Too many people equals chaos; too few means decisions have to be put on hold—bad news.

If you're trying to solve a problem or make a decision on a controversial issue, make sure you have adequate representation from all groups who have a voice in the decision. If you only invite people with one point of view, your meeting will run smoothly, but your decision may not stand up later.

4. DECIDE WHERE AND WHEN THE MEETING SHOULD OCCUR

Check the schedules of any heavy hitters. Often the scheduling of your meetings will be determined by the schedule of any key personnel that will be attending. If you're briefing a three-star, odds are that his secretary will be telling you when the meeting can occur, not the other way around.

If possible, pick the time of day to meet your purpose. If you've got flexibility, you might select the time of day for your meeting to help you meet your objectives. If you want your meeting attendees fired up and eager to contribute, you may want to schedule a meeting in the morning, the time when most people have more energy. If you want them inpatient and anxious to get done, try just before lunch. If you want them agreeable, try right after lunch. If you want them asleep, try midafternoon. Finally, you might try the just-before-quitting time tactic. If you're lucky, you might have any opposition collapse just so they can catch the car pool.

Try to avoid meetings the first thing Monday morning. Give folks some time to read e-mail and prioritize the days and week ahead. If your organization is a service organization, don't schedule meetings during customer service hours. The "Closed for Training" sign on the door does not fare well in the customer satisfaction department!

Keep it under an hour, or plan for breaks. Keep in mind that after 20 minutes or so, our minds tend to get lazy and wander. Try to keep the flow of the meeting going so that no agenda item goes longer than 20 minutes. Watch the clock (a timekeeper comes in handy). Try to keep the meeting to 90 minutes or less, and plan for breaks if the meeting goes over an hour.

Reserve the room. Follow established procedures to reserve a conference room that can handle the attendee list.

5. PLAN FOR CAPTURING MEETING INFORMATION

If this is not a routine meeting with an appointed recorder, take a moment to think about how you will capture the meeting information, both in the meeting itself and afterwards. Capturing the information is critical, but how best to do it will depend on the nature of the meeting.

Think about how you're going to capture information during the meeting. Will you use a white board? (If so, make sure your conference room has one ... or try to borrow a portable whiteboard that can print out what is written on them.) Will you capture brainstorming efforts using flip charts and "butcher paper"—oversized paper that can be taped up on a wall and looked at by a group? Do you have an administrative support person that can act as a recorder? Can you ask a coworker to take notes?

Meeting minutes capture the process and outcome of the meeting. They "close the loop" on the meeting, and let the attendees know what was decided. If you have administrative personnel that can prepare minutes, consider having them attend the meeting to act as the recorder. On the other hand, drafting the minutes yourself has one clear advantage: you can make sure they include what they should.

Explaining the objectives, reviewing the agenda, going over the ground rules and making sure everyone knows each other and understands why they're in attendance.

6. SEND OUT AN AGENDA

Create an agenda and send it to attendees no later than 1 or 2 days prior to the meeting. If this is not a time critical issue, if the attendees don't work for you, or if you are asking people to present material or review long documents prior to the meeting, try to give at least a week's notice. Also tell presenters to bring a copy of their material to leave with the recorder.

The agenda should include the date, time, location and purpose of the meeting. This advance notice gives everyone an opportunity to prepare their thoughts and know where the meeting is going before they get there.

If you are asking people to present material or prepare their thoughts, make sure they know how much time they have been allocated and when they will present. You may also want to use a detailed agenda on the day of the meeting to keep the group on track and stay focused.

If you have trouble coming up with a solid agenda, chances are you really don't have a reason to meet in the first place (see tip one)!

7. REMEMBER: MEETINGS CAN BE CANCELLED

> **Presenting material at a meeting**
>
> At some time or another, you may be asked to *present* information at a meeting, possibly in a briefing format. Whether or not you have time to prepare, it's always important to remember the "Seven Steps for Effective Communication" introduced in Chapter 2. A meeting is just another communication platform and it's still important to analyze your purpose and audience, research your topic, support your ideas, organize, draft, edit and seek feedback. If you don't, you may fail to hit the target, lead the audience to a place where no man has gone before, or worst case scenario, waste everyone's time.

Before you take 5, 10 or 20 people away from productive work time ask yourself these questions:

- Is there a real need for this meeting? You might receive feedback from your agenda that indicates that the business can be accomplished by other means. Consider sending an e-mail or a memo to disseminate information.

- Will key decisionmakers (if a decision is the goal) and/or the majority of the group be present? If key decision makers can't be there, why bother? If the key decision makers are the entire group, make sure the majority can attend. Otherwise you spend precious time updating those who missed the meeting before you can come to consensus on the issue. It's better to put the meeting on hold.

RUNNING YOUR MEETING

You won't peg out anyone's fun meter by dragging a meeting out unnecessarily. It's your job to keep it focused, and separate the "wheat from the chaff."

START ON TIME

Start on time with an upbeat note, and don't wait for tardy attendees. State your desired outcome.

FOLLOW YOUR AGENDA

One of the worst things you can do is to ignore your own agenda. It drives people nuts! Consider a review of the detailed agenda in the opening minutes of the meeting to remind people of the goals and plan for the meeting.

Regular meetings usually follow a fixed order of business. Below is a typical example:

1. Introduction/call to order.
2. Attendance/roll call of members present.
3. Acknowledgment/correction of previous meeting minutes.
4. Reports from committees/persons [these items were previously identified in the Agenda].
5. Old Business from previous minutes/Action Items.
6. New Business previously designated for consideration at this meeting [identified in Agenda].
7. Round table/New Business not previously identified for consideration at this meeting.
8. Appointments/Assignments [new committees or personnel changes].

Some suggest the following allocation of time:

- ¼ of meeting—past agenda items and follow-up (old business)
- ½ of meeting—current agenda items
- ¼ of meeting—future agenda items

Establish the date, time, and location of the next meeting, if needed and if ready. Evaluate the meeting before adjourning. Are there things that need to be improved?

A word about Robert's Rules of Order

Formal meetings sometimes follow the sequence, procedures and terminology documented in *Robert's Rules of Order,* a set of rules used in Parliamentary procedure. General Henry M. Robert first published the rules in 1876, and there are several updated versions and summaries available in print or on the web. (For example, the complete 1915 version is available on line at http://www.bartleby.com/176/

Most Air Force meetings do not use *Robert's Rules of Order*, and we won't go into them here. The terminology is somewhat formal, and the structure seems overly constraining to some. Even so, you should be aware that they exist as agreed-upon set of guidelines on how to organize a potentially chaotic situation.

Also, some day you may find yourself in a meeting where there are a lot of rules, where people read out the minutes of the previous meeting, and where parts of the meeting seem scripted ("With a quorum present, the meeting is called to order…"). If this happens to you, don't panic. They're probably following some form of *Robert's Rules of Order*. Ask about them, or look them up yourself. Everything will become clear.

UNDERSTAND GROUP DYNAMICS AND LOOK FOR WAYS TO MAKE MEETINGS FUN

If you're in charge of a group that will be meeting over a period of time, it pays to learn the basics about group dynamics. Also, if you dare, you might want to try to inject some fun into your meetings. (As always—know your audience. Some believe meetings should be miserable … anything else is unnatural!) We'll talk more about both these topics at the end of the chapter.

FOLLOWING UP: PREPARING MEETING MINUTES

If you did your homework, following up is a snap—it involves sending out meeting minutes and starting the whole cycle over again. Prepare meeting minutes in the official memo format. Minutes are a clear summary of the participants' comments. They document planned or completed action. Date the minutes the day they are distributed. You may list names of members present in two columns to save space.

Place information regarding a future meeting in the last paragraph. When a person signs a paper as a member of a board or committee, the signature element indicates that person's status on that board or committee, not any other position the person may hold. Type "Approved as written" two lines below the recorder's signature block, followed by the approving authority's signature block.

Minutes are typed either single or double-spaced, with additional space between items of business and paragraphs.

No erasures should appear in the minutes. All typing should be neat and orderly, paying particular attention to uniformity of margins and text.

All names should be spelled correctly; acceptable grammar should be used; sentences should be well constructed and correctly punctuated. All verbs should be in past tense.

The order of the minutes usually coincides with the order of the agenda, and the following items are generally included:

- Kind of meeting (regular, special, etc.)
- Day, date, time and place of meeting
- The word "Minutes" in the heading
- Name of the meeting body
- Opening paragraph; i.e., The Executive Committee met for _____ meeting on day, date and time
- Members present
- Members absent
- Action taken on last meeting's minutes
- Reports
- Current business, with complete discussions and conclusions
- Old Business, with discussions, and follow-up, as recommended
- New Business, with discussions and recommendations
- Adjournment

MEETINGS AND GROUP DYNAMICS

All meetings, teams or groups move through predictable stages. You will save yourself a lot of frustration if you are familiar with these stages. Let's look at a snapshot of these stages first and then discuss them in more detail.

FORMING: a period of uncertainty in which members try to determine their place in the team and the procedures and rules of the team.

STORMING: conflicts begin to arise as members resist the influence of the team and rebel against accomplishing the task.

NORMING: the team establishes cohesiveness and commitment, discovering new ways to work together and setting norms for appropriate behavior.

PERFORMING: the team develops proficiency in achieving its goals and becomes more flexible in its patterns of working together.

Because the forming, storming and norming stages result in minimal output, it is tempting to try to rush through or short circuit these stages and hope the team can thereby achieve peak productivity. You may want to stand up and say, "Can't we just all be friends?" Although seductive, this idea is dysfunctional. Just as individuals go through predictable stages of growth (depending on age, experience, maturity and other factors), teams go through predictable stages. The duration of these stages depends on factors such as individual and team maturity, task complexity, leadership, organizational climate and external climate. Teams can fixate at various stages. Some teams (like some people) are never fully functioning. How can you reduce the nonproductive time commonly spent in the forming and storming stages? Given that these stages are inevitable, try sharing rumors, concerns and expectations of the team to minimize their tensions, fears or anxiety. Also, encourage the team members to contact one another so that there will be no "surprises." Therefore, an atmosphere of trust will be achieved early on (norming stage), allowing for interpersonal issues to be put aside in favor of task issues and for the team to move on to the performing stage.

Read on for more detailed descriptions of these stages from *The Team Handbook* written by Peter R. Scholtes.

FORMING. When a team is forming, members cautiously explore the boundaries of acceptable group behavior. This is a stage of transition from individual to member status, and of testing the leader's guidance both formally and informally. Because there is so much going on to distract the members' attention in the beginning, the team accomplishes little, if anything, that concerns its project goals. Don't despair and flush your project down the toilet! This is perfectly normal!

STORMING. Storming is probably the most difficult stage for the team. You may ask yourself, "What was I thinking?" The team members begin to realize the task is different and more difficult than they imagined, becoming testy, blameful, or overzealous. Impatient about the lack of progress, but still too inexperienced to know much about decision making or the scientific approach, members argue about just what actions the team should take. They try to rely solely on their personal and professional experience, resisting any need for collaborating with other team members. Their behavior means team members have little energy to spend on progressing towards the team's goal. Still, they are beginning to understand one another.

NORMING. During this stage, members reconcile competing loyalties and responsibilities. They accept the team, team ground rules (or "norms"), their roles in the team, and the individuality of fellow members. Emotional conflict is reduced as previously competitive relationships become more cooperative. As team members begin to work out their differences, they now have more time and energy to spend on the project. Thus, they are able to at last start making significant strides.

PERFORMING. By this stage, the team has settled its relationships and expectations. They can begin performing—diagnosing and solving problems, and choosing and implementing changes. At last team members have discovered and accepted each other's strengths and weaknesses, and learned what their roles are. The team is now an effective, cohesive unit. You can tell when your team has reached this stage because you start getting a lot of work done—finally!

MEETINGS: CAN THEY BE FUN?

Have you ever wondered why pagers miraculously go off during meetings or "scheduling conflicts" come up at the last minute? It's because many people avoid meetings like the plague. Let's face it … they're boring! Does that mean you should hold meetings like a standup comedian? No! In fact, you need to really know your audience before you go out on this limb. But you can interject some fun into meetings to encourage participation and creativity. Here are just a few ideas to get you going.

- Hold your meeting off-site. We military types are creatures of habit. This does not change at meetings. Everyone usually sits in the same seat, brings the same coffee mug, brings up the same questions and takes breaks at the same time … you get the picture. Sometimes a change of scenery is all that is needed to rejuvenate a brain-dead group. **Bonus:** Change the meeting time to coincide with lunch. This gives the group a chance to talk about "non-meeting issues" and connect in a more nonthreatening environment over a meal.

- Have a contest to generate ideas for projects. A little friendly competition can be a boost for all and bring great results.

- Tone down the conversation dominators with "fees." Attendees are required to pay each time they interject, and they are limited to inputs. Be careful with this one … you don't want to shut folks down either.

- Appoint a "Director of Creativity" for each meeting to come up with ideas like these! Appoint the next meeting's director at the conclusion of the current meeting to give them time to plan. Have the group vote on the favorite idea at the end of the year and give out a humorous award.

SUMMARY

Although many of us don't really look forward to meetings they are indeed a fact of life in today's Air Force and they certainly serve a purpose in our mission and how we get things done on a daily basis. This chapter gave you the nuts and bolts for holding more successful and productive meetings. Remember the "Seven Steps for Effective Communication" and the tips introduced in this chapter as you think about and prepare your next meeting. And as always, pay attention to the local guidelines for conducting meetings in your organization. Good luck!

Never invite people who don't need to be there. Never meet without an agenda. Never start late. Don't get off the subject. Don't tolerate excessive interruptions.

PART VI:
THE QUILL

This section expands upon the functions and formats for written communications with the US Air Force as outlined in Air Force Manual (AFMAN) 33-326, *Preparing Official Communications*. We've included the most frequently used formats, which are only a fraction of the total used throughout the Department of Defense. Although functions and formats may differ somewhat among major commands, the information that follows should be useful at any command or staff level. However, if your organization publishes a supplement to AFMAN 33-326, an OI (operating instruction), or its own administrative style guide, be sure to check those sources for command-unique guidance on preparing staff work.

AIR FORCE WRITING PRODUCTS AND TEMPLATES

WHEN YOU WRITE FOR "THE BOSS"

There may be a time when you will be tasked to prepare a staff instrument for a general officer's signature. Is there an added dimension when we write for a general officer? From discussions with numerous generals and their executive officers, the answer is "yes." These are facts and hints to help you through this added dimension!

✪ **First ... see if the boss has preferences!** Run to see the general's secretary! It will ease your pain. For instance, when one particular general writes personal letters to higher ranking individuals, he uses "Respectfully" as the salutation, and for lower ranking individuals, "Sincerely." Use "Sincerely" on all personal memos. Now having said that ... it's up to your commander's preference. Some generals write "Very Respectfully" above "Sincerely" when addressing higher ranking officers.

✪ **Read step one of the seven-step checklist.** Learn all you can about the general's views on the subject and the relationship with the addressee. Try to capture the general's wider perspective before you pick up your pen. What peripheral issues facing the general could be directly or indirectly affected by your words? What is the desired *purpose*? What tone (pages 23-24), pattern (pages 59-62) and correspondence style (pages 167-169) are most appropriate?

✪ **Keep it simple or face frustration.** A general's time is spread over many issues. Get to the point, make it, and move on. If the addressee needs only the time, don't send instructions on how to build a clock. Your first draft will probably be twice as long as needed. If you must include details, use attachments.

✪ **Go easy on the modifiers.** A general doesn't need to be *very* interested in something—being interested is sufficient. Also, avoid emotionalism.

✪ **Quality check, quality check, quality check!** From your logic to your grammar, from your facts and figures to your format, triple check your work. You have nothing to lose but your credibility! HINT: Ensure you use the *current* address since military members move around so often.

✪ **Go one step further.** Look efficient when doing a personal letter and provide the general with the "go-by name" of the addressee. Try using a yellow sticky!

✪ **Don't expect your glistening product to fly the first time.** Not even the best staffers are clairvoyant.

Why should we write differently for general officers? Why should that added dimension apply only to *the boss*? There's no good reason. We'd become better communicators if we assume **all** of our correspondence were "starbound."

PERSONAL LETTER

DEPARTMENT OF THE AIR FORCE
HEADQUARTERS UNITED STATES AIR FORCE

15 Jan 04

SMSgt Robin Edwards
Assistant Executive
1700 Air Force Pentagon
Washington DC 20330-1700

Mr. Peter Overall
ECI/EDECT
50 South Turner Boulevard
Maxwell AFB-Gunter Annex AL 36118-5643

Dear Mr. Overall

　　Thanks for your recent efforts to analyze the feasibility of creating a world wide web home page to educate customers on intelligence capabilities. The extensive research you performed during your off-duty time will assist Air Force Intelligence in determining the most effective methods to advertise our capabilities to support warfighting.

　　We plan to use the background papers and prototype hypertext computer program you developed to brief the Technology Tiger Team in our upcoming general officer summit. Your work and demonstrations will add credibility to our concepts and visions. Customers worldwide will be able to instantaneously access up-to-date information to understand how to integrate and use intelligence as an integral part of warfighting.

　　We look forward to continuing our work with you. We found your input invaluable in our efforts to disseminate and communicate intelligence capabilities to our many customers throughout the world.

Sincerely

Robin Edwards
ROBIN EDWARDS, SMSgt, USAF

cc:
Mrs. Elizabeth Adams, ECI/EDECT

DEPARTMENT OF THE AIR FORCE
AIR UNIVERSITY (AETC)

[1.75 inches or 10 lines from top of page] 5 Sep 04¶

¶
¶
Chaplain, Lieutenant Colonel David Davis¶
Staff Chaplain¶
105 South Hansell¶
Maxwell AFB AL 36112-6332¶
¶
¶
Chaplain, Lieutenant Colonel Fred Walker, CAP¶
824 Sunrise Court¶
Marshall MO 65340-2846¶
¶
Dear Fred¶
¶
      Thanks for your fax. Here is the information you requested concerning the Air Force's use of the personal letter. This letter also functions as a cover letter for the draft copy of the new CAP Chaplain Handbook I promised you.¶
¶
      Personal letters are usually prepared on letterhead stationery with the sender's address element one to two lines below the date. Font size should be Times New Roman 12 point. The date is placed 10 lines from the top of the page on the right side. The salutation is normally in the format "Dear Xxxxx" and complimentary close is normally "Sincerely." Begin the salutation one line space under the receiver's address. Type the complimentary close element "Sincerely" one line space below the text of the letter three spaces to the right of page center. Notice that you do not use punctuation after either. The signature element begins five line spaces below and aligned with the complimentary close element. Place your list of attachments one line below the signature element and flush with the left margin.¶
¶
      Personal letters are really official memos prepared in a personal style. They are appropriate for welcome letters, letters of appreciation, letters of condolence, or any other occasion when a situation might be better handled in a personal manner. Attachments, if any, are listed the same way as in an official memorandum.¶
¶
      Hope this answers your questions. I look forward to your input on the handbook. My e-mail address is david.davis@maxwell.af.mil. Look at my homepage at http://chapel.maxwell.af.mil/.¶
¶
                                              Sincerely [3 spaces to the right of page center]¶

¶
¶
¶
¶
¶ [Start signature block on fifth line]   *David Davis*
                                              DAVID DAVIS, Chaplain
                                              Lieutenant Colonel, USAF¶
¶
Attachment:¶
Draft CAP Chaplain Handbook¶

> **NOTE: Triple space equal 2 blank lines**
> **Double space equal 1 blank line**

2-inch top margin and 1-inch bottom margin

1-inch left and right margins

DEPARTMENT OF THE AIR FORCE
AIR UNIVERSITY (AETC)

23 May 04

Colonel Jacob R. Bradley
Director of Plans and Programs
550 McDonald Street
Maxwell AFB-Gunter Annex AL 36118-5643

Colonel William J. Nash
Program Director
75 South Butler Avenue
Patrick AFB FL 85469-6357

Dear Colonel Nash

    Thank you for your outstanding presentation to the Air Command and Staff College Class of 2003. Your briefing was right on target and expertly integrated many aspects of our curriculum into a focused leadership perspective. Our students face increasingly complex challenges, and your keen insights were invaluable in preparing them for the future.

    We appreciate your support and look forward to future visits.

                Sincerely

                Jacob R. Bradley
                JACOB R. BRADLEY, Colonel, USAF

Letter of Appreciation

FORMS OF ADDRESS, SALUTATION, AND CLOSE

In the *Office of the Secretary of Defense Manual for Written Material* there are some general, convenient forms of address for both letters and envelopes. We think these are important enough to include here. Using titles correctly in the salutation of a letter makes a good impression. We suggest a word of caution … avoid the use of gender distinctive titles. This can be generally done by rewriting or varying the layout.

To summarize some of the key rules to follow when addressing a letter to an individual by name or title, we have provided these points:

- Do not use punctuation with the address element. Do not use punctuation after the last word of salutation and complimentary close.

- Spell out all titles of address (except Dr., Mr. and Mrs.), always using only one title. (*Dr. James Norrix or James Norrix, MD; not Dr. James Norrix, MD*).

- Address correspondence by professional or organizational title when the name or the gender, or both are not known. (*Dear Resource Manager, Dear Department Head, etc.*).

- Use Ms. with the surname or first and last name when a woman's marital status is unknown. Using the title "Ms." is okay as a general rule, unless the person prefers Miss or Mrs. (*Ms. Shroyer or Maxine Shroyer*). In business use a married woman's first name (*Mrs. Lee Reising*); socially, the custom has been to use her husband's first name (*Mrs. Stephen Reising*).

- Respond to a letter written jointly by more than three people by preparing single replies, or replying to the person signing and mentioning the others early in the letter.

- Address all Presidential appointees and federal and state elected officials as "The Honorable," even when addressing social correspondence. As a general rule, do not address county and city officials, except mayors, as "The Honorable."

- Use "Judge," "General," "The Honorable," etc., if appropriate. People, once entitled, may keep the title throughout their lifetime. Some dignitaries holding doctoral degrees may prefer the use of "Dr."

- Use two lines for a couple with different last names with the man's name on the bottom.

- Use Miss, Mrs. or Ms. with a woman's surname, but Madam with formal or position titles. (*Dear Mrs. Carolyn Brown, but Dear Madam Justice or Dear Madam Secretary*)

To keep down the following samples, we've substituted an * for Mr., Mrs., Miss, Ms., or Madam.

The White House:

The President	The President The White House Washington DC 20500-0001	Dear * President Respectfully yours
Spouse of the President	* (full name) The White House Washington DC 20500-0001	Dear * (surname) Sincerely
Assistant to the President	The Honorable (full name) Assistant to the President The White House Washington DC 20500-0001	Dear * (surname) Sincerely
Secretary to the President	The Honorable (full name) Secretary to the President The White House Washington DC 20500-0001	Dear * (surname) Sincerely
Secretary to the President (with military rank)	(full rank) (full name) Secretary to the President The White House Washington DC 20500-0001	Dear (rank) (surname) Sincerely

The Vice President:

The Vice President	The Vice President The White House Washington DC 20500-0001	Dear * Vice President Respectfully yours
The President of the Senate	The Honorable (full name) President of the Senate Washington DC 20510-0001	Dear * President Sincerely

The Federal Judiciary:

The Chief Justice	The Chief Justice of the United States The Supreme Court of the United States Washington DC 20543-0001	Dear Chief Justice (surname) Sincerely
Associate Justice	Justice (surname) The Supreme Court of the United States Washington DC 20543-0001	Dear Justice (surname) Sincerely
Retired Associate Justice	The Honorable (full name) (local address)	Dear Justice (surname) Sincerely
Presiding Justice	The Honorable (full name) Presiding Justice (name of court) (local address)	Dear Justice (surname) Sincerely
Chief Judge of a Court	The Honorable (full name) Chief Judge of the (court; if a US district court, give district) (local address)	Dear Judge (surname) Sincerely
Clerk of a Lower Court	* (full name) Clerk of the (court; if a US district court, give district) (local address)	Dear * (surname) Sincerely

The Congress:

President pro Tempore of the Senate	The Honorable (full name) President pro Tempore of the Senate United States Senate Washington DC 20510-0001	Dear Senator (surname) Sincerely
Committee Chairman, US Senate	The Honorable (full name) Chairman, Committee on (name) United States Senate Washington DC 20510-0001	Dear Madam Chair or Mr. Chairman Sincerely
Subcommittee Chairman, US Senate	The Honorable (full name) Chairman, Subcommittee on (name) (parent committee) United Sates Senate Washington DC 20510-0001	Dear Senate (surname) Sincerely
Senator (Washington DC office)	The Honorable (full name) United States Senate Washington DC 20510-0001	Dear Senator (surname) Sincerely
(Away from Washington DC)	The Honorable (full name) United States Senator (local address)	Dear Senator (surname) Sincerely
Senate Majority (or Minority) Leader (Washington DC office)	The Honorable (full name) Majority (or Minority) Leader United States Senate Washington DC 20510-0001	Dear Senator (surname) Sincerely
(Away from Washington DC)	The Honorable (full name) Majority (or Minority) Leader United States Senate (local address)	Dear Senator (surname) Sincerely
Senator-elect	The Honorable (full name) United States Senator-elect United States Senate Washington DC 20510-0001	Dear * (surname) Sincerely
Office of a deceased Senator	* (Secretary's full name, if known) Office of the late Senator (full name) United States Senate Washington DC 20510-0001	Dear * (surname) or Dear Sir Sincerely
Speaker of the House of Representatives	The Honorable (full name) Speaker of the House of Representatives Washington DC 20510-0001	Dear * Speaker Sincerely
Committee Chairman, House of Representatives	The Honorable (full name) Chairman, Committee on (name) House of Representatives Washington DC 20515-0001	Dear Mr. Chairman or Madam Chair Sincerely
Subcommittee Chairman, House of Representatives	The Honorable (full name) Chairman, Subcommittee on (name) (parent committee) House of Representatives Washington DC 20515-0001	Dear * (surname) Sincerely or Dear Mr. Chairman or Madam Chair Sincerely (when incoming correspondence is so signed and it pertains to subcommittee affairs)

Representative (Washington DC office)	The Honorable (full name) House of Representatives Washington DC 20515-0001	Dear * (surname) Sincerely
(Away from Washington DC)	The Honorable (full name) (local address)	Dear * (surname) Sincerely
House Majority (or Minority) Leader (Washington DC office)	The Honorable (full name) Majority (or Minority) Leader House of Representatives Washington DC 20515-0001	Dear * (surname) Sincerely
(Away from Washington DC)	The Honorable (full name) (local address)	Dear * (surname) Sincerely
Representative-elect	The Honorable (full name) Representative-elect House of Representatives Washington DC 20515-0001	Dear * (surname) Sincerely
Office of a decreased Representative	* (secretary's full name, if known) Office of the late (full name) House of Representatives Washington DC 20515-0001	Dear * (surname) *or* Dear Sir *or* Madam Sincerely
Resident Commissioner	The Honorable (full name) Resident Commissioner from (area) Washington DC 20515-0001	Dear * (surname) Sincerely
Delegate of the District of Columbia	The Honorable (full name) House of Representatives Washington DC 20515-0001	Dear * (surname) Sincerely

Legislative agencies:

Comptroller General (Head of the General Accounting Office)	The Honorable (full name) Comptroller General of the United States Washington DC 20548-0001	Dear * (surname) Sincerely
Public Printer (Head of the US Government Printing Office)	The Honorable (full name) Public Printer US Government Printing Office Washington DC 20541-0001	Dear * (surname) Sincerely
Librarian of Congress	The Honorable (full name) Librarian of Congress Washington DC 20540-0001	Dear * (surname) Sincerely

Executive departments:

Members of the Cabinet addressed as *Secretary*	The Honorable (full name) Secretary of (department) Washington DC (ZIP + 4)	Dear * Secretary Sincerely
Attorney General (Head of the Department of Justice)	The Honorable (full name) Attorney General Washington DC 20530-0001	Dear * Attorney General Sincerely
Under Secretary of a Department	The Honorable (full name) Under Secretary of (department) Washington DC (ZIP + 4)	Dear * (surname) Sincerely
Deputy Secretary of a Department	The Honorable (full name) Deputy Secretary of (department) Washington DC (ZIP + 4)	Dear * (surname) Sincerely

Assistant Secretary of a Department	The Honorable (full name) Assistant Secretary of (department) Washington DC (ZIP + 4)	Dear (surname) Sincerely

Titles for Cabinet Secretaries are: Secretary of State, Secretary of the Treasury, Secretary of Defense, Secretary of the Interior, Secretary of Agriculture, Secretary of Commerce, Secretary of Labor, Secretary of Education, Secretary of Housing and Urban Development, Secretary of Transportation, Secretary of Energy, Secretary of Health and Human Services and Secretary of Veterans Affairs.

Military departments:

Address and/or sign off with current rank, not to be promoted rank

The Secretary	The Honorable (full name) Secretary of (department) Washington DC (ZIP + 4)	Dear * Secretary Sincerely
Under Secretary of a Department	The Honorable (full name) Under Secretary of (department) Washington DC (ZIP + 4)	Dear * Sincerely
Assistant Secretary of a Department	The Honorable (full name) Assistant Secretary of (department) Washington DC (ZIP + 4)	Dear * (surname) Sincerely

Military personnel:

Army, Air Force, and Marine Corps officers:

General of the Army	General of the Army (full name) (local address)	Dear General (surname) Sincerely
General, Lieutenant General, Major General, Brigadier General	(full rank) (full name), (abbreviation of service designation) (post office address of organization and station)	Dear General (surname) Sincerely
Colonel, Lieutenant Colonel	(full rank) (full name), (abbreviation of service designation) (post office address of organization and station)	Dear Colonel (surname) Sincerely
Major, Captain	(same as above)	Dear (rank) (surname) Sincerely
First Lieutenant, Second Lieutenant	(same as above)	Dear Lieutenant (surname) Sincerely
Chief Warrant Officer, Warrant Officer	(same as above)	Dear * (surname) Sincerely

Navy officers:

Fleet Admiral, Admiral Vice Admiral, Rear Admiral, Commodore	(same as above)	Dear Admiral *or* Commodore (surname) Sincerely
Captain, Commander, Lieutenant Commander	(same as above)	Dear (rank) (surname) Sincerely
Lieutenant	(same as above)	Dear Lieutenant (surname) Sincerely
Lieutenant (junior grade)	(same as above)	Dear Lieutenant JG (surname) Sincerely

Ensign	(same as previous page)	Dear Ensign (surname) Sincerely
Chief Warrant Officer	(same as previous page)	Dear Chief Warrant Officer (surname) Sincerely

Academy members:

Cadet, Midshipman, Air Cadet	(rank) (full name) (local address)	Dear (rank) (surname) Sincerely

Army enlisted personnel:

Sergeant Major of the Army	Sergeant Major (full name) (local address)	Dear Sergeant Major (surname) Sincerely
Command Sergeant Major, Sergeant Major	(rank) (full name), (abbreviation of service designation) (post office address of organization and station)	Dear Sergeant Major (surname) Sincerely
First Sergeant	(same as above)	Dear First Sergeant (surname) Sincerely
Master Sergeant, Platoon Sergeant, Sergeant First Class, Staff Sergeant, Sergeant	(same as above)	Dear Sergeant (surname) Sincerely
Corporal	(same as above)	Dear Corporal (surname) Sincerely
Specialist (all grades)	(same as above)	Dear Specialist (surname) Sincerely
Private First Class, Private	(same as above)	Dear Private (surname) Sincerely

Navy enlisted personnel:

Master Chief Petty Officer	(rank) (full name), (abbreviation of service designation) (post office address of organization and station)	Dear Master Chief Petty Officer (surname) Sincerely
Senior Chief Petty Officer, Chief Petty Officer	(same as above)	Dear (rank) (surname) Sincerely
Petty Officer First Class, Second Class, Third Class	(same as above)	Dear Petty Officer (surname) Sincerely
Seaman, Seaman Apprentice, Seaman Recruit	(same as above)	Dear Seaman (surname) Sincerely
Fireman, Fireman Apprentice, Fireman Recruit	(same as above)	Dear Fireman (surname) Sincerely
Airman, Airman Apprentice, Airman Recruit	(same as above)	Dear Airman (surname) Sincerely
Construction Man, Construction Man Apprentice, Construction Man Recruit	(same as above)	Dear Construction Man (surname) Sincerely

Hospitalman, Hospitalman Apprentice, Hospitalman Recruit	(same as above)	Dear Hospitalman (surname) Sincerely
Dentalman, Dentalman Apprentice, Dentalman Recruit	(same as above)	Dear Dentalman (surname) Sincerely
Stewardsman, Stewardsman Apprentice, Stewardsman Recruit	(same as above)	Dear Stewardsmand (surname) Sincerely

Marine Corps enlisted personnel:

Sergeant Major of the Marine Corps	Sergeant Major (full name) (local address)	Dear Sergeant Major (surname) Sincerely
Sergeant Major	(rank) (full name), (abbreviation of service designation) (post office address of organization or station)	Dear Sergeant Major (surname) Sincerely
Master Gunnery Sergeant, First Sergeant, Gunnery Sergeant, Staff Sergeant, Sergeant	(same as above)	Dear (rank) (surname) Sincerely
Corporal, Lance Corporal	(same as above)	Dear Corporal (surname) Sincerely
Private First Class, Private	(same as above)	Dear Private (surname) Sincerely

Air Force enlisted personnel:

Chief Master Sergeant of the Air Force	Chief Master Sergeant of the Air Force (full name) (local address)	Dear Chief (surname) Sincerely
Chief Master Sergeant	Chief Master Sergeant (full name) (local address)	Dear Chief (surname) Sincerely
Senior Master Sergeant, Master Sergeant, Technical Sergeant, Staff Sergeant	(rank) (full name), (abbreviation of service designation) (post office address of organization and station)	Dear Sergeant (surname) Sincerely
Senior Airman, Airman First Class, Airman, Airman Basic	(same as above)	Dear Airman (surname) Sincerely

Retired military personnel:

| All retired military personnel | (rank) (full name), (abbreviated service designation), Retired
(local address) | Dear (rank) (surname)
Sincerely |

Independent agencies:

| Director, Office of Management and Budget | The Honorable (full name)
Director, Office of Management and Budget
Washington DC 20503-0001 | Dear * (surname)
Sincerely |
| Head of Federal Agency Authority or Board | The Honorable (full name)
(title), (organization)
(agency)
Washington DC (ZIP + 4) | Dear * (surname)
Sincerely |

Head of a major organization within an agency (if the official is appointed by the President)	The Honorable (full name) (title), (organization) (agency) Washington DC (ZIP + 4)	Dear * (surname) Sincerely
President of a Commission	The Honorable (full name) President, (commission) Washington DC (ZIP + 4)	Dear * (surname) Sincerely
Chairman of a Commission	The Honorable (full name) Chairman, (commission) Washington DC (ZIP + 4)	Dear Madam Chair *or* Mr. Chairman Sincerely
Chairman of a Board	The Honorable (full name) Chairman, (board) Washington DC (ZIP + 4)	Dear Mr. Chairman *or* Madam Chair Sincerely
Postmaster General	The Honorable (full name) Postmaster General Washington DC (ZIP + 4)	Dear * Postmaster General Sincerely

American Mission:

American Ambassador	The Honorable (full name) American Ambassador (city), (country) (local address)	Sir *or* Madam (formal) Dear * Ambassador (informal) Sincerely
American Ambassador (with military rank)	(full rank) (full name) American Ambassador (city), (country) (local address)	Sir or Madam (formal) Dear * Ambassador or Dear (rank) (surname) (informal) Sincerely
American Minister	The Honorable (full name) American Minister (city), (country) (local address)	Sir or Madam (formal) Dear * Minister (informal) Sincerely
American Minister (with military rank)	(full rank) (full name) American Minister (city), (country) (local address)	Sir or Madam (formal) Dear * Minister (surname) (informal) Sincerely

Foreign government officials:

Foreign Ambassador in the United States	His/Her Excellency (full name) Ambassador of (country) (local address)	Excellency (formal) Dear * Ambassador (informal) Sincerely
Foreign Minister in the United States	The Honorable (full name) Minister of (country) (local address)	Sir *or* Madam (formal) Dear * Minister (informal) Sincerely

| Foreign Chargé d'Affaires in the United States | * (full name)
Chargé d'Affaires of (country)
(local address) | Sir *or* Madam (formal)
Dear * Chargé d'Affaires
(informal)
Sincerely |

The Organization of American States:

Secretary General of the Organization of American States	The Honorable (full name) Secretary General, the Organization of American States Pan American Union Washington DC 20006-0001	Sir *or* Madam (formal) Dear * Secretary General *or* Dear * (Dr.) (surname) (informal) Sincerely
Assistant Secretary General of the Organization of American States	The Honorable (full name) Assistant Secretary General, the Organization of American States Pan American Union Washington DC 20006-0001	Sir *or* Madam (formal) Dear * (Dr.) (surname) (informal) Sincerely
United States Representative on the Council of the Organization of American States	The Honorable (full name) United States Representative on the Council of the Organization of American States Department of State Washington DC (ZIP + 4)	Sir *or* Madam (formal) Dear * (Dr.) (surname) (informal) Sincerely

United Nations:

Communications to the United Nations are addressed to the United States Representative to the United Nations, through the Department of State. Exceptions, which are sent direct to the United States Representative, include those intended for the Economic and Social Council, the Disarmament Commission, the Trusteeship Council and the delegation to the General Assembly (when it is in session). Direct communications with the United Nations is inappropriate unless exceptions arise. Where it is necessary, the communication should be sent to the Secretary General of the United Nations through the United States Representative by means of a cover letter.

Secretary General of the United Nations	His/Her Excellency (full name) Secretary General of the United Nations New York NY 10017-3582	Excellency (formal) Dear * Secretary General (informal) Sincerely
United States Representative to the United Nations	The Honorable (full name) United States Representative to the United Nations New York 10017-3582	Sir or Madam (formal) Dear * (surname) (informal) Sincerely
Chairman, United States Delegation to the United Nations Military Staff Committee	The Honorable (full name) Chairman, United States Delegation to the United Nations Military Staff Committee New York 10017-3582	Sir or Madam (formal) Dear * (surname) (informal) Sincerely
Senior Military Adviser to the United States Delegation to the United Nations General Assembly	(full rank) (full name) Senior Military Adviser United States Delegation to the United Nations General Assembly New York NY 10017-3582	Dear (rank) (surname) Sincerely
United States Representative to the Economic and Social Council	The Honorable (full name) United States Representative on the Economic and Social Council of the United Nations New York NY 10017-3582	Sir *or* Madam (formal) Dear * (surname) (informal) Sincerely
United States Representative to the United Nations Disarmament Commission	The Honorable (full name) United States Representative on the Disarmament Commission of the United Nations	Sir *or* Madam (formal) Dear * (surname) (informal) Sincerely

New York NY 10017-3582

United States Representative to the Trusteeship Council	The Honorable (full name) United States Representative on the Trusteeship Council New York NY 10017-3582	Sir *or* Madam (formal) Dear * (surname) (informal) Sincerely

State and local governments:

In most states, the lower branch of the legislature is the House of Representatives. In some states, such as California and New York, the lower house is known as the Assembly. In others, such as Maryland, Virginia and West Virginia, it is known as the House of Delegates. Nebraska has a one-house legislature; its members are classed as senators.

Governor of a State	The Honorable (full name) Governor of (state) (city) (state) (ZIP + 4)	Dear Governor (surname) Sincerely
Acting Governor of a State	The Honorable (full name) Acting Governor of (state) (city) (state) (ZIP + 4)	Dear * (surname) Sincerely
Lieutenant Governor of a State	The Honorable (full name) Lieutenant Governor of a State (city) (state) (ZIP + 4)	Dear * Secretary Sincerely
Secretary of State of a State	The Honorable (full name) Secretary of State of (state) (city) (state) (ZIP + 4)	Dear * Secretary Sincerely
Chief Justice of Supreme Court of a State	The Honorable (full name) Chief Justice Supreme Court of the State of (state) (city) (state) (ZIP + 4)	Dear * Chief Justice Sincerely
Attorney General of a State	The Honorable (full name) Attorney General State of (state) (city) (state) (ZIP + 4)	Dear * Attorney General Sincerely
Treasurer, Comptroller, Auditor of a State	The Honorable (full name) State Treasurer State of (state) (city) (state) (ZIP + 4)	Dear * (surname) Sincerely
President of the Senate of a State	The Honorable (full name) President of the Senate of the State of (state) (city) (state) (ZIP + 4)	Dear * (surname) Sincerely
State Senator	The Honorable (full name) (state) Senate (city) (state) (ZIP + 4)	Dear * (surname) Sincerely

Speaker of the House of Representatives or the Assembly or the House of Delegates of a State	The Honorable (full name)
	Speaker of the House of Representatives
	  (or Assembly or House of Delegates)
	  of the States of (state)
	(city) (state) (ZIP + 4)
		Dear * (surname)
		Sincerely

State Representative, Assemblyman, Delegate	The Honorable (full name) (state) House of Representatives (or Assembly of House of Delegates) (city) (state) (ZIP + 4)	Dear * (surname) Sincerely
Mayor	The Honorable (full name) Mayor of (city) (city) (state) (ZIP + 4)	Dear Mayor (surname) Sincerely
President of a Board of Commissioners	The Honorable (full name) President, Board of Commissioners of (city) (city) (state) (ZIP + 4)	Dear * (surname) Sincerely

Ecclesiastical organizations:

Protestant Minister, Pastor, Rector
(with doctoral degree)
	The Reverend (full name)
	(title), (church)
	(local address)
		Dear Dr. (surname)
		  or Reverend, Sir *or* Madam
		   (formal)
		Sincerely

Protestant Minister, Pastor, Rector (without doctoral degree)	The Reverend (full name) (title), (church) (local address)	Dear * (surname) *or* Reverend, Sir *or* Madam (formal) Sincerely
Rabbi (with doctoral degree)	Rabbi (full name) (local address)	Dear Dr. (surname) *or* Dear Rabbi (surname) Sincerely
Rabbi (without doctoral degree)	Rabbi (full name) (local address)	Dear Rabbi (surname) Sincerely
Catholic Cardinal	His Eminence (Christian name) Cardinal (surname) (local address)	Your Eminence (formal) Dear Cardinal (surname) (informal) Sincerely
Catholic Archbishop	The Most Reverend (full name) Archbishop of (province) (local address)	Your Excellency (formal) Dear Archbishop (surname) (informal) Sincerely
Catholic Bishop	The Most Reverend (full name) Bishop of (province) (local address)	Your Excellency (formal) Dear Bishop (surname) (informal) Sincerely

Catholic Monsignor
(higher rank)
	The Right Reverend Monsignor (full name)
	(local address)
		Right Reverend Monsignor
		  (formal)
		Dear Monsignor (surname)
		  (informal)
		Sincerely

Catholic Monsignor
(lower rank)
	The Very Reverend Monsignor (full name)
	(local address)
		Very Reverend Monsignor
		  (formal)
		Dear Monsignor (surname)
		  (informal)
		Sincerely

Catholic Priest	The Reverend (full name)
	  (add initials of order, if any)
	(local address)
		Reverend Sir (formal)
		Dear Father (surname)
		  (informal)
		Sincerely

Catholic Mother Superior of an Institution	Mother (name) (initials of order, if any) Superior, (institution) (local address)	Dear Mother (name) Sincerely
Catholic Sister	Sister (full name) (organization) (local address)	Dear Sister (full name) Sincerely
Catholic Brother	Brother (full name) (organization) (local address)	Dear Brother (full name) Sincerely
Mormon Bishop	Mr. (full name) Church of Jesus Christ of Latter-day Saints (local address)	Sir (formal) Dear Mr. (surname) (informal) Sincerely
Protestant Episcopal Bishop	The Right Reverend (full name) Bishop of (name) (local address)	Right Reverend Sir (formal) Dear Bishop (surname) (informal) Sincerely
Protestant Episcopal Dean	The Very Reverend (full name) Dean of (church) (local address)	Very Reverend Sir (formal) Dear Dean (surname) (informal) Sincerely
Methodist Bishop	The Reverend (full name) Methodist Bishop (local address)	Reverend Sir (formal) Reverend Madam (informal) Sincerely
Chaplain (military service)	Chaplain (rank) (name) (local address)	Dear Chaplain (surname) Sincerely

Corporations, companies, and federations:

A company or corporation	(company or corporation) (local address)	Mesdames, Gentlemen *or* Sirs Sincerely
A federation	(name of official) (title), (federation) (local address)	Dear * (surname) Sincerely

Private citizens:

President of a university or college (with doctoral degree)	Dr. (full name) President, (institution) (local address)	Dear Dr. (surname) Sincerely
President of a university or college (without doctoral degree)	* (full name) President, (institution) (local address)	Dear * (surname) Sincerely
Dean of a school (with doctoral degree)	Dr. (full name) Dean, School of (name) (institution) (local address)	Dear Dr. (surname) Sincerely

Dean of a school (without doctoral degree)	Professor (full name) School of (name) (institution) (local address)	Dear Professor (surname) Sincerely
Professor (with doctoral degree)	Dr. or Professor (full name) Department of (name) 　(institution) (local address)	Dear Dr. (surname) Dean Professor (surname) Sincerely
Professor (without doctoral degree)	Professor (full name) Department of (name) (institution) (local address)	Dear Professor (surname) Sincerely
Associate Professor, Assistant Professor	* (full name) Associate (or Assistant) Professor Department of (name) 　(institution) (local address)	Dear Professor (surname) Sincerely
Physician	(full name), MD (local address)	Dear Dr. (surname) Sincerely
Lawyer	* (full name) Attorney at Law (local address)	Dear * (surname) Sincerely
One individual	Mr., Mrs., Miss, or Ms. (full name) (local address)	Dear Mr., Miss, Ms. or Mrs. 　(surname) Sincerely
Two individuals	Mr. and Mrs. (full name) Mr. (full name) Mrs. (full name) and Miss (full name) Mrs. (full name) and Mr. (full name) (local address)	Dear Mr. and Mrs. 　(surname) Dear Mr. (surname) and 　Mr. (surname) Dear Mrs. (surname) and 　Miss (surname) Dear Mrs. (surname) and 　Mr. (surname) Sincerely
Three or four individuals	Messieurs (surnames) Mesdames (surnames) Misses (full names or given 　names and surnames) or The Misses (surnames) (local address)	Gentlemen, Sirs, or Mesdames or Dear Misses 　(surnames) Sincerely

Former government officials:

Former President	The Honorable (full name) (local address)	Dear President (surname) Respectfully yours
Former Vice President	(same as above)	Dear * (surname) Sincerely
Former Member of the Cabinet addressed as "Secretary"	(same as above)	Dear Mr. (surname) Sincerely
Former Attorney General	(same as above)	Dear Mr. (surname) General (surname) Sincerely

OFFICIAL
MEMORANDUM

DEPARTMENT OF THE AIR FORCE
AIR UNIVERSITY (AETC)

24 Nov 04

MEMORANDUM FOR CATEGORY 1

FROM: AU/CV
      55 LeMay
      Maxwell AFB AL 36112-6335

SUBJECT: Responsible Use of Electronic Communications (S/S Memo, 10 Nov 04)

1. Access to the Internet provides each person at Air University information sources that rival any library. From the comfort of your desk, in about the time it takes you to pour that second cup of coffee, your computer can connect you with other computers halfway around the world. You can view volumes of text and images, all neatly organized for your convenience. If used correctly, the Internet can be an incredibly powerful tool. However, this new capability must be used responsibly. The computer and communications lines you use to connect the Internet are government resources. Similarly, the information you review or transfer while at work becomes property of the Air Force, and must serve the purpose of helping you to do your job. You would not consider using the phone on your desk for inappropriate and unofficial calls; that same logic applies to your computer and accessing Internet. *Remember it is an abuse of government resources to use government equipment or duty time to conduct unofficial business.*

2. Electronic mail provides an immediate and convenient means of communication. A single message can be programmed to send to any number of persons, at any time of the day. Before we hit that "Send" button, we should consider where our intended messages may end up. The negative repercussion of unintentional distribution was illustrated recently when a pilot involved in the rescue of Captain Scott O'Grady in Bosnia sent a full and graphic account of the rescue mission to a number of his friends. These friends forwarded the message to others, and soon the entire text of the message—including misclassified information—found its way to an electronic bulletin board, available for anyone to access. This story shows a lack of discretion on the part of the original writer, who didn't consider the sensitive nature of his message or the potential scope of his audience. We should learn from his mistakes.

3. Another important point with regard to e-mail, especially when you communicate with outside agencies, is that these messages are official correspondence. They reflect your professionalism and present an image of Air University. Let's make certain that image is a positive one.

Bill Grant
BILL GRANT, Colonel, USAF
Vice Commander

DEPARTMENT OF THE AIR FORCE
HEADQUARTERS UNITED STATES AIR FORCE

[1.75 inches or 10 lines from top of page] 2 February 2004¶

¶
MEMORANDUM FOR HQ AETC/IM¶
                        HQ AU/IM¶

¶
FROM: HQ USAF/SCMV¶
         1250 Air Force Pentagon¶
         Washington DC 20330-1250¶

¶
SUBJECT: Format of the Air Force Official Memorandum (Your Memo, 15 Jan 03)¶

¶
References: (a) AFMAN 33-326, *Preparing Official Communications*¶
              (b) AFH 33-337, *The Tongue and Quill*¶
              [One reference is listed in the Subject line; two or more are listed as shown.]¶

¶
1. This is a prepared letterhead format of the Air Force official memorandum. The heading will be generic consisting of two or three lines: Department of the Air Force, etc., and the organization (Headquarters Air Combat Command, Headquarters United States Air Force, etc.), and the location with the ZIP code (optional). Printed letterhead stationery for wing level is normally used if the quantity needed justifies the printing cost. Any unit without its own letterhead may use its parent unit's and identify its organization and office symbol in the FROM caption. Be sure to include the 9-digit ZIP code with the full address in the FROM caption.¶

¶
2. The style of writing is yours. For some helpful guidelines see pages 17-24. However, when writing for someone else's signature, try to write as though that person were speaking. Be succinct, use active voice and keep it short (one page, if possible, see Atch 1). Include extensive background material as an attachment rather than within the memorandum itself (see Atch 2).¶

¶
3. Even though most signers want their signatures on a perfect product, minor errors may be neatly corrected in ink.¶

¶
4. If you want a response directed to a project officer rather than the signer, include that person's name and rank (if applicable), office symbol, telephone number and e-mail in the body of the memo.¶

¶
¶
¶
¶
¶ **[Start signature block on fifth line]**   *Shirley Collins*
                                      SHIRLEY COLLINS ¶
                                      Management Analyst¶
                                      Visual Information/Publishing Division¶

¶
2 Attachments:¶
1. Minutes, 28 Dec 03¶
2. HQ AF/ILCXE Memo, 15 Dec 03¶
¶
cc:¶
HQ AETC/DP¶

NOTE: Triple space equal 2 blank lines
Double space equal 1 blank line

2-inch top margin and 1-inch bottom margin

1-inch left and right margins

DEPARTMENT OF THE AIR FORCE
UNITED STATES AIR FORCE
RANDOLPH AIR FORCE BASE

21 October 2004¶

¶
MEMORANDUM FOR ADMINISTRATORS¶
¶
FROM: ACSC/DEO¶
    225 Chennault Circle¶
    Maxwell AFB AL 36112-6426¶
¶
SUBJECT: Subdividing Paragraphs in the Official Memo¶
¶
1. Introduction paragraph with purpose, statement, and overview.¶
¶
2. First main idea.¶
¶
 a. Fact and reasoning supporting this idea.¶
¶
  (1) Fact and reasoning to support a.¶
¶
  (2) Additional fact and reasoning to support a.¶
¶
   (a) Support for (2).¶
¶
   (b) Additional support for (2).¶
¶
    <u>1</u> Support for (b).¶
¶
     <u>a</u> Support for <u>1</u>.¶
¶
     <u>b</u> Additional Support for <u>1</u>.¶
¶
      [1] Support for <u>a</u>.¶
¶
      [2] Additional Support for <u>a</u>.¶
¶
    <u>2</u> Additional support for (b).¶
¶
 b. Additional fact for reasoning to support 2 (First main idea).¶
¶
  (1) Support for b.¶
¶
  (2) Additional support for b.¶
¶
3. Second main idea.¶
¶
 a. Fact and reasoning supporting this idea.¶
¶
 b. Etc.¶
¶
4. Optional closure and point of contact information.¶
¶
¶
¶
¶

> **REMEMBER: When subdividing paragraphs, never use a *1* without a *2* or an *a* without a *b*.**

¶   **[Start signature block on fifth blank line]**

Edward Clinton

EDWARD CLINTON, Lt Col, USAF¶
Publishing Editor¶

As director of communications, I was asked to prepare a memo reviewing our company's training programs and materials. In the body of the memo, one of the sentences I mentioned the "pedagogical approach" used by one of the training manuals. The day after I routed the memo to the executive committee, I was called into the HR director's office, and told that the executive vice president wanted me out of the building by lunch. When I asked why, I was told that she couldn't stand for "perverts" (pedophilia?) working in her company. Finally, he showed me her copy of the memo, with her demand that I be fired, and the word "pedagogical" circled in red. The HR manager was fairly reasonable, and once he looked the word up in his dictionary and made a copy of the definition to send back to her, he told me not to worry. He would take care of it. Two days later, a memo to the entire staff came out, directing us that no words, which could not be found in the local Sunday newspaper, could be used in company memos. A month later, I resigned. In accordance with company policy, I created my resignation memo by pasting words together from the Sunday paper.

IN TURN MEMO

DEPARTMENT OF THE AIR FORCE
AIR UNIVERSITY (AETC)

19 Feb 04

MEMORANDUM FOR AU/CC
                                    ACSC/CC
                                    IN TURN

FROM: AETC/CV
         1200 Ash Street, Room 123
         Randolph AFB TX 78236-2550

SUBJECT: Format of the IN TURN Memo

1. RADEX 2003 was the best conference we have offered. Every detail was planned and executed professionally and expertly. Everyone who took part in making the conference such a success is to be commended.

2. Through the support of ACSC, the plans for this event progressed from mere concept to reality. Majors Tom Childress and Keith Tonnies and Ms. Carrie Long were especially helpful in their expert handling of the communications requirements.

3. On behalf of the Commander, Air Education and Training Command, please thank everyone who worked so hard to make the conference a success.

Dean Porter
DEAN PORTER, JR.
Major General, USAF
Vice Commander

DEPARTMENT OF THE AIR FORCE
HEADQUARTERS UNITED STATES AIR FORCE
WASHINGTON DC

¶

[1.75 inches or 10 lines from top of page] 2 January 2004¶

¶
MEMORANDUM FOR   HQ AETC/PA¶
                                        LG¶
                                        XP¶
                        IN TURN¶
¶
FROM: HQ USAF/SCMV¶
              1250 Air Force Pentagon¶
              Washington DC 20330-1250¶
¶
SUBJECT: IN TURN Memo Format¶
¶
1. The IN TURN memo is an official memo addressed to two or more individuals or offices.
It's primarily used when you want the final addressee to see the coordination or action of
the other addressees as the memo moves along the chain.¶
¶
2. The above address element shows the proper way to address an IN TURN memo <u>within</u> an
organization. Here's how to address it to several different organizations.¶
¶
MEMORANDUM FOR   ACSC/CC¶
                                       AWC/CC¶
                                       SOS/CC¶
                                       AU/CC¶
                         IN TURN¶
¶
3. When you receive an IN TURN memo strike through your office symbol, put your initials and
the date next to your office symbol (shown below) and send the memo on to the next addressee.
When you have comments, write a note next to your office symbol and either attach your
comments to the package or send a separate letter directly to the final addressee.¶
¶
MEMORANDUM FOR   ACSC/~~DE~~¶ – *Concur, JN, 3 Jan*
                                      ~~DS~~¶ *Comments encl, 5 Jan*
                                      ~~AS~~¶ *PM – 7 Jan*
                        IN TURN¶
¶
¶
¶
¶
¶   **[Start signature block on fifth line]**   *Brian Magers*
                                                                BRIAN MAGERS, Lt Col, USAF¶
                                                                Chief of Publishing Policy¶
¶
cc:¶
HQ PACAF/DP¶
¶
DISTRIBUTION:¶
HQ USAF/SCM¶

NOTE: Triple space equal 2 blank lines
Double space equal 1 blank line

2-inch top margin and 1-inch bottom margin

1-inch left and right margins

INDORSEMENT MEMO

DEPARTMENT OF THE AIR FORCE
AIR UNIVERSITY (AETC)

1 Apr 04

MEMORANDUM FOR ACSC/DEOT
                    THROUGH: MAJOR BOWE

                 ACSC/DEC
                    THROUGH: MAJOR JONES

FROM: ACSC/DL

SUBJECT: ACSC Research Internet Homepage Guidelines

1. All student research teams who want to place a homepage on the Air University Internet server must follow the guidelines in the attachment. Before submitting your completed homepage to ACSC/DL, ensure you:

    a. Provide a hard copy of your layout and design. Have your research advisor review and approve the homepage.

    b. Test your hypertext links to be sure they work correctly.

2. If you have any questions or suggestions on these guidelines, please let me know. My e-mail address is robin.jones@maxwell.af.mil.

Robin Jones
ROBIN JONES, Major, USAF
Chief, Internet Technology

Attachment:
Internet Homepage Guidelines

cc:
Lt Col McCann

1st Ind, ACSC/DEO   5 Apr 04

MEMORANDUM FOR ACSC/DL

Good guidelines. Please include a reminder that all information must be unclassified.

Rickey Bowe
RICKEY BOWE, Major, USAF
Chief, Scheduling

DEPARTMENT OF THE AIR FORCE
AIR UNIVERSITY (AETC)

[1.75 inches or 10 lines from top of page] 2 November 2004¶

¶
MEMORANDUM FOR HQ AU/CFA
ATTENTION: MS. SHELLY DIKE

¶
FROM: ACSC/DEOP
¶
SUBJECT: Indorsement Memo Format
¶
1. Indorse official memos only, not personal letters. Use within or between US military organizations or between US military organizations and civilian organizations under contract with the Air Force.
¶
2. Type indorsements on the same page as the original memo or previous indorsement when space is available. Begin typing on the left margin two lines below the basic memo or previous indorsement. Number each indorsement in sequence (lst Ind, 2d Ind, 3d Ind). Follow the number with your abbreviated functional address symbols.
¶
¶
¶
¶
¶   **[Start signature block on fifth line]**

Sharon McBride
SHARON MCBRIDE
The Tongue and Quill OPR¶

¶
3 Attachments:¶
1. AFMAN 33-326, 1 Nov 99 ¶
2. HQ USAF/IM ltr, 28 Aug 04¶
3. AFCA/CC msg, 28 Sep 04¶
¶
cc: ¶
SAF/AAIQ¶
¶
1st Ind, HQ AU/CFA 8 November 04¶
¶
MEMORANDUM FOR ACSC/DEOP¶
¶
Use a separate-page indorsement when there isn't space remaining on the original memorandum or previous indorsement. The separate-page indorsement is basically the same as the one for the same page except the top line always cites the indorsement number with the office of origin date and subject of the original communication; the second line reflects the functional address symbols of the indorsing office with the date. An example is on the next page.¶
¶
¶
¶
¶   **[Start signature block on fifth line]**

Bart R. Kessler
BART R. KESSLER, Lt Col, USAF¶
Vice Dean of Operations¶

¶
3 Attachments:¶
1. & 2. nc¶
3. wd¶

2-inch top margin and 1-inch bottom margin

1-inch left and right margins

2 [.5 inches from top of page or 4 blank line]

2d Ind to ACSC/DEOP, 2 Nov 04, Indorsement Memo Format

ACSC/DEOP   11 December 2004

MEMORANDUM FOR HQ AU/CFA

The first line of the attachment element on indorsements should indicate the total number of attachments being forwarded. Succeeding lines should indicate the action regarding all the attachments listed (see previous indorsements and attachment listing below).

Sharon McBride
SHARON MCBRIDE
The Tongue and Quill OPR

3 Attachments:
1. & 2. nc
3. wd
4. (added) 9 AF ltr, 8 Nov 04

NOTE: *Tongue and Quill* recommends the second and following lines of indented text began at the left margin, see example 1a on page 189.

Are your memos cut and dry?
Perhaps you need English that's alive!

(See pages 65-89)

SHORT-NOTE REPY

DEPARTMENT OF THE AIR FORCE
AIR FORCE SPACE COMMAND

15 Feb 04

MEMORANDUM FOR ACSC/DEOP

FROM: 1CCS
250 South Ramp Road
Peterson AFB CO 80914-3050

SUBJECT: Request for *The Tongue and Quill*

1. I work for the Air Force, slinging ink at paper, pounding a computer, giving briefings, pushing packages and opening my mouth quite frequently in the conduct of today's mission. I need a personal copy of *The Tongue and Quill*.

2. This copy would help tremendously to improve my communications techniques and those of the people who work for me. My personal opinion is that everyone who works in the Air Force, civilian or military, should have a personal copy of *The Tongue and Quill*.

Ralph M. Phillips
RALPH M. PHILLIPS
Supervisor, Printing Specialist

Memorandum for Mr. Phillips
Here's your T&Q~check the
"Communication Basics" section.
I couldn't agree with you more,
everyone should have their own copy!

Sharon McBride
Attachment:
T&Q

DEPARTMENT OF THE AIR FORCE
HEADQUARTERS PACIFIC AIR FORCES

[1.75 inches or 10 lines from top of page] 6 Jan 04¶

¶
MEMORANDUM FOR ACSC/DEOP¶

¶
FROM: HQ PACAF/XP¶
       25 East Street STE G214¶
       Hickam AFB HI 96853-5417¶

¶
SUBJECT: The Short-Note Reply Memo Format¶

¶
Please explain the function and format of the short-note reply. I understand
it saves time, as well as typing, and can be used to acknowledge, provide a brief
routine reply, or forward correspondence.¶

¶
¶
¶
¶

Raymond E. Daniels

¶ **[Start signature block on fifth line]**   RAYMOND E. DANIELS, Lt Col, USAF¶
                                Director of Plans and Administration¶

¶
Memorandum For HQ PACAF/XP¶

¶
You just explained it! Write (or type) it on the letter if space allows;
if not, put it on a separate page and attach it. Make a copy for your files.¶

¶
Sharon McBride¶
T&Q OPR¶

> All too often a penciled question on a staff package
> generates an explosive blast of paperwork. (Rather
> like the commander sneezing and the staff catching
> pneumonia!) Why not answer the question with a
> penciled note? Heresy? Hogwash! "Reply in kind" is
> a concept at least as old as powered flight. Remember,
> tearing off the paper tiger is everybody's business—
> from the Chief of Staff down to the lowest ballpoint
> jockey. The short-note reply isn't always appropriate,
> but it is used much less than it could be. Try it; you'll
> like it!

NOTE: Triple space equal 2 blank lines
Double space equal 1 blank line

2-inch top margin and 1-inch bottom margin

1-inch left and right margins

MEMORANDUM FOR RECORD

The **separate-page memorandum for record (commonly referred to as memo for record, MR or MFR)** is used as an in-house document. It records info that is generally not recorded in writing (e.g., a phone call or meeting results) and informally passes it to others. People working together everyday generally pass info back and forth verbally, but sometimes it needs to be recorded and filed—memo for record is perfect. A "MEMO FOR RECORD" line can be added to target a specific addressee.

The **explanatory memo for record** gives you a quick synopsis of the purpose of the memo, tells who got involved and provides additional background info not included in the basic memo. By reading both the basic memo and the memo for record, readers should understand enough about the subject to coordinate or sign the memo without having to call or ask for more info.

If the basic memo really does say it all, an explanatory memo for record may not be needed. However, some organizations require you to acknowledge it by including "MR: Self-explanatory."

Turn to the next page for memo for record examples.

"A compilation of what outstanding people said
or wrote at the age of 20 would make a
collection of asinine pronouncements."
                                    – Eric Hoffer

"Eric wrote that when he was 19?"
                                    – The Quill

MEMO FOR RECORD   2 January 2004

SUBJECT: Preparing a Separate-Page Memorandum for Record (MR)

1. Use a separate-page MR to fulfill the functions discussed on the preceding page.

2. Type or write the MR on a sheet of paper in this format. Use 1-inch margins all around and number the paragraphs if there is more than one. A full signature block is not necessary, but the MR should be signed.

Carolyn R. Brown

CAROLYN R. BROWN
ASCS/DE

MEMO FOR RECORD   2 Jan 04

Omit the subject when typing the explanatory MR on the record copy. If space permits, type the MR and date two lines below the signature block. When there isn't, type "MR ATTACHED" or "MR ON REVERSE" and put the MR on a separate sheet or on the back of the record copy if it can be read clearly. Number the paragraphs when there is more than one. No signature block is required; merely sign your last name after the last word of the MR.

Brown

MR: When you have a very brief MR and not much space on the bottom of your correspondence, use this tighter format. Sign your last name and put the date following the last word.

Brown
2 Jan

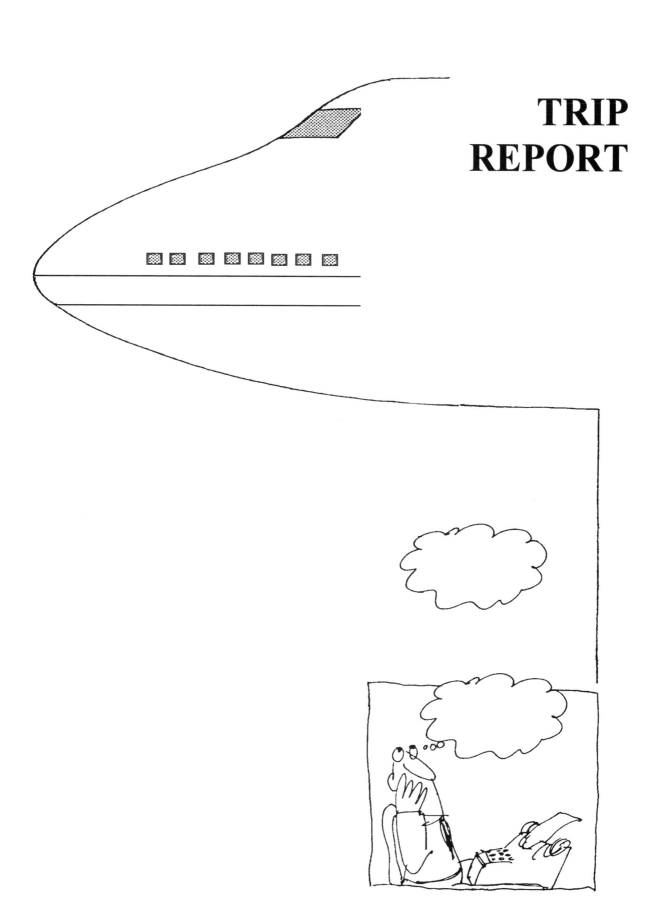

TRIP REPORT

DEPARTMENT OF THE AIR FORCE
HEADQUARTERS UNITED STATES AIR FORCE

29 January 2004

MEMORANDUM FOR HQ USAF/SCM
　　　　　　　　　　　　SC
　　　　　　IN TURN

FROM: HQ USAF/SCMV
　　　　　1250 Air Force Pentagon
　　　　　Washington DC 20330-1250

SUBJECT: The Trip Report Format

1. PURPOSE: Briefly state the reason for your trip. The report should answer the questions who, what, when, where, why, and how much, and then provide recommendations and conclusions. Attach meeting minutes or any other background documents that provide more detailed information, if needed. The format for the report is not particularly important. The official memorandum shown here is a good example; however, if another format better suits your needs or your organization has a preferred format, use it.

2. TRAVELER(S): Include rank, first name or initial, and surname. Provide position titles if travelers are from different offices or organizations. You may list names of members present in two columns to save space.

3. ITINERARY: List location(s) visited, inclusive dates, and key personnel contacted.

4. DISCUSSION: Base the amount of detailed information you include here on the knowledge level of your intended readers. Always include the trip objective, problems encountered, findings, future commitments made and your contribution to the event.

5. CONCLUSIONS/RECOMMENDATIONS: Summarize your findings and/or recommended actions.

Debby Clow
DEBBY CLOW
Committee Member
Visual Information/Publishing Committee

Attachment:
Minutes, 20 Dec 03

STAFF STUDY REPORT

The staff study report is used to analyze a clearly defined problem, identify conclusions and make recommendations. Not all organizations routinely use it, but it is an accepted format for a problem-solution report in both Air Force and Joint Staffs. (It's one of several staff formats identified in JFSC Pub 1—*The Joint Staff Officer's Guide*.)

This section provides information on how to analyze a staff problem ("Actions Before Writing Your Report") and guidance on writing a formal staff study ("Writing Your Report"). You may never write up a problem-solution in the staff study format. However, if you understand and apply the essential elements of problem analysis, you'll be better prepared for any staff communication. The staff study, as a thought process, is far more important than what you call it or what precise format you use to communicate your problem-solution explanation—letters, estimates of the situation, operational plans and orders, or the staff study report itself.

"Since the purpose of a staff is to assist the commander in the exercise of command, the work of the staff revolves around the solution of problems...
The staff study is one of the more flexible problem-solving procedures available to a staff."

— JFSC Pub 1

ACTIONS BEFORE WRITING YOUR REPORT

Before you can report on a problem, you must mentally solve it. Here's a logical sequence of essential elements:

1. ANALYZE THE AUDIENCE. You usually solve problems dropped on you by the hierarchy. Sometimes you generate your own areas or subjects that call for analysis. In any case, there will be political and operational constraints that affect your problem-solving process. Do some reflective thinking about the "environment" in which you're operating.

2. LIMIT THE PROBLEM. Restrict it to manageable size by fixing the *who, what, when, why* and *how* of the situation. Eliminate unnecessary concerns. Narrow the problem statement to exactly what you will be discussing—a common error is a fuzzy or inaccurate problem statement. For example, if the problem is the use of amphetamines and barbiturates among junior airmen, the problem statement "To reduce the crime rate on base" would be too broad. So would "How to detect and limit the use of dangerous drugs on base." More to the point would be "To detect and end the causes of amphetamine and barbiturate use among the junior Airmen at Wright-Patterson."

The problem should eventually be stated in one of three ways:

As a question:
What should we do to detect and end the causes of amphetamine and barbiturate use among junior airmen on this base?

As a statement of need or purpose:
This base needs to develop ways to detect and end the …

As an infinitive phrase:
To detect and end the causes …

3. ANALYZE THE WHOLE PROBLEM. Do the parts suggest other problems that need separate handling? Or do the parts relate so closely to the whole situation you need only one approach?

4. GATHER DATA. Collect all information pertinent to the problem. (Tips on how and where to conduct staff research can be found on Chapter 4).

5. EVALUATE YOUR INFORMATION. Is the information from reliable witnesses? Is it from qualified authorities? Does it qualify as solid support?

"Education should be as gradual as the moonrise, perceptual not in progress but in result."

– George John Whyte-Melville

6. ORGANIZE YOUR INFORMATION. One way to organize information is to place it under headings titled "Facts," "Assumptions" and "Criteria."

- **Facts** should be just that, not opinions or assertions. Identify only those facts that directly bear on the problem.

- **Assumptions** are important because they are always necessary. To reduce a research project to manageable size, it is usually necessary to accept certain things as being true, even if you are not absolutely sure. The validity of your assumptions usually has a great deal to do with the validity of your conclusions. Sometimes desired conclusions can be supported with certain unrealistic assumptions. In evaluating research, seek out the assumptions and make some judgment as to how reasonable they are. If you feel they are unrealistic, make whatever assumptions you feel are correct and try to judge their effect on the conclusions of the study. Sometimes a perfectly logical study explodes in your face because your assumptions were incredibly weak or simply not supportable.

- **Criteria** are those standards, requirements or limitations used to test possible solutions. The criteria for a problem-solution are sometimes provided in complete form by your boss when you are assigned the problem. Sometimes criteria are inherent in the nature of the obstacle causing the problem. The obstacle can only be overcome within certain physical limits, and these limits will establish the criteria for the problem-solution. In most cases, however, criteria are usually inherent in your own frame of reference and in the goal you are trying to attain. This goal and this frame of reference will tolerate only certain problem-solutions, and the limits of this tolerance will establish the criteria for the problem-solution.

  Remember this: **The criteria will not be very useful if you cannot clearly test the possible solutions against them!** Since weak or even lousy criteria are often seen in problem-solution reports, let's examine three examples of criteria and assess their value.

  ❶ "The total solution must not cost more than $6,000 annually."

  ❷ "The solution must result in a 75 percent operationally ready (OR) rate."

  ❸ "The solution must be consistent with the boss' philosophy on personnel management."

  Criterion one is fine; you could easily "bump" your proposed solutions against a specific cost. Criterion two looks good on the surface, but OR rates result from numerous and complex variables. You probably could not guarantee the decisionmaker your "solution" would lead to a 75 percent OR rate. It might **improve** the OR rate or actually lead to a rate **higher** than 75 percent, but before your boss actually **implements** your solution, how would you **know** that? If a criterion cannot be used to test solutions **before** implementation, it is not an acceptable criterion. Criterion three isn't bad, but it's fuzzy. Perhaps it could be written more precisely or left off the formal report altogether. You could still use it intuitively to check your solutions, but realize when you use "hidden" criteria, your report will be less objective.

7. LIST POSSIBLE SOLUTIONS. Approach the task of creating solutions with an open mind. Develop as many solutions as possible. The "brainstorming" technique using several knowledgeable people is a popular approach to generating possible solutions.

8. TEST POSSIBLE SOLUTIONS. Test each solution by using criteria formed while gathering data. Weigh one solution against another after testing each. Be sensitive to your personal biases and prejudices. Strive for professional objectivity.

9. SELECT FINAL SOLUTION. Select the best possible solution—or a combination of the best solutions—to fit the mission. Most Air Force problem-solutions fall into one of the three patterns listed below. Do not try to force your report into one of these patterns if it doesn't appear to fit.

① **Single best possible solution.** This one is basic and the most commonly used. You select the best solution from several possible ones.

② **Combination of possible solutions.** You may need to combine two or more possible solutions for your best possibilities.

③ **Single possible solution.** At times, you may want to report on only one possible solution.

10. ACT. Jot down the actions required for the final solution. Your comments here will eventually lead to the **specific action**(s) your boss should take to implement the solution (this will eventually appear in the "Action Recommended" portion when you write the report). If there is no implementing document for the decisionmaker to sign, you need to state clearly what other specific action the boss must take to implement your proposal. No military problem is complete until action has been planned and executed.

NOTE: In actual practice, the steps of problem solving do not always follow a definite and orderly sequence. The steps may overlap, more than one step may be considered at one time, or developments at one step may cause you to reconsider a previous step. For example, the data you collect may force you to redefine your problem. Similarly, while testing solutions, you may think of a new solution or, in the process of selecting a final solution, you may discover you need additional information. The steps just outlined can serve as a checklist to bring order to your mental processes.

WRITING YOUR REPORT

Here is the suggested format for a staff study report. Use only those portions of this format necessary for your particular report. If you omit certain paragraphs, renumber subsequent paragraphs accordingly.

DEPARTMENT OF THE AIR FORCE
AIR UNIVERSITY (AETC)

4 February 2004

MEMORANDUM FOR

FROM: ACSC/DEO
225 Chennault Circle
Maxwell AFB AL 36112-6426

SUBJECT: Preparing a Staff Study Report

PROBLEM

1. Clearly and concisely state the problem you are trying to solve.

FACTORS BEARING ON THE PROBLEM

2. Facts. Limit your facts to only those directly relating to the problem.

3. Assumptions. Should be realistic and support your study.

4. Criteria. Give standards, requirements, or limitations you will use to test possible solutions. Ensure you can use standards to measure or test solutions.

5. Definitions. Describe or define terms that may confuse your audience.

DISCUSSION

6. This section shows the logic used in solving the problem. Introduce the problem and give some background, if necessary. Then explain your solution or possible solution.

CONCLUSION

7. State your conclusion as a workable, complete solution to the problem you described previously in "Discussion."

ACTION RECOMMENDED

8. Tell the reader the action necessary to implement the solution. This should be worked so the boss only needs to sign to make the solution happen.

David J. Tanthorey
DAVID J. TANTHOREY, Major, USAF
Computer Operations

2 Attachments:
(listed on next page)

By now you probably realize the staff study is a problem-solution report that presents data collected, discusses possible solutions to the problem, and indicates the best solution. **It is not a style to solve a problem.** You should *mentally* solve your problem, and then *report* the solution in writing. The format of the staff study report includes a **heading,** a **body,** an **ending** and, when necessary, the **attachments.**

1. **HEADING.** Leave after MEMORANDUM FOR blank. This allows the report to seek its own level. After FROM, enter your complete office address (see page 184, paragraph 4). After SUBJECT, state the report's subject as briefly and concisely as possible. However, use a few extra words if this will add meaning to your subject.

2. **BODY.** The body of the report contains five parts: (1) **Problem,** (2) **Factors Bearing on the Problem,** (3) **Discussion,** (4) **Conclusion**, and (5) **Action Recommended**. These parts coincide with the steps of problem solving. That's why the staff study report is a convenient way to report your problem-solution.

Steps of problem solving	Body of staff study
1. Recognize the problem	1. Problem
2. Gather data	2. Factors Bearing on the Problem
3. List possible solutions	3. Discussion
4. Test possible solutions	4. Conclusion (a brief restatement of final solution)
5. Select final solution	5. Action Recommended
6. Act	

Problem. The statement of the problem tells the reader what you are trying to solve. No discussion is necessary at this point; a simple statement of the problem is sufficient. You have ample opportunity to discuss all aspects of the problem later in the report.

Factors Bearing on the Problem. This part contains the facts, assumptions, criteria, and definitions you used to build possible solutions to your problem. Devote separate paragraphs to facts, assumptions, criteria, and definitions as shown in the sample study report. Obviously, if you write a report in which you have no assumptions or definitions, omit either or both. Include only those important factors you used to solve your problem. Briefly state whatever you include. Put lengthy support material in attachments. Write each sentence completely so you don't force the reader to refer to the attachments to understand what you've written.

Discussion. This part of the report is crucial because it shows the logic used to solve the problem. Generally, some background information is necessary to properly introduce your problem. The introduction may be one paragraph or several paragraphs, depending on the detail required. Once the intro is complete, use one of the following outlines to discuss your thought process.

➤➤➤ When using the single best possible solution:

❶ List all possible solutions you think will interest the decisionmaker.

❷ Show how you tested each possible solution against the criteria, listing both the advantages and disadvantages. Use the same criteria to test each possible solution.

❸ Show how you weighed each possible solution against the others to select the best possible solution.

❹ Clearly indicate the best possible solution.

➤➤➤ When using a combination of possible solutions:

❶ List all the possible solutions you think will interest the decisionmaker.

❷ Show how you tested each possible solution against the criteria, listing both the advantages and disadvantages. Use the same criteria to test each possible solution.

❸ Show how you weighed each possible solution against the other possible solutions and why you retained certain ones as a partial solution to the problem.

❹ Show how and why you combined the retained possible solutions.

➤➤➤ When using the single possible solution:

❶ List your single solution.

❷ Test it against the criteria.

❸ Show how and why this solution will solve the problem.

No matter how you organize your report, these points are important: (1) make it brief, (2) maintain a sequence of thought throughout, (3) show the reader how you reasoned the problem through, and (4) use attachments for support, but include enough information in the body of the report to make sense without referring to the attachments.

Conclusion. After showing how you reasoned the problem through, state your conclusion. The conclusion must provide a complete, workable solution to the problem. The conclusion is nothing more than a brief restatement of the best possible solution or solutions. The conclusion must not continue the discussion. It should completely satisfy the requirements of the problem; it should never introduce new material.

Word the recommendations so your boss need only sign for action. Do not recommend alternatives. This does not mean you cannot consider alternative solutions in "Discussion." It means you commit yourself to the line of action you judge best.

You must relieve the decisionmaker of the research and study necessary to decide from several alternatives. Give precise guidance on what you want the decisionmaker to do; i.e., "Sign the implementing letter at attachment 1." (Normally, implementing documents should be the first attachment.) Don't submit a rubber turkey. Recommendations like "Recommend further study" or "Either solution A or B should be implemented" indicate the decisionmaker picked the wrong person to do the study.

3. ENDING. Follow the format shown on the sample report. The ending contains the name, rank, and title of the person or persons responsible for the report and a listing of attachments. Do not use an identification line.

4. ATTACHMENTS. Since the body of the staff study report must be brief, relegate as much of the detail as possible to the attachments. Although seldom required, identify material needed to support an attachment as an appendix to the attachment.

- Include, as attachments, the directives necessary to support the recommended actions.

- The body may reference the authority directing the study. An attachment may contain an actual copy of the directive.

- The body may contain an extract or a condensed version of a quotation. An attachment may contain a copy of the complete quotation.

- The body may contain a statement that requires support. An attachment may state the source and include the material that verifies that statement.

- The body may refer to a chart or information in a chart. An attachment may include the complete chart. (Design the chart to fit the overall proportions of the report or fold the chart to fit these proportions.)

- If directives or detailed instructions are required to implement the recommended action, include the drafts as attachments.

5. TABS. Number tabs (paper or plastic indicator) to help the reader locate attachments or appendices. Affix each tab to a blank sheet of paper and insert immediately preceding the attachment. If it is not practical to extract the supporting material from a long or complex document used as an attachment, affix the tab to that page within the attachment or appendix where the supporting material is located.

Position the tab for attachment 1 to the lower right corner of a sheet of paper. Position the tab for each succeeding attachment slightly higher on a separate sheet so all tabs can be seen.

COMPLETED STAFF WORK

A staff study report should represent completed staff work. This means the staff member has solved a problem and presented a complete solution to the boss. The solution should be complete enough that the decisionmaker has only to approve or disapprove.

The impulse to ask the chief what to do occurs more often when the problem is difficult. This impulse often comes to the inexperienced staff member frustrated over a hard job. It's easy to ask the chief what to do, and it appears easy for the chief to answer, but you should resist the urge. Your job is to advise your boss what should be done—provide answers, not questions. Of course, it's okay to inquire at any point in the problem-solving procedure if you need to find out whether you are on the right track. This coordination often saves untold hours.

Some final thoughts on completed staff work and problem-solution reporting:

★ Completed staff work provides the creative staffer a better chance to get a hearing. Unleash your latent creativity!

★ Schedule time to work the problem. Most problems worthy of analysis require considerable study and reflection.

★ There's usually no school solution—no "hidden cause" that will jump up and bite your kneecap. That's life. Avoid simplistic solutions; e.g., "Fire the idiots and get on with the program."

★ Don't assume that the heavier and fancier the study, the better it is. A smart decisionmaker focuses on the relevance and accuracy of your supporting material and the logic of your argument.

★ Don't work a study in isolation. If you point your finger at someone or some unit, or if the solution requires a change in someone's operation, you'd better get their reaction before you drop the bomb. You can look mighty foolish if you find out later they were operating under a constraint of which you were unaware.

★ Remember the final test for completed staff work: If you were the boss, would you be willing to stake your professional reputation on this problem-solution report? If the answer is "no, go directly to Go. Do not collect $200." It's time to start over.

"No man is fully able to command unless he has first learned to obey."
— Latin Proverb

The Latest, Greatest or Not So Greatest English

Audi = good-bye, I'm leaving, I'm out of here: "I'm audi."

Baldwin = attractive guy, a male Betty

Barney = unattractive guy, not a Baldwin

Betty = beautiful woman

Big time = totally, very

Buggin' = irritated, perturbed, flipping out: "I'm bugging"

Clueless = lost, stupid, mental state of people who are not yet your friends, uncool

Furiously = very, extremely, majorly

Hang = get tight with, ally with

I'm all ... = I was saying such things as "He's all 'Where were you?'" and "I'm all 'What's it your business?'"

Majorly = very, totally, furiously

Monet = looks fine from a distance but a mess up close, not a babe, really

Monster = much too big and loud, very good

Postal = a state of irritation, psychotic anger and disorientation

TB = true blue, loyal, faithful

Toast = in trouble, doomed, exhausted, towed up, history

Tow up = tore up, in bad condition, trashed, toast

Wass up? = see "zup"

Wig, wigged or wiggin' = become irritated, freak out, go postal

Zup? = Question: Is anything new? What's up

—from the film *Clueless*

TALKING
PAPER

TALKING PAPER

ON

USE OF COPYRIGHTED MATERIAL

- A copyright is the exclusive right granted under Title 17, United States Code, to the owner of an original work to reproduce and to distribute copies, to make derivative works, and to perform or display certain types of the works publicly.

- Use of copyrighted material in works prepared by or for the Air Force is governed by AFI 51-303, *Intellectual Property - Patents, Patent Related Matters, Trademarks and Copyrights*.

  -- No Air Force personnel should incorporate copyrighted material into works prepared by or for the Air Force to an extent that would clearly infringe a copyright without the written permission of the copyright owner.

  -- Each Air Force activity may seek permission in the form of a license or release to make limited use of copyrighted material without charge. The request should:

    --- Be for no greater rights than are actually needed.

    --- Identify fully the material for which permission to use is requested.

    --- Explain the proposed use and state conditions of use, so that the copyright proprietor or agent need only give affirmative consent of the proposed use.

    --- Be submitted in two copies to the copyright proprietor so that the proprietor may retain one copy and return the other copy after it is signed.

    --- Include a self-addressed return envelope.

- The Judge Advocate General controls and coordinates all copyright activities of the Air Force.

  -- The patents division directs those activities within The Judge Advocate General.

  -- Forward a copy of each license or release or any permission obtained without charge to the patents division.

Capt Jones/CPD/JA/3-3426/ada/30 Aug 04

TALKING PAPER

¶

ON

¶

WRITING TALKING AND POINT PAPERS

¶
¶

- Talking paper: quick-reference outline on key points, facts, positions, questions
  to use during oral presentations

¶

- Point paper: memory tickler or quick-reference outline to use during meetings
  or to informally pass information quickly to another person or office

¶

  -- No standard format; this illustrates space-saving format by eliminating headings
   (PURPOSE, DISCUSSION, RECOMMENDATION)

¶

  -- Usually formatted to conform to user's desires

¶

   --- Both papers assume reader has knowledge of subject

¶

   --- Prepare separate talker for each subject

¶

  -- Prepared in short statements; telegraphic wording

¶

  -- Center title (all cpital letters). Use 1-inch margin all around

¶

   --- Single dashes before major thoughts; multiple dashes for subordinate
    thoughts

¶

   --- Single space each item; double space between items

¶

  -- Use open punctuation; ending punctuation not required

¶

  -- Avoid lengthy details or chronologies, limiting to one page when possible

¶

  -- See DoD 5200.1-R/AFI 31-401 to prepare classified papers

¶

  -- Include writer's identification line as shown below on first page only if
   multiple pages

¶

- Include recommendations, if any, as last item

¶

- Include supporting information in an FYI (for your information) note in parenthesis
  at the appropriate place in the text or in attached background paper. EXAMPLE:
  (FYI: This is an FYI note. END FYI)

Ms. Adams/ACSC/DEC/3-5043/lcm/7 Apr 04

> **NOTE: Triple space equal 2 blank lines**
> **Double space equal 1 blank line**

1-inch top and bottom margins

1-inch left and right margins

BULLET
BACKGROUND
PAPER

BULLET BACKGROUND PAPER

ON

BULLET STATEMENTS

PURPOSE

The Bullet Background Paper is an excellent tool designed to present concisely written statements centered around a single idea or to present a collection of accomplishments with their respective impacts.

DISCUSSION

- All bullet statements must be accurate, brief and specific (ABS)

- Types of bullet statements

  -- Single Idea Bullets

    --- Definition: A concise written statement of a single idea or concept

    --- Cannot delineate

    --- Must serve purpose

  -- Accomplishment-Impact Bullets

    --- Definition: A concise written statement of a person's single accomplishment and its impact on the unit mission, organization, etc.

    --- Must clearly state a single accomplishment

    --- Must have an impact

      ---- Impact can be implied (Specific Achievement) or expressly stated

      ---- Strive to relate impact to the unit mission, organization, etc.

      ---- Make impact clear to those not familiar with career field, unique terminology, jargon, etc.

      ---- Expressly stated impact should strengthen bullet statement

      ---- Expressly stated impact should put accomplishment into perspective

    --- Accomplishment element must always precede impact element

    --- Types of Accomplishment-Impact Bullets

      ---- Action Verb: Accomplishment begins with strong, descriptive action verb

      ---- Modified Verb: Accomplishment begins with modifier (typically adverb)

      ---- Specific Achievement: Impact is implied, not expressly stated

SUMMARY

This paper summarizes characteristics of Single Idea and Accomplishment-Impact bullet statements.

MSgt McCarty/CEPME/EDC/6-4309/dcm/18 Aug 04

¶

¶

¶
¶

BULLET BACKGROUND PAPER

ON

THE BULLET BACKGROUND PAPER

An increasingly popular version of the background paper is the "bullet" background paper. The bullet format provides a concise, chronological evolution of a problem, a complete summary of an attached staff package or a more detailed explanation of what appears in an attached talking paper.

Use the first paragraph to identify the main thrust of the paper.

Main ideas follow the intro paragraph and may be as long as several sentences or as short as one word (such as "Advantages").

- Secondary items follow with a single dash and tertiary items follow with multiple indented dashes. Secondary and tertiary items can be as short as a word or as long as several sentences.

- Format varies.

  -- Center title (all capital letters); use 1-inch margins all around; single space the text; double space between items—except double space title and triple space to text; use appropriate punctuation in paragraphs and complete thoughts.

  -- Headings such as SUBJECT, PROBLEM, BACKGROUND, DISCUSSION, CONCLUSION, or RECOMMENDATION are optional.

Keys to developing a good backgrounder:

- Write the paper according to the knowledge level of the user; i.e., a person who is very knowledgeable on the subject won't require as much detail as one who knows very little.

- Emphasize main points.

- Attach additional support data; refer to it in the backgrounder.

- Require minimum length to achieve brevity with short transitions.

- End with concluding remarks or recommendations.

Include an identification line (author's rank and name, organization, office symbol, phone number, typist's initials, and date) on the first page 1 inch from the bottom of the page.

1-inch top and bottom margins

> **NOTE:** Triple space equal 2 blank lines
> Double space equal 1 blank line

> **REMEMBER: When subordinating bullets never use a double-dash (--) without a single dash (-), or a triple-dash (---) without a double-dash (--), etc.**
>
> - Main bullet
>   -- Secondary bullet
>   --- Tertiary bullet
>   --- Tertiary bullet
>   -- Secondary bullet
> - Main bullet

Mrs. Wilson/ACSC/CCA/3-2295/cab/7 Apr 03

1-inch left and right margins

TYPES OF BULLET STATEMENTS

1. Single Idea Bullets ~ This type of bullet is typically used for Talking Papers, Point Papers and Bullet Background Papers. Various examples can be found on the sample BBPs and in the Display Dot punctuation guidelines on pages 289-290.

2. Accomplishment-Impact Bullets ~ This type of bullet is typically used in performance reports (EPR, OPR, etc.), recommendations, and award submissions (Air Force IMT 1206). These bullets typically describe someone's work performance (good or bad) and/or noteworthy off-duty pursuits. For further help on how to develop accomplishment-impact bullet statements, see pages 226-235.

- **The Accomplishment Element:** This portion briefly describes the person's actions or behavior. Keep the following questions in mind as you construct the accomplishment element of your bullet statement:

    - What did the person (or group) do?
    - What was the success (or, less often, the failure)?
    - Might some terms be unfamiliar to readers outside this career field?

- **The Impact Element:** The impact element describes the results of the accomplishment and it may be either expressly stated or implied. The impact element is vital to describing the relative importance of the action. Keep the following questions in mind as you construct the impact element of your bullet statement:

    - What is the impact on the mission of the organization or the Air Force?
    - Is this impact statement accurate in scope and strength?
    - Does it put things into perspective?

There are three variations of the accomplishment-impact bullet. In most performance related documents (i.e., performance reports, award submissions, etc.), the Action Verb and Modified Verb variations are predominantly used, whereas the Specific Achievement variation is used sparingly.

- **Action Verb:** The accomplishment element begins with a strong action verb and ends with an expressly stated impact element.

    - **Developed** new customer sign-in log reducing customer complaints by 35 percent
    - **Implemented** a schedule to pick up customer equipment items; reduced delays by 5 days
    - **Conducted** 10 staff assistance visits this year
        -- **Ensured** all units received an EXCELLENT rating on their IG inspections
        -- **Selected** two units for Air Force-wide recognition
- **Modified Verb:** The accomplishment element begins with a modifier in front of the action verb and ends with an expressly stated impact element.

    - **Consistently** exceeded all standards of ... [*Consistently* modifies the verb *exceeded*]
    - **Solely** responsible for production increases in ... [*Solely* modifies the verb *responsible*]
- **Specific Achievement:** The accomplishment element may begin with a noun, verb or modifier; the impact of the accomplishment is implied. Specific Achievement bullets are primarily used to describe professional development or personal attainment.
    - Wing NCO of the Quarter for Jul-Sep 2004
    - CCAF degree in Aeronautical Systems Technology
    - Tirelessly maintained 4.0 GPA towards Bachelor's degree in Civil Engineering

Spell Chequer

Eye halve a spelling chequer
It came with my pea sea
It plainly marques four my revue
Miss Steaks eye kin knot sea.

Eye strike a key and type a word
And weight four it two say
Weather eye am wrong oar write
It shows me strait a weigh.
As soon as a mist ache is maid
It nose bee fore two long
And eye can put the error rite
Its rarely ever wrong.

Eye have run this poem threw it
I am shore your pleased two no.
Its letter perfect in it's weigh
My chequer tolled me sew.

— Sauce Unknown

BACKGROUND PAPER

BACKGROUND PAPER

ON

JOINT COMMUNICATIONS PLANNING AND MANAGEMENT

1. Joint Communications Planning and Management System (JCPMS) is an open-system, UNIX-based, automated communications network planning and management tool. It gives all the services and commanders in chief an automated capability to plan and manage a joint task force communications network. The requirement for a JCPMS capability comes from a requirement submitted by the United States Atlantic Command. The Joint Staff and the JCS Publication 6-05 working group validated the requirement and rewrote it into a joint mission needs statement. The Defense Information Systems Agency (DISA) evaluated the services' current and under development communications management programs to determine suitability for the joint requirement. DISA selected the Army's Integrated Systems Control (ISYSCON) as the best candidate to be developed into a JCPMS.

2. Joint Staff failed to obtain funding from The Office of the Secretary of Defense. The Joint Staff requested the Air Force, Army and Navy share equally in the cost to develop JCPMS from ISYSCON. The services attempted to write an input to the Program Object Memorandum for this requirement, but the initiative fell short within each service.

3. JCPMS will be based on commercial off-the-shelf (COTS) and government off-the-shelf (GOTS) software. This is prudent, cost-effective, timely, and in-line with COTS network management and control products already planned for other Air Force and DoD programs (e.g., theater deployable communications equipment, Base Network Control Center System and Defense Message System). A COTS/GOTS-based implementation of JCPMS will ease compatibility and interoperability issues likely to arise when the different network management systems are ultimately tied together.

4. The Air Force supports a joint solution that functions as a "manager of managers" to unite both strategic and deployable individual equipment managers in a seamless fashion. This will ease compatibility and/or interoperability issues likely to arise when the different network management systems are ultimately tied together. The Air Force agrees a JCPMS is needed and supports adapting a COTS-based ISYSCON to meet JCPMS requirements with the understanding the Air Force has no additional funding to contribute toward this effort.

Maj Don Duckett/ACSC/PI/6-5585/fv/28 Feb 04

BACKGROUND PAPER

ON

BACKGROUND PAPERS

¶
¶
¶
¶

1. The background paper is a multipurpose staff communications instrument that transmits ideas or concepts from one agency or person to another. It is an excellent way to express ideas on specific topics and to describe conditions that require a particular staff action. This background paper outlines its basic function and format with concluding comments covering the style and length.

2. The most common function of the background paper is to present the background (chronological, problem-solution, etc.) underlying an issue or subject, but it also has other purposes. One command uses it to condense and summarize important or complex matters into a single, easily read document limited to three pages. Another command uses it to inform and prepare senior officers to talk on a subject. Still another command specifies it as a summary of background material to provide a speaker with historical, technical, or statistical data.

3. A good background paper contains a brief *introductory paragraph* that provides the reader with a clear statement of purpose and an outline or "road map" for the paper, the *basic discussion* that comprises the bulk of the paper (cohesive, single-idea paragraphs), and leads the reader logically to the *conclusion.* Generally, write in the third person, although writing in first person, active voice will sometimes be more appropriate. The concise, telegraphic style of the talking paper is sometimes used and is encouraged. The bullet background paper shown next illustrates a tighter format. If it will do the job, use it.

4. The specific format (including style and length) of the background paper also varies. It may be longer than one page; the main point is to make it as short or as long as necessary to cover the topic adequately. Some agencies specify no particular format. The bullet format and this one illustrate the formats used in most organizations. General guidelines follow:

    a. Begin the header "BACKGROUND PAPER" six lines (1 inch) from the top of the first page and three lines above text.

    b. Use a 1-inch margin all around, double space.

    c. Type an identification line (originator's rank and name, organization, office symbol, phone number, typist's initials, and date) on the first page 1 inch from the bottom of the page.

    d. Number the pages consecutively, starting with page 2. Enter the page number on the left margin 1 inch from the bottom of the page or at least two lines below the last line of the text.

5. The key to an effective background paper, like any well-written document, is to get to the point quickly cover all aspects of the issue in sufficient detail to meet your objective, and close the paper with a sense of finality.

> **NOTE: Triple space equal 2 blank lines**
> **Double space equal 1 blank line**

MSgt Reed/ACSC/DPS/3-2620/jk/1 Oct 04

1-inch left and right margins

1-inch top and bottom margins

POSITION PAPER

POSITION PAPER

ON

QUALITY AIR FORCE

1. This paper addresses the importance of embracing and implementing Quality Air Force (QAF) tools into the Air Force. QAF tools are proven management techniques that allow better decision analysis and decision-making, but many do not recognize their usefulness. Several of the tools are based on supported measurement (metrics) and, like tools in a master carpenter's toolbox, they work well when properly used.

2. Traditionally, we have been taught to depend on our leader to analyze and solve problems. Based on the leader's experience and "analysis," many of those decisions were excellent, but not all. Today, we can better analyze problems using a quality tool called a flow chart. This tool asks people involved in, and thoroughly aware of, a process to examine and display the process steps so it can be analyzed and improvements can be made. This allows a logical and more "fact-based" improvement decision. Still, people say the process takes too long, it is unwieldy, and they aren't comfortable with it. So all the "improvement" decisions are made without it.

3. We must use QAF tools better. We have all bent a nail using a hammer, but how many of us blame the hammer? The same philosophy applies to quality tools. If we apply them properly, they work. We can certainly use a wrong tool to solve a problem. (We've all hammered in a nail with something other than a hammer with mixed results—QAF tools are the same.) But consider the master carpenters; they not only know and use tools more effectively they also add unique tools of their own. If we can add more "quality" tools to our toolkits, we must.

4. Using QAF tools in this time of dramatic, fast-paced change makes a lot of sense. If you haven't tried using them, do it. If you have tried once and a tool didn't work, think of all the nails you bent with a perfectly good hammer. The tools are proven performers, but we must use them and use them properly.

Lt Col Miles/ACSC/DE/3-7070/ly/14 Mar 04

POSITION PAPER

¶

ON

¶

THE POSITION PAPER

¶
¶

1. When you must evaluate a proposal, raise a new idea for consideration, advocate a current situation or proposal, or "take a stand" on an issue, you'll find the position paper format ideally suited for the task. This position paper builds a case for that assertion by describing the function and format of the Air Force position paper.

2. The opening statement or introductory paragraph must contain a "clear statement" of your purpose in presenting the issue and "your position" on that issue. The remainder of the paper should consist of integrated paragraphs or statements that logically support or defend that position. Adequate, accurate and relevant support material is a must for the position paper.

3. Since a position paper is comparable to an advocacy briefing in written format, the concluding paragraph must contain a specific recommendation or a clear restatement of your position. This is where you reemphasize your bottom line.

4. The format for the position paper is the same as for the background paper—number the paragraphs and double space the text. For more specifics, refer to the background paper on pages 215-216.

> **NOTE: Triple space equal 2 blank line**
> **Double space equal 1 blank line**

Capt Suarez/ACSC/DPS/3-7911/yb/2 Mar 04

1-inch top and bottom margins

1-inch left and right margins

STAFF SUMMARY SHEET

	TO	ACTION	SIGNATURE *(SURNAME)*, GRADE AND DATE		TO	ACTION	SIGNATURE *(SURNAME)*, GRADE AND DATE
1	AU/ES	Coord	*Cooper, Lt Col, 6 Apr 04*	6			
2	AU/CV	Coord	*Wright, Col, 6 Apr 04*	7			
3	AU/CC	Sig	*Woods, Col, 6 Apr 04*	8			
4				9			
5				10			

SURNAME OF ACTION OFFICER AND GRADE	SYMBOL	PHONE	TYPIST'S INITIALS	SUSPENSE DATE
Major Robinson	ACSC/CCE	3-2224	gst	

SUBJECT	DATE
Proposed Letter of Invitation to General Colin Powell, USA, Retired	2 Apr 04

SUMMARY

1. The proposed letter at Tab 1 invites General Powell to be the guest speaker at Air Command and Staff College's (ACSC) graduation on 12 Jun 04. This year, ACSC has added emphasis on leadership throughout the curriculum. General Powell's presentation will provide the capstone to the school's Leadership Focus guest speaker series.

2. RECOMMENDATION. CC sign proposed letter at Tab 1.

Michael E. Harris
MICHAEL E. HARRIS, Colonel, USAF
Dean of Distance Learning, ACSC
Ext 3-2456

1 Tab
Proposed Invitation Ltr

SAMPLE

AF FORM 1768, 19840901 *(IMT-V1)*   PREVIOUS EDITION WILL BE USED

STAFF SUMMARY SHEET

	TO	ACTION	SIGNATURE (SURNAME), GRADE AND DATE		TO	ACTION	SIGNATURE (SURNAME), GRADE AND DATE
1	ACSC/DEO	Coord	*Herron, Lt Col, 5 Apr 04*	6			
2	ACSC/DE	Coord	*Graham, Dr., 6 Apr 04*	7			*Sign your surname, rank or grade, and date on the bottom line if you are the addressee; sign on the top line if you aren't the addressee. If more than 10 coordinators, use another form, renumber and fill in all info through Subject line.*
3	ACSC/CV	Coord	*German, Col, 6 Apr 04*	8			
4	ACSC/CC	Sig	*Ladnier, Brig Gen, 7 Apr 04*	9			
5				10			

SURNAME OF ACTION OFFICER AND GRADE	SYMBOL	PHONE	TYPIST'S INITIALS	SUSPENSE DATE
SMSgt Tyndale	DPS	3-2290	jv	11 Apr 04

SUBJECT	DATE
Preparing the Staff Summary Sheet (SSS)	1 Apr 04

SUMMARY

1. The SSS introduces, summarizes, coordinates, or obtains approval or signature on a staff package. It should be a concise (preferably one page) summary of the package. It states the purpose, pertinent background information, rationale, and discussion necessary to justify the action desired. Show the action desired (Coord [Coordination], Appr [Approval], Sig [Signature]). Use Info (Information), when the SSS is submitted for information only. (**NOTE:** Usually show only one Appr entry and one Sig entry.) Use complete address when coordinating with outside organizations.

2. The SSS is attached to the front of the correspondence package. If an additional page is necessary, prepare it on plain bond paper. Use the same margins you see here. Summarize complicated or lengthy correspondence or document attached, or any appropriate portion of any document you reference. Do not use in place of a memo; use with a package it summarizes.

3. List attachments to the SSS as tabs. List the documents for action as Tab 1. List incoming letter, directive or other paper—if any—that prompted you to prepare the SSS as Tab 2. (If you have more than one document for action, list and tab with as many numbers as needed and list the material you're responding to as the next number: Tabs 1, 2, and 3 for signature, Tab 4 incoming document.) List supplemental documents as additional tabs followed by the record or coordination copy and information copies. If nonconcurrence is involved, list it and the letter or rebuttal as the last tab.

4. VIEW OF OTHER. Explain concerns of others external to the staff (i.e., OSD, Army, Navy, State, etc.) For example: "OSD may disapprove of the approach." Use a period or colon after VIEW OF OTHER, OPTION, etc. Either is acceptable; just be consistent.

5. OPTIONS. If there are significant alternative solutions, explain. For example, "Buying off-the-shelf hardware will reduce the cost by 25 percent, but will meet only 80 percent of the requirements."

6. RECOMMENDATION. Use this caption when SSS is routed for action. State the recommendation, including action necessary to implement it, in such a way the official need only sign an attachment or coordinate, approve, or disapprove the recommended action. Do not recommend alternatives or use this caption when submitted for info only.

Carrie H. Long
CARRIE H. LONG
Student Services, ACSC
Ext 3-7901

2 Tabs
1. Proposed Ltr
2. HQ AETC/CC Ltr, 25 Mar 04 w/1 Atch

AF FORM 1768, 19840901 (IMT-V1)   PREVIOUS EDITION WILL BE USED

ARRANGING ATTACHMENTS TO AF FORM 1768

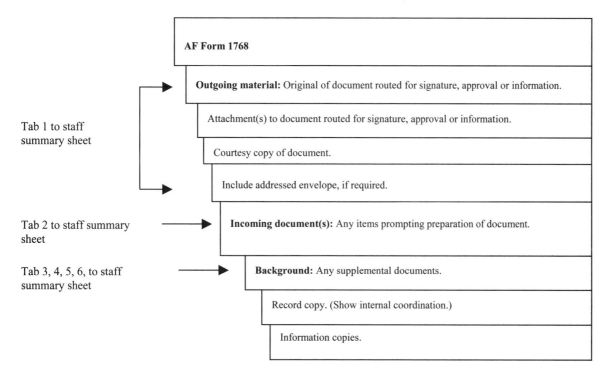

AF Form 1768

Tab 1 to staff summary sheet

Outgoing material: Original of document routed for signature, approval or information.

Attachment(s) to document routed for signature, approval or information.

Courtesy copy of document.

Include addressed envelope, if required.

Tab 2 to staff summary sheet

Incoming document(s): Any items prompting preparation of document.

Tab 3, 4, 5, 6, to staff summary sheet

Background: Any supplemental documents.

Record copy. (Show internal coordination.)

Information copies.

FORMAT TO INDICATE TABS

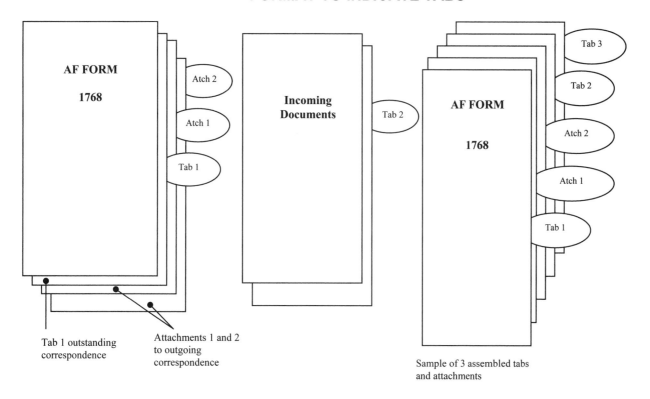

Tab 1 outstanding correspondence

Attachments 1 and 2 to outgoing correspondence

Sample of 3 assembled tabs and attachments

PREPARATION OF AN ELECTRONIC STAFF SUMMARY SHEET (SSS)

Staff Summary Sheets requiring your Group or Wing Commander's signature should be sent through your internal channels via e-mail. Office of Primary Responsibility (OPR) transmits package via e-mail to first reviewer to coordinate/comment. The first reviewer should forward (never reply) package with comments (if any) to the next reviewer. This procedure is repeated until last reviewer has coordinated on the package. The last reviewer forwards the entire package back to the OPR.

E-mail Message

From:

Sent:

To: 42 SPTG/CCE

Cc:

Subject: Electronic Staff Summary (ESS) Format

COORDINATION

Office	Action	Last Name/Rank/Date
AU/ES	Coord	Humphries, GS-11, 11 Sep 03
AU/DSS	Review	Gordon, GS-07, 11 Sep 03
AU/DS	Coord	Phillips, Col, 12 Sep 03
AU/CV	Coord	Copy rovided/mal
AU/CCS	Review/Print	Thomas, GS-09, 13 Sep 03
AU/CCE	Review	Tonnies, Maj, 16 Sep 03
AU/CC	Sign	[AU/CC igned s o CC Secretary for mailing. Prh]
42 CS/SCB	Info/File	Smith, GS-11, 16 Sep 03

*Note: Options for Action are Sign, Coord, Appr, Review, Info, or File

---STAFF SUMMARY

AO: Mr. John J. Smith, GS-11, 42 CS/SCB, 3-3467

SUPENSE: N/A

Summary: (If applicable)

1. PURPOSE. Please FORWARD (never REPLY) to the next agency after coordination. (If you require that the package be forwarded back to the originator for closer tracking, then so state.) The only attachment should be the actual letter for signature, or any extremely lengthy document, nothing else. All pertinent information should be included in this section. The reader should not have to look anywhere else for an explanation of the package's contents.

2. DISCUSSION. State the reason for the needed signature/coordination and any pertinent information needed to inform them why they are receiving the package. Include all information in the E-mail itself (i.e., tabs, talking papers, etc.).

3. RECOMMENDATION. In this section include the document you need signed (attachment) or action required. State whether you require an electronic or original signature in this section.

//Signed/mal/30 Sep 03//
MARTHA A. LONG
Lieutenant Colonel, USAF
Commander, 42d Communications Squadron
Ext 3-4289

Tab:
Example Electronic Staff Summary (ESS) Format

---Tab – ESS Format

ESS Format.DOC (attach document here)

RULES FOR WRITERS:

🖎 *Verbs HAS to agree with their subjects.*

🖎 *Prepositions are not words to end sentences with.*

🖎 *And don't start a sentence with a conjunction.*

🖎 *It is wrong to ever split an infinitive.*

🖎 *Avoid clichés like the plague. (They're old hat.)*

🖎 *Also, always avoid annoying alliteration.*

🖎 *Be more or less specific.*

🖎 *Parenthetical remarks (however relevant) are (usually) unnecessary.*

🖎 *Also too, never, ever use repetitive redundancies.*

🖎 *No sentence fragments.*

🖎 *Foreign words and phrases are not apropos.*

🖎 *Do not be redundant; do not use more words than necessary; it's highly superfluous.*

🖎 *One should NEVER generalize.*

🖎 *Comparisons are as bad as clichés.*

🖎 *Don't use no double negatives.*

🖎 *Eschew ampersands & abbreviations, etc.*

🖎 *One-word sentences? Eliminate.*

PERFORMANCE REPORT

Officer Performance Reports (OPR), Enlisted Performance Reports (EPR), and civilian reports (appraisals), are used by commanders, managers and supervisors to document an individual's performance over a specific period of time, and to identify top performers. As a supervisor, you play a vital role in your employee's career. You are the single most important person in a dynamic process that, through mentoring and effective writing, ensures each individual is afforded the opportunity for success and increased responsibility.

Performance reports provide a permanent, long-term record of an individual's performance and potential based on their performance. Performance reports should follow appropriate instructions/regulations and command guidance, if required. For more information, consult writing guides that support your organization's needs.

"My guidelines are simple. Be selective. Be concise, don't tell someone what you know; tell them what they need to know, what it means, and why it matters."
— General David C. Jones

GETTING STARTED

1 GETTING ORGANIZED. Keep records of all employee's accomplishments, awards and recommendations.

2 FORMAT. Refer to bullet format on page 213 and 226-235 for writing style. Use the font specified by your organization—in most cases it's built into the software used to produce the report. Limit each statement to three lines in length.

3 EDITING PRACTICES. Use the *Mechanics of Writing* section for grammatical terms, punctuation, abbreviations, capitalization, hyphens, and numbers. Avoid misspellings, typos, badly smudged documents, and misaligned bullet statements, which make a poor impression.

4 WRITE EFFECTIVELY. Get the reader's attention. Positive words and phrases leave a lasting impression with readers, so check out the word list on pages 228-229. Neutral or negative words and phrases give the impression that the person you are writing about is average or below average. The heart of effective writing in a performance report involves writing effective accomplishment-impact statements.

WRITING ACCOMPLISHMENT-IMPACT BULLET STATEMENTS

Like many Air Force writers, you've likely stared down the barrel of a loaded bullet statement tasking and wondered, "How in the world do I even *START* to write effective bullet statements?" Perhaps you're in that situation right now, or perhaps you're there for the 40th time! Well, it's time to stop this endless cycle of pain and torment. Over the next few pages you'll learn a tried and true system for building excellent bullet statements—plus it will save you lots of time and eliminate all of that frustration. Oh, got your attention, huh? Well OK, let's get you started!

STEP 1: EXTRACT THE FACTS

The first leg of this journey begins by tackling the hardest part of bullet statement writing—getting started! Most distressed bullet statement writers get into trouble early because they fail to identify all the information needed. It's a common mistake!

Gather the Information

Collect all of the information you can find that is relevant to the actual accomplishment. When identifying this information, be sure to capture everything on paper or in a computer. At this stage, capture information that may seem even remotely related to the accomplishment; you can cut stuff out later. You are looking for the following types of information:

Isolate one specific action the person performed (don't generalize).

Try to select the proper "power verb" that best describes this action (i.e., repaired, installed, designed, etc.).

Look for as much numerical information as possible that is related to this action (number of items fixed, dollars generated, man-hours saved, people served, pages written, etc.).

Track down information about how this accomplishment impacted the bigger picture, the larger scope (How did it help the work center? How did it support the unit's mission? How did it benefit the entire Air Force?).

Be persistent when collecting your information, and start by talking to the person you're writing about. They are in the best position to clarify and give exact details about what they've done and how they went about doing it.

Another place to draw information from is coworkers and other supervisors who may have seen this person in action. But don't stop there! You may need to consult Technical Orders, customers served, letters of appreciation, automated work production documents, or other sources to get all the information you need.

It's a great idea to track your subordinate's accomplishments as they happen. Keep a record of significant work performance (both good and bad) by jotting them down on a document somewhere. This habit will help you be prepared when it's time for a performance report or feedback.

In most cases, gathering this information should not require a great deal of time. But be prepared to schedule ample time with various people and jot down things you need to include (or verify). Keep in mind that if you fail to collect the information needed up front, putting together strong bullet statements will be nearly impossible. A lack of research is something to avoid ... it will force you to rely less on convincing facts and more upon loose generalizations and incorrect guesses.

Sort the Information

Once you have the information captured, don't be surprised if it looks like a host of different things all thrown together. Your next step is to review each item you've gathered and test it to see if it is truly associated with the single accomplishment you identified earlier or if these facts are completely unrelated. You can do this by applying the million-dollar question to each item:

$$$**   **"Is this bit of information solidly connected to this single accomplishment?"**   **$$$

If the answer is yes, flag the information as a keeper. If the answer is no, line through or flag the information as a nonkeeper ... but never throw it away or delete it! Although it may not be a good fit for the current bullet statement, it may be just what you need for another bullet you will soon be framing.

Continue applying this question to all of the items you've collected for this bullet statement. Once you're done with this process of sorting, you should have a neat little stack of information that pertains precisely to the target bullet statement. You've just eliminated any doubt concerning what should be included in this bullet, and what should not be included.

STEP 2: BUILD THE STRUCTURE

The next leg of our journey involves taking the information from Step 1 and organizing it into the proper structure of an accomplishment-impact bullet. There are essentially two major components: the accomplishment element and the impact element.

The Accomplishment Element

The accomplishment element should always begin with an action. Most of the time, this action takes the form of a strong action verb. Below you'll find a short list of action verbs that can be used to start bullet statements. It's nice to keep a list of verbs like this handy in preparation for writing that next EPR, OPR, or quarterly/annual award submission package.

SAMPLE ACTION VERB

Accomplished	Compared	Ensured	Produced
Achieved	Compelled	Escalated	Projected
Acquired	Competed	Established	Promoted
Acted	Compiled	Exceeded	Prompted
Activated	Completed	Excelled	Propagated
Actuated	Composed	Expanded	Propelled
Adapts	Comprehend	Expedited	Quantified
Adhered	Computed	Exploited	Rallied
Adjusted	Conceived	Explored	Recognized
Administered	Concentrated	Fabricated	Rectified
Advised	Conducted	Facilitated	Refined
Agitated	Conformed	Focused	Reformed
Analyzed	Confronted	Forced	Regenerated
Anticipated	Considered	Formulated	Rehabilitated
Applied	Consolidated	Generated	Rejuvenated
Appraised	Consulted	Grasped	Renewed
Approved	Contacted	Helped	Renovated
Aroused	Continued	Honed	Reorganized
Arranged	Contract	Identified	Required
Articulated	Contributed	Ignited	Resolved
Assembled	Controlled	Impassioned	Revived
Asserted	Cooperate	Implemented	Sacrificed
Assessed	Coordinated	Improved	Scrutinized
Assigned	Created	Initiated	Sought
Assisted	Cultivated	Inspired	Solved
Assured	Delegated	Insured	Sparked
Attained	Demonstrated	Invigorated	Spearheaded
Attend	Deterred	Kindled	Stimulated
Authorized	Developed	Launched	Strengthened
Averted	Devised	Maintained	Strove
Bolstered	Displayed	Manipulated	Supervised
Brought	Dominated	Motivated	Supported
Build	Drove	Organized	Surpassed
Calculated	Elicited	Originated	Sustained
Capitalized	Embodied	Overcame	Transformed
Catalyzed	Emerged	Oversaw	Utilize
Chaired	Emulated	Performed	
Challenged	Encouraged	Perpetuated	
Clarified	Endeavored	Persevered	
Collaborate	Energized	Persuaded	
Collected	Enforced	Planned	
Commanded	Enhanced	Practiced	
Communicated	Enriched	Prepared	

In some cases, action verbs alone just cannot fully stress the strength or depth of someone's accomplishment. If you need to give action verbs an added boost, use an adverb to modify the verb. Most adverbs are real easy to pick out … they end with the last two letters "ly." Try connecting some of the adverbs listed below to the verbs listed above to get a feel for how the adverb/verb combination can intensify the accomplishment element.

Sample Adverbs

Actively	Creatively	Forcefully	Quickly
Aggressively	Decisively	Frantically	Relentlessly
Anxiously	Eagerly	Impulsively	Restlessly
Ardently	Energetically	Incisively	Spiritedly
Articulately	Enterprisingly	Innovatively	Spontaneously
Assertively	Enthusiastically	Intensely	Swiftly
Avidly	Expeditiously	Powerfully	Tenaciously
Boldly	Exuberantly	Promptly	Vigorously
Competitively	Feverishly	Prosperously	Vigilant
Compulsively	Fiercely	Provocatively	

Now that you get the general idea about how to begin the accomplishment element, let's look at the rest of this critical part of the bullet statement. Broadly speaking, the accomplishment element contains all the words that describe a single action performed by a person. Yes, I know that sounds pretty simple, but you'd be surprised at how many writers violate this basic rule. If two or more actions are combined together in the same bullet, each of the actions is forced to share the strength of that entire statement. So rather than combining two or more actions to strengthen a single bullet, bullet statement writers must ensure bullets focus on only one single accomplishment. Two examples of an accomplishment element are listed below; one uses simply an action verb while the other uses a modifier (adverb) for added emphasis:

- **Processed** over 300 records with no errors as part of the 42 ABW Mobility Exercise
- **Tenaciously processed** over 300 records with no errors as part of the 42 ABW Mobility Exercise

Quick review: the accomplishment element begins with some form of action (action verb only or modifier plus action verb) and contains a description of one single action or accomplishment. With that established, let's look at the second portion of the bullet statement: the impact element.

The Impact Element

This part of the bullet statement explains how the person's actions have had an effect on the organization. The impact element can show varying levels of influence. That is, the person's actions may be connected to significant improvements to a work center's mission, an entire unit mission, or as broad as the entire Air Force. One thing to keep in mind about the scope of the impact is that it should be consistent with the person's accomplishment. For example, if the accomplishment explains how a person processed a large number of records during a base exercise, the impact should not be stretched or exaggerated to show how the Air Force will save millions of dollars. We'll chat more about "accuracy" later. For the time being, be extremely careful not to stretch the truth when rendering full credit for someone's accomplishment. Check out the suggested impact element below:

    …all wing personnel met their scheduled clock times

Connecting the Elements Together

Connecting the accomplishment and impact elements together can be accomplished several ways. One of these ways is to use the "ing" form of words. See how the word "ensuring" connects our two elements in the example below:

- Processed over 300 records with no errors as part of the 42 ABW Mobility Exercise **ensuring** all wing personnel met their scheduled clock times

Another way to connect these two elements together is to use conjunctive (just a fancy word meaning "joining things together") punctuation. The most common form of conjunctive punctuation in bullet statements is the semicolon. Let's set off our previous example with a semicolon to see this approach in action:

- Processed over 300 records with no errors as part of the 42 ABW Mobility Exercise; all wing personnel met their scheduled clock times

You may have noticed that in the two examples above, each bullet statement is set off with a single dash (-). If you have a situation where a person's single accomplishment has more than one significant impact, you may be better off showing each impact element separately. To do this properly, cut the impact completely off the main bullet and make each impact element a separate subbullet (subordinate to the main bullet) with double-dashes (--). When making an impact serve as a subordinate bullet, try to start each one with an action verb. It's much easier to comprehend this option by seeing it in action rather than by reading about it. So check out the example below:

- Processed over 300 records with no errors as part of the 42 ABW Mobility Exercise

    -- **Ensured all wing personnel met their scheduled clock times**

    -- **Helped the wing garner an overall "OUTSTANDING" rating from the IG team**

There is just one more thing concerning structure we want to nail down before moving on to our third and final step. All Air Force bullet statement writers need to adhere to several rules that standardize the appearance and format of bullets on various documents. Take a few moments to get familiar with these rules.

BULLET STATEMENT MECHANICS

- Start main bullets with a single dash (-)

- Indent bullets so that the first character of the second (and subsequent) line(s) aligns directly under the first character (not dash) of the line above

- Start subordinate bullets with additional dashes (-- for secondary, --- for tertiary)

- Indent subordinate bullets so that the first dash of the secondary (and tertiary) bullet(s) aligns directly under the first character (not dash) of the bullet above

  -- Secondary bullet is subordinate to main bullet

     --- Tertiary bullet is subordinate to secondary bullet

- Never use ending punctuation in accomplishment-impact bullet statements

- Always start an accomplishment-impact bullet statement with some form of action (action verb or modified verb)

- Never start an accomplishment-impact bullet with a proper noun or pronoun

- Minimize the use of the individual's name in bullet statement when it is printed elsewhere on the document

- Avoid using personal pronouns (he, she, his, her, etc.) in accomplishment-impact bullet statements; these devices typically serve to form complete sentences

STEP 3: STREAMLINE THE FINAL PRODUCT

Time for another quick review. After first extracting mounds of data to be used in drafting the bullets, we used a process to separate our extracted information into two different components: the accomplishment element and the impact element. Finally, we connected the two elements together and applied a handful of standardized rules to make the bullets appear strong and healthy looking.

So where do we go from here? Now we need to **Streamline the Final Product** and make the bullet statement. It is here that we'll refine the bullet statement by making it Accurate, Brief, and Specific (ABS). Did you catch that acronym? Similar to doing "crunches" in the gym to shape our abdominal muscles, we're going to crunch our bullet statements to give them some strong, healthy looking ABS. We'll start this section by first defining each of these attributes in A-B-S order.

Accuracy

For anything to be "accurate" it must be "correct." People who have the most trouble with accuracy are those who try to stretch the truth. You may remember that we touched on this briefly in our description of the impact element earlier. Mild exaggeration or embellishment of the facts is poison and is only a half-step away from outright lying and falsifying the facts. Don't ever let this happen to you. Let honesty, integrity, and accuracy be your watchwords.

Brevity

In editing for brevity you are actually accomplishing two separate activities. First, you should select words (replacing as necessary) that are shortest, clearest, yet most descriptive to the readers. That means that long, confusing words or phrases get swapped out with shorter, clearer, more common terms. Second, you need to reduce the number of unnecessary words. Some of the words that all bullet writers should be looking to eliminate (or at least sharply reduce) are:

- Articles: a, an, the
- Helping verbs: can, could, may, might, must, ought, shall, should, will, would
  -- Also forms of *be*: be, am, is, are, was, were, been, being
  -- Also forms of *have*: have, has, had, having
  -- Also forms of *do*: do, does, did
- Linking verbs: forms of verbs associated with five senses: look, sound, smell, feel, taste
- Name of the person when their name is printed elsewhere on the document
- Personal pronouns
- Prepositions (use them sparingly): over, under, in, during, within, etc.

Specificity

Specific bullet statements contain detailed facts. To write them, you'll need to be familiar with the people and systems involved. Fight the urge to estimate or generalize. Don't be satisfied with 10-20 of this, or about $1000 of that … you need to make sure of the exact numbers … right down to the penny if possible!

Knowing the meaning of ABS (also known as "the critical attributes of bullet statements") should help you make some "sweeping" changes to all the bullet statements you write. When you properly apply these attributes to your bullet statements, you will clean them **up**, trim them **down**, and give them the **all-around** scrubbing they need before they become a permanent part of someone's official record.

Sculpting Your ABS

Working these attributes into your bullet statements is like climbing into the ring for a few rounds of boxing. Oh yes, it may be a tough fight, but you're going to pulverize that challenger!

Ding! Ding! Ding! Ladies and gentlemen, in that corner over there, the challenger. Weighing in at "less than intimidating" and wearing the sagging trunks: the half-baked bullet statement!

In this corner right here, sporting an impressive set of ABS, the reigning bullet statement editor, and undisputed champion: the Air Force supervisor!

All right, come to the center of the ring and take a closer look at your opponent. You're going to put your ABS into action against the bullet statement below. No contest!

> - **Repaired 17 seriously corroded broken or missing Log Periodic (LP) antenna elements in the Atlantic Gateway Antenna System within 3 days by using elements from decommissioned antenna parts saving an estimated $3500 in procurement cost and 4 weeks of expected delivery time**

Back to your corner for a moment to formulate a strategy. In an effort to save time, we'll combine two of the attributes together during our first round of editing. Accuracy and specificity are going in first, while brevity will wait patiently at ringside for round two.

ACCURACY and SPECIFICITY

Ding! Round One. To make the bullet statement **ACCURATE**, you need to ensure the facts are correct. In a real-world setting we would verify the facts by simply asking a few questions. To simulate a realistic question and answer process that might occur in this situation, we'll pose a few obvious questions and also list some answers that would be reasonably collected.

- **Were there truly 17 antenna elements repaired?** Well, actually no ... not exactly. Seventeen elements were actually replaced, but an additional 23 were either sanded and painted or simply rescued.

- **Four weeks of delivery time seems like a long time ... why so much time to receive some antenna parts?** The antenna elements come from only one vendor and the estimated delivery time is based on the relatively low priority of the work order.

In order to make the facts in the bullet statement as **SPECIFIC** as possible, follow the same question and answer method we used to guarantee accuracy ... that's why we combined these two steps together!! Let's see what kind of information a few questions can scrape up.

- **Exactly how many antennas (end items) in the system were fixed?** The 40 antenna elements that were repaired or replaced were distributed among all 6 LP antennas in the system.

- **Can we be more specific than $3500 about the estimated cost savings?** Yes, the exact cost charged to the unit for purchasing the 17 replacement antenna elements would have been $3479.

- **What would the $3500 be spent on (item names, quantity, etc.)?** The estimated cost of $3500 was limited to varying sizes of antenna elements for an AN/GRA-4(V)4 Log Periodic Antenna System.

- **Who does the Atlantic Gateway Antenna System service?** The Atlantic Gateway connects Air Force and other DOD users from the eastern seaboard of the US to military personnel in Western Europe.

- **How has the repair of the antennas improved service to their customers?** Transmit and receive signal strength was improved; static and cross-talk was reduced.

Before we move on to the next round of editing, let's work in some of the answers to our questions into the bullet statement. You'll notice that not every tidbit of information we found could be added, but the items we did integrate contribute significantly to the message being sent. Items added or altered are represented in bold print.

- **Restored 40** seriously corroded, broken, or missing elements **on 6 AN/GRA-4 Log Periodic Antenna Systems** in the Atlantic Gateway Antenna System within 3 days by using elements from decommissioned antenna parts

  -- **Saved** an estimated **$3479** in procurement cost and 4 weeks of expected delivery time

  -- **Sharply improved clarity of voice signal for operators in US and Europe**

Ding! Ding! Ding! End of Round One. The added information enabled us to build an additional impact element and sharpen the exacting details of the preexisting bullet. You may have noticed that the bullet has grown in length again. Don't sweat it! Because we are rested and ready to get back in the ring and wear down that lengthy bullet! Time to employ the secret weapon: brevity!

BREVITY

Ding! Round Two. Let's apply what we know about brevity to knock this bullet out! Keep in mind that keeping your *editing for shorter words* separate from your *editing for a reduction of words* may be difficult. Therefore, give each bullet statement the old "one-two" and do both at the same time! Changes made to improve one aspect often promote the other.

Since we cannot boldface words or sections that have been removed or reduced, we need another way to indicate deletions, reductions, and occurrences of rephrasing. Carefully contrast the two versions of the bullet statements to see how certain terms and sections were reduced and consolidated.

BEFORE Editing for Brevity

- Restored 40 ~~seriously corroded, broken, or missing~~ elements on 6 AN/GRA-4 Log Periodic Antenna ~~Systems in the Atlantic Gateway Antenna System~~ ~~with~~in 3 days ~~by~~ using ~~elements from~~ decommissioned antenna parts
  -- Saved ~~an estimated~~ $3479 ~~in procurement cost~~ and 4 weeks of expected delivery time
  -- Sharply improved clarity ~~of~~ voice ~~signal~~ for operators in US and Europe

AFTER Editing for Brevity

- Restored 40 damaged elements on six AN/GRA-4 Log Periodic Antennas in three days using decommissioned antenna parts
  -- Saved $3479 and 4 weeks of expected delivery time
  -- Sharply improved voice clarity for Atlantic Gateway Antenna System operators in US and Europe

Ding! Ding! Ding! End of Round Two and the Match. The winner and still undisputed champion by KNOCK OUT! Yep, that's you … way to go champ!

Well, I guess congratulations are in order! You've just successfully completed the process of drafting effective accomplishment-impact bullet statements. Not as difficult as you made it out to be, is it? With a little more practice, you can Extract the Facts, Build the Structure, and Streamline the Final Product with less and less effort. Writing accomplishment-impact bullets will become almost second nature to you. And the next time you're hit with a bullet statement tasking (who knows … maybe tomorrow), you'll be prepared, fully armed and eager to climb back into that ring. Good luck!

"The way we communicate with others and with ourselves ultimately determines the quality of our lives."

— Anthony Robbins, motivation speaker

RULES FOR WRITERS continue:

✍ *Analogies in writing are like feathers on a snake.*

✍ *Eliminate commas, that are, not necessary. Parenthetical words however should be enclosed in commas.*

✍ *Never use a big word when substituting a diminutive one would suffice.*

✍ *Kill all exclamation points!!!*

✍ *Use words correctly, irregardless of how others use them.*

✍ *Understatement is always the absolute best way to put forth earth-shaking ideas.*

✍ *Use the apostrophe in it's proper place and omit it when its not needed.*

✍ *If you've heard it once, you've heard it a thousand times: Resist hyperbole; not one writer in a million can use it correctly.*

✍ *Puns are for children, not groan readers.*

✍ *Go around the barn at high noon to avoid colloquialisms.*

✍ *Even IF a mixed metaphor sings, it should be derailed.*

✍ *Who needs rhetorical questions?*

✍ *Exaggeration is a billion times worse than understatement.*

And finally...

✍ *Proofread carefully to see if you any words out.*

AWARDS AND DECORATIONS

MILITARY DECORATIONS

Awarded to an individual in recognition of heroism, or meritorious or outstanding service or achievement. Awarded for exceptionally distinguished service and accomplishments having significant Air Force-wide scope and impact covering a period of at least one year.

> Decorations are recommended to those who are PCSing, PCAing, separating, retiring, or members who have served more than 3 years on station.

MILITARY AWARDS

A decoration, medal, badge, ribbon or appurtenance bestowed on an individual or a unit.

CIVILIAN DECORATIONS

Awarded for exceptionally meritorious service. Performed their assigned duties for at least one year that resulted in an Air Force-wide impact to programs or projects that benefited the government.

CIVILIAN AWARDS

Special recognition for superior accomplishments for a person in the form of a cash award, certificate or medal.

For more in-depth guidance refer to Air Force Pamphlet 36-2861, *Civilian Recognition Guide* and/or Air Force Instruction 36-2803, *Air Force Awards and Decorations Program.*

"For a few yards of ribbon (medals), I could conquer the world."
– Napoleon

AWARD PACKAGE CHECKLIST

DÉCOR 6 CHECKLIST: The DÉCOR 6 is the computer form used to recommend a decoration. Check to ensure it's filled out completely (preferably typed, but can be neatly handwritten).

_____Section 1A (NAME OF DECORATION), 1B (CLUSTER) and 1C (INCLUSIVE DATES)

_____Section 1D and 1E—use a typewriter to X out the items not needed or line through with a black pen.

_____SECTION 1F (PRESENTATION DATE DESIRED): Include date award is to be presented or N/A if no date is needed. Dates are mandatory for retirements and separations.

_____1H (NEXT DUTY ASSIGNMENT OR FUTURE ADDRESS): This must be filled in if the person is separating or retiring. Personnel will send it back if an address is not included.

_____SECTION 5. Make sure the duty title matches the title used in the citation. The date on the citation cannot be before the DATE ARRIVED STATION. Must be one day or more after. (See AFI 36-2803 for information about including more than one assignment in an award/citation.)

_____SECTION 6 DECORATION HISTORY. This is the section where you will see the different decorations a person has received. When a person receives their first AFAM, this will be considered a basic award. When they receive a second AFAM, this will become their first oak leaf cluster.

NOTE: If there is another decoration within the inclusive dates, ensure that a copy of the certificate is provided with the package; this helps to ensure two awards are not given for the same achievement.

_____BLOCK 8. Make sure the supervisor has marked through either the "I RECOMMEND/DO NOT RECOMMEND" statement, as appropriate. Also be sure the supervisor and the commander have signed the printout.

CITATION CHECKLIST: The citation is read at the ceremony and summarizes the achievement or service for which the award is given.

Check the body of the citation

_____ Use 10- or 12-point font; borders should be top 1 to 1 1/2 inch, side 1 to 2 inch and bottom 3 inches.

_____ Opening and closing statements should be in accordance with AFI 36-2803.

_____ Length of citation should not exceed guidelines in AFI 36-2803.

      -- MSM and AFCM citations: 14 lines maximum

      -- AFAM citations: 11 or 12 lines maximum

_____ Make sure the following information is not on separate lines:

      -- Rank and name

      -- Unit and organization

      -- Date and month

_____ Do not include inclusive dates in the text portion of the citation

_____ Numbers 10 and above should be expressed in figures, and with exceptions, numbers "zero" through "nine" should be expressed in words. If both categories of numbers are used in the same related series, use figures for all. "Million" and "billion" should be spelled out ($20 million). Use numbers for organizations; they would not detract from the readability or the professionalism of a decoration citation/certificate. The use of the dollar ($) and percent (%) signs are authorized for use in decoration citations/certificates.

_____ No acronyms/symbols in body of citation. Rank should be spelled out the first time and shortened thereafter (e.g., Technical Sergeant Brian S. Bowers will be used in the first sentence. When referred to again, it should be "Sergeant Bowers.")

Check to make sure these items on the citation match the Décor 6:

_____ Dates of accomplishment.

_____ Oak leaf clusters (OLC). Use parentheses around the OLC on the citation.

_____ Duty Title.

Is the signature block correct? Different medals are signed at different levels of command. Senior commanders sign the higher level medals.

JUSTIFICATION/APPROVAL CHECKLIST:

_____ Attach all OPR/EPRs that cover the inclusive dates.

_____ For awards for specific achievement, if the period is too short to supply a report, a letter or memorandum must be provided with descriptive justification.

_____ Check with your organization to see if additional forms are required to staff an award through the chain of command. (For example, Air Education and Training Command uses AETC Form 114; check your unit for any specific requirements.)

WIN WITH WORDSMANSHIP

After years of hacking through etymological thickets at the US Public Health Service, Philip Broughton, a 69-year old official, created a surefire method to convert frustration into fulfillment (jargonwise). Euphemistically called the *Systematic Buzz Phrase Projector,* Broughton's system employs a lexicon of 30 carefully chosen **"buzzwords"**:

COLUMN 1	COLUMN 2	COLUMN 3
0. integrated	0. management	0. options
1. total	1. organizational	1. flexibility
2. systematized	2. monitored	2. capability
3. parallel	3. reciprocal	3. mobility
4. functional	4. digital	4. programming
5. responsive	5. logistical	5. concept
6. optional	6. transitional	6. time-phase
7. synchronized	7. incremental	7. projection
8. compatible	8. third-generation	8. hardware
9. balanced	9. policy	9. contingency

The procedure is simple. Think of any three-digit number, then select the corresponding buzzword from each column. For instance, numbers 2, 5, and 7 produce "systematized logistical projection," a phrase that can be dropped into virtually any report with that ring of decisive, knowledgeable authority. "No one will have the remotest idea of what you're talking about," says Broughton, "but the important thing is they're not about to admit it."

This is definitely a four eighty-five.

AIR FORCE PUBLICATIONS

This section expands on the functions and formats for Air Force publications—doctrine documents, policy directives, instructions, manuals, supplements, pamphlets, indexes, directories, handbooks, catalogs, joint publications, and changes (page and message) to those publications—as in the Air Force Instruction 33-360, Volume 1, *Air Force Content Management Program—Publications*. There may be a time in your military career when you will put on your action officer hat and write or develop an Air Force publication, form or Information Management Tool (IMT), regardless of your command level.

Publications serve a vital purpose—they direct and explain the policies and procedures governing Air Force functions. You, the action officer who writes these publications, are the expert in the various functions; you are the specialist in the different kinds of work done in the Air Force. When you write a publication, your role is to explain what work must be done and how it must be done by the people who must do it. The action officer's role is a demanding one. Preparing a publication is, by itself, a complex process. Additionally, you have two important responsibilities to meet.

- To provide clear, accurate guidance that others understand and follow.

- To get your guidance out quickly to those who need it.

These responsibilities are equally important, but at times will seem to conflict. A well-prepared publication takes less time to publish and reach its users than a poorly prepared one. Always take the time to prepare a well-written publication. To do this, you ...

- Plan carefully and become familiar with AFI 33-360, Volume 1, procedures to prepare, to submit and to coordinate a publication.

Follow the Seven Steps to Effective Communication and concentrate on organizing material logically, drafting a comprehensive outline, and writing clear and concise sentences.

GETTING STARTED...

1 GETTING ORGANIZED. For more information, go to page 56 without collecting $200 or contact your local publishing office for local procedures and guidance.

2 WRITING EFFECTIVELY. Use plain English with active voice; plain, concrete words; correct verb tenses (see pages 98-99) and simple sentences in a formal tone and style.

3 EDITING PRACTICES.

✎ Avoid contractions (flip to page 310), and *except* and other qualifiers detailed on page 77.

✎ Use the *Mechanics of Writing* section for punctuation, abbreviation, capitalization and numbers.

✎ Use functional address symbols sparingly. If needed, spell out symbols with the abbreviation and include a complete mailing address. [Visual Information/Publishing Division (HQ USAF/SCMV), 1250 AF Pentagon, Washington DC 20330-1250, see the *Envelope* section]

4 DEVELOPING TITLES. Writing good titles requires some thought ... a few guidelines to get you moving are given below:

Do ..

👍 Keep titles short and descriptive. The title of each paragraph describes the subject of the entire paragraph; the title of a section describes the overall subject of all the paragraphs within the section; the title of a chapter describes the overall subject of all the sections within that chapter.

👍 Use "General" sparingly—it's too broad. It may describe the material in a division but not the subject of that division. If your first chapter contains intro material, try "Introduction."

👍 Use only form/IMT numbers, or figure and table numbers. But, when using in the text, ensure its title follows.

👍 Include the report control symbol only in the title of the division it prescribes.

Do not ...

👎 Repeat the titles of main divisions or the titles of subdivisions within one division.

👎 Repeat the titles to a chapter and a section within that chapter, or paragraph titles within a section.

👎 Use abbreviations and acronyms. If you cannot avoid, include both the abbreviation and its meaning, or include it by itself after it has been explained in a preceding title.

👎 Use contractions in titles!

BIOGRAPHY

A current biography can be a helpful personal and management tool. You may want to send your biography to organizations you may be visiting or groups you may be speaking before. It may be helpful to your organization for key personnel to maintain biographies so newcomers can have a better understanding of the backgrounds of unit leadership. Your public affairs office may use your biography to assist in preparation of articles for publication.

A biography gives the reader some insight into who you are, where you've been assigned and the jobs you have handled. It should also include your education, awards, and if rated, flying hours and the type of aircraft experience. Ideally, it would also include an official Air Force photograph.

The Air Force electronically publishes official biographies for active-duty general officers, senior-level civilian executives and the chief master sergeant of the Air Force. Air Force public affairs prepare these biographies following the instructions found in AFI 35-101, *Public Affairs Policies and Procedures*. See http://www.af.mil/lib/bio/index.html for examples.

Applying the procedures from the Air Force instruction, on pages 244-245 is an outline of the parts of an Air Force biography. It is important to follow the recommended format so that there is consistency of presentation, style, and length of information. A "corporate look" needs to be maintained so readers are not confused going from one biography to another. Also, it makes for ease of distribution whether the biography is printed, posted on a web site, or sent by e-mail.

(**NOTE:** Air Force general officers and the Senior Executive Service must follow the outline in AFI 35-101 and the *Associated Press* style of news writing.)

FORMAT FOR BIOGRAPHY

Rank and Name (ALL CAPS)

First paragraph. Rank/Mr.; Mrs.; Ms. and Name, Duty Title, Unit/Organization, Base/City and State. *(Include a short statement of current job responsibilities. If included, the statement must conform to current policies concerning Internet security.)*

Second paragraph. *Courtesy title* (i.e.: Major Smith; Mr. Smith; Ms. Smith) is a (year) graduate of (college). Begin a brief nontechnical narrative (two short paragraphs) in chronological order of military/civilian career, including more prestigious tours of duty and locations. Avoid listing every assignment.

Duty titles, not job descriptions or designations, will be capitalized. (Example 1: His staff tours include duty as Manager, name of Section/Directorate. Example 2: Major Smith served as (Duty Title or job title.) He managed the logistics group in Europe, including duty in special interest projects (name project) for the commander.

Third paragraph. If needed, continue job information. DO NOT include family information.

EDUCATION

(In chronological order, list year, type of degree earned, field in which earned, school attended, and location. Executive courses should include the year, full name of course, school attended, and the location [city and state].)

1965 Bachelor of Science degree in (degree), Syracuse University, Syracuse, NY

1976 Air Command and Staff College, Maxwell Air Force Base, AL

1976 Master's Degree in Business Administration, Auburn University, AL

ASSIGNMENTS (CAREER CHRONOLOGY FOR CIVILIANS)

Show service from beginning month/year assigned to ending month/year, position, unit assigned, location. (Civilians may list year to year only. Career Chronology will include military and civilian assignments for civilians who have served in the Armed Forces. For their military assignments include year-to-year, position, unit assigned, and location. Duty titles will be capitalized; job descriptions or job designations will not.)

1. Month and Year – Month and Year, job description, organization, Base/City, State.

2. Continue entries to present assignment showing dates, duties, and locations. The last entry in the assignments section should match the first line of the biography. **NOTE:** Significant temporary duty assignments within a normal duty assignment should be included in the same entry with the main assignment. Example: 3. Month and year – Month and year, job duty, unit, Base/City, State (Month and year – Month and year, job duty, unit, Base/City, State). The assignment in parenthesis is the temporary assignment.

MAJOR AWARDS AND DECORATIONS (AWARDS AND HONORS FOR CIVILIANS)

If an award needs to be explained, it should not be included in this section. Instead, include the information as part of the narrative in paragraphs 2 or 3.

Vietnam Service Medal

Republic of Vietnam Campaign Medal

Kuwait Liberation Medal *(specify Government of Saudi Arabia or Government of Kuwait)*

PROFESSIONAL MEMBERSHIPS AND AFFILIATIONS

(List memberships or former memberships within last 5 years only)

Air Force Association

Order of Daedalians

Women's Bar Association

OTHER ACHIEVEMENTS

2000 "Who's Who in America"

PUBLICATIONS

"Titles of Books/Published Articles," Name of publication and the year published

EFFECTIVE DATES OF PROMOTION: (CHRONOLOGICAL ORDER)

(Current as of January 2003)

ADDITIONAL INFORMATION:

1. When preparing a biography, list the name and address of the organization owning the biography on the top portion of the first page.

2. Only official Air Force photographs may be used on the biography. Post the photo on the *right* side of the narrative.

Reasons why the English language is so hard to learn:

1) The bandage was wound around the wound.

2) The farm was used to produce produce.

3) The dump was so full that it had to refuse more refuse.

4) We must polish the Polish furniture.

5) He could lead if he would get the lead out.

6) The soldier decided to desert his dessert in the desert.

7) Since there is no time like the present, he thought it was time to present the present.

8) A bass was painted on the head of a bass drum.

9) When shot at, the dove dove into the bushes.

10) I did not object to the object.

11) The insurance was invalid for the invalid.

12) There was a row among the oarsmen about how to row.

13) They were too close to the door to close it.

14) The buck does funny things when the does are present.

15) A seamstress and a sewer fell down into a sewer line.

16) To help with planting, the farmer taught his sow to sow.

17) The wind was too strong to wind the sail.

18) After a number of Novocain injections, my jaw got number.

19) Upon seeing the tear in the painting I shed a tear.

20) I had to subject the subject to a series of tests.

21) How can I intimate this to my most intimate friend?

22) I spent last evening evening out a pile of dirt.

RÉSUMÉ

No, the résumé is not an official Air Force communication, but it is included for two reasons: (1) to help you when you are considered for a "special" assignment and need to give "someone" a summary of your experience and qualifications, and (2) to help you when you start job searching after your military or civil service career.

Getting interviewed for positions in which you are interested isn't always easy, but a good résumé can help. In many cases, your résumé is the first impression a potential employer has of you. The way it is written and how it looks are a direct reflection of you and your communication and organization ability. Your résumé must arouse the interest of a potential employer enough to make him want to meet you from a stack of hundreds of applicant résumés for a single job. Since no one is likely to interview a large number of applicants, "someone" will scan the résumés (for about 20 seconds each), choose those that will be read further, and trash the remainder. Therefore, your résumé has to make an impression quickly to make it pass the hiring official.

FUNCTION AND FORMAT

In some cases, the format you use will depend upon why you need a résumé. Is it a "feeler" to send to multiple companies? A response to a particular announcement? A specific request from a person or company? Generally speaking, a format that works best for you is the right one.

① **Chronological Résumé.** An outline of your work experience and periods of employment (in reverse chronological order—most recent information first) that shows steady employment. Titles and organizations are emphasized as are duties and accomplishments. This format is used most often by those with steady employment and/or who want to remain in a current career field. It's also excellent for those who have shown advancement within a specific career field. Detail a 10-year period and summarize earlier experience that is relevant to the position you are seeking.

② **Functional Résumé.** Emphasizes your qualifications (skills, knowledge, abilities, achievements) as opposed to specific dates and places of employment, and allows you to group them into functional areas such as training, sales, procurement, and accounting. List the functional areas in the order of importance as related to the job objective and stress your accomplishments within these functional areas. This format is used most often by people who are reentering the work force or those who are seeking a career change.

③ **Combination Résumé.** Combines the best of the chronological and functional résumés because you can group relevant skills and abilities into functional areas and then provide your work history, dates and places of employment, and education. This format allows you to cover a wider variety of subjects and qualifications, thereby showing skills that are transferable from one career to another. It works well for those "special" assignment requirements, for military retirees (those who have frequently switched career fields), and is ideal for people whose career paths have been somewhat erratic.

④ **Targeted Résumé.** As the name implies, this format focuses on your skills, knowledge, abilities, achievements, experience, and education that relate to the targeted position. It features a series of bullet statements regarding your capabilities and achievements related to the targeted job. Experience is listed to support statements, but it does not need to be emphasized. Education is listed after achievements. This format is probably the easiest to write, but keep in mind it must be completely reaccomplished for each position you are seeking.

⑤ **Federal Résumé.** An outline for the Federal Government and private industry to apply for Federal job searches. This résumé is commonly five pages by the time all your information is included so it does *not* follow the "keep it to two pages max" rule. Once you submit your résumé, there may be further forms to complete. When in doubt as to what to include, read your announcement carefully. Still in doubt? Call and ask questions. Generally include this: *Job info:* announcement number, title and grade of job for which you're applying; *personal info:* full name, address, day and evening phone numbers, social security account number, country or citizenship, reinstatement eligibility, veteran's preference, highest Federal civilian grade held; *education:* high school and college name and address, type and year of degree, major; *work experience:* job title, duties and accomplishments, employer's address, supervisor's name and phone number, starting and ending dates, hours per week, salary; *job related info:* training (with title, date), skills, certificates, honors, awards, accomplishments.

⑥   **Military Résumé.** If you are applying for a competitive assignment, a résumé may be a useful tool to summarize your qualifications for the position. You can use whatever format you think showcases your abilities, but make sure you include the following:

☑ **Security clearance and date of investigation.** This is critical for some jobs—and in some cases, you'll need a higher level clearance just to get in the building!

☑ **Date of rank.** This helps potential supervisors know how your seniority relates to the incumbents in the office … and helps them know when you'll be up for promotion.

☑ **Professional Military Education.** Make sure you identify any schools completed while in residence and any distinguished graduate (DG) recognition.

☑ **Service Status.** Air Force Reserve or National Guard personnel should identify their service status.

☑ **Availability.** When will you be eligible for reassignment? Include this in the "Job Objective" section.

☑ **Special Qualifications.** As appropriate, identify any special workplace qualifications such as foreign language skills, Acquisition Profession Development Program (APDP) Certification or Joint Officer credentials.

☑ **Flying Data.** Even if the position is not a "flying job," rated officers may want to include rated information.

"I lost my last three jobs because of clumsiness. I got fired from the optical supply co. because I was an eye-dropper, from the sign co. because I was a name-dropper, and from the gutter co. because I was an eavesdropper."

– from a Frank and Earnest cartoon
by Bob Thaves

PUTTING IT TOGETHER

Let's march through the six-step checklist again to write your résumé.

1 ANALYZE PURPOSE AND AUDIENCE. Your purpose … get a meeting with a potential employer.

2 CONDUCT THE RESEARCH. Know yourself: your needs and wants (type and level of the position), what you can offer and what you can do for them. Now take time to list your skills and accomplishments. And research the prospective company to learn about the job as well as to "speak their language."

3 SUPPORT YOUR IDEAS. Your "ideas" … the qualifications for the job. The "support" … your knowledge, skills, abilities, experience and education that support those qualifications.

4 GET ORGANIZED. Gather your documents: job descriptions, certificates, licenses, transcripts. If you worked for DOD, "civilianize" those job titles and descriptions to those used in the private sector … consult a *Dictionary of Occupational Titles*, published by the US Department of Labor. Learning the company's terminology could mean the difference between a 20-second scan and a "meeting."

5 DRAFT AND EDIT. Type your draft and edit, edit, edit for typos, extraneous information, action words (see page 228), plain language (businesses typically write on an 8th grade level), neatness, accuracy, and consistency in format. It must be long enough to cover relevant information but brief enough not to bore a potential employer (two pages max). Use only key phrases and words appropriate to the job you're seeking. To help you develop your accomplishment statements, review the advice on writing "accomplishment-impact" bullets on pages 226-235.

6 FIGHT FOR FEEDBACK. Have someone you trust read the résumé and suggest changes and recommendations. Are all the t's crossed and i's dotted? Does it look professional? Visually appealing? Is it easy to read with the strongest points quickly apparent? Is there good use of spacing, margins, indentions, capitalization and underlining?

TO INCLUDE OR NOT TO INCLUDE

As a minimum, include the following on all résumés:

☑ **Name, address, and phone number** (including the area code) centered at the top of the first page.

☑ **Job objective and/or summary statement.**

☑ **Qualifications and work experience** relevant to the job you are seeking.

☑ **Education and training** (anything acquired during military service or through workshops, seminars, and continuing education classes relevant to the position you're seeking).

The following are other topic areas to include *but* choose only those pertinent to the job you're seeking. CAUTION: More is not necessarily better. **Keep your résumé to one page, if possible, or two pages max.**

- Special skills or capabilities
- Career accomplishments
- Languages studied
- Honors and awards
- Military service
- E-mail address

- Professional experience and memberships
- Leadership activities
- Credentials, licenses
- Papers, presentations, published works
- Security clearance

The following information can sometimes be detrimental and takes up valuable space on a résumé, so we recommend you don't include it unless a potential employer specifically asks for it.

- Personal data (age, marital status, number and ages of children)
- Photograph
- Salary history or requirement
- Reason for leaving a job and names of bosses

- Religious affiliation
- Irrelevant information
- Hobbies or personal interests
- References
- Months, days—use "years" only
- Specific security information

COVER MEMO

A cover letter is a personal communication written to a specific person in an organization. But your same résumé can be sent to many potential employers. Write your cover memo so that the person reading it will want to read the résumé you've attached. To help out here's a checklist:

☑ Address the letter, if possible, to a person (use a title) with whom you wish to meet. Avoid Dear Sir, Gentlemen—clearly these are not gender-neutral terms.

☑ Use a positive tone and stress how your association with them will benefit you both. To sound genuine, research the organization and the position to learn all you can. Make it sound natural, relaxed, and not self-conscious. The letter needs action verbs, personal pronouns, life, conviction, humor, assurance, and confidence, ensuring you exclude irrelevant and negative information. Try saying "changing careers" or "looking for new challenges" instead of *retired*; and "single" instead of *divorced*.

☑ Limit it, to hold interest, to one page with about three paragraphs.

☑ Develop a strong first sentence to grab the reader's attention and, when possible, use a name of a mutual contact.

☑ Devote the next paragraph to brief facts by highlighting your relevant experience, skills, and accomplishments that make you unique for the job. It needs to entice the reader to call you. Consider using "bullets" to create eye appeal. And place the most relevant information in the first (or top) position.

☑ Close with a bid for a brief meeting (don't use *interview*) and write as though you expect it to occur. Indicate *you* will follow up with a call to arrange a time.

☑ Make it look professional—use the computer to make a good first impression! And edit, edit, edit!

☑ Use 8½- by 11-inch high-quality paper that is white or off-white.

☑ Maintain electronic file copies and borrow from past letters to save your brain power when submitting future letters!

"Writing is an adventure. To begin with, it is a toy and an amusement. Then it becomes a mistress, then it becomes a master, then it becomes a tyrant. The last phase is that just as you are about to be reconciled to your servitude, you kill the monster and fling him to the public."

— Winston Churchill

BODY STYLE

The examples that follow on pages 254-258 are provided primarily to give you a starting point and to illustrate style. More detailed information can be found in libraries and book stores or gotten from professional résumé writers. When writing your résumé, remember, the best qualified person doesn't always get the job—sometimes it's the person who knows **how** to get the job.

Cover Letter

PO Box 9553
Seattle, Washington 54321-9553
July 7, 2004

Ms. Rena Mitchell
Government Contracts Advisor
Bowe and Burke International Corporation
1472 South 303d Street
Seattle, Washington 98003-1472

Dear Ms. Mitchell,

I recently read about Bowe and Burke in the *Seattle Tribune*. I talked to Nancy Herron today and she mentioned you need an additional staff person for your government contract work.

I am interested in this position and believe you will find these particular aspects of my background relevant to this position:

- As project manager for a statewide fund-raising campaign, I recruited and supervised personnel, administered the budget, and oversaw the development of promotional materials.

- As manager of 300 volunteers for an organization, I handled diverse management functions—scheduled work, evaluated personnel, organized supply and equipment resources, and controlled cash receipts and other assets.

- I have extensive knowledge of your company's computer software and have developed personnel and accounting programs to expedite management functions.

- My bachelor's degree in business administration includes course work in personnel administration and accounting.

Since I am difficult to reach, I will telephone you on Wednesday, July 17, to arrange for a meeting where I can learn more about your requirements and tell you more about my background as it relates to you. Thank you for your consideration and I look forward to being in touch.

Sincerely,

Mary Middleton
MARY MIDDLETON

Attachment:
Résumé

Chronological résumé

KAREN CALLOWAY
173 Brunnen Street
Germantown, Iowa 54321-6652
(319) 386-0464

JOB OBJECTIVE

Challenging writer-editor position.

EXPERIENCE

<u>Writer-Editor</u>, Middletown College, Middletown, Iowa, 1993-Present

Researched and wrote curriculum materials for ... Edited curriculum materials written by ... ensuring ... and verifying facts and figures. Researched, designed, wrote, typeset, and distributed a 55-page administrative handbook and a 21-page textbook preparation guide for ... Supervised three ...

<u>Editorial Assistant</u>, The Middletown <u>Journal</u>, Middletown, Iowa, 1990-1993

Typed, edited, and proofread all articles prepared by reporters and staffers for the weekly newspaper. Produced ... Recommended ... Prepared ...

EDUCATION

Enrolled in Education Specialist Program, Troy State College (Anticipated completion December 1997)

MAG in Adult Education, Troy State College, Middletown, Iowa, 1993

BS in English, Troy State College, Middletown, Iowa, 1992

PROFESSIONAL MEMBERSHIPS

American Writers Guild

Association of Professional Editors

AWARDS

Best Book Award, American Council of Teachers of English, 1991

PERSONAL INTERESTS

Free-lance writing, photography, publishing

Functional résumé

CHARLES CATO
3941 Graff Road
Millbrook, Alabama 36054-0001
(205) 285-4333

JOB OBJECTIVE

Information systems resource manager specializing in microcomputers with emphasis on training and development.

QUALIFICATIONS

Resource Management: Managed $300 million inventory of hardware and software resources for 13 individual computer systems, 5 aircraft simulators, and 40 microcomputers. Reorganized ... Identified ..., formulated new policies, updated ... and revised ... Researched and developed ... Planned and supervised ... Reduced computer supply acquisition costs by $150K through ...

Systems Analysis: Coordinated weekly ... Organized, developed, and supervised the ... Designed training ... and developed self-study course ... implemented data base ... that resulted in ...

Quality Control: Developed, coordinated, and managed ... Assessed contractor ... Provided technical analysis of ...

Programming: Developed and maintained ...

EDUCATION

Enrolled in MA in Business Administration with emphasis in Information Systems, University of Alabama, Tuscaloosa, Alabama (Expected graduation December 1997)

BS, Mathematics and Computer Science, Tulane University, New Orleans, Louisiana, 1996 (GPA 3.1)

Combination (chronological and functional) résumé

FELICIA VINSON
317 Rue Du Congo
Vienna, Virginia 23230-4590
(703) 236-7052

OBJECTIVE	A position using skills and education that offers growth with increasing responsibility.
SUMMARY	Experienced in financial and retail organizations emphasizing office administration, accounting and supervisory skills.
QUALIFICATIONS	<u>Accounting and Finance</u>. Performed accounting and administrative procedures for large banking institution and retail sales organization. Completed … Balanced … Recorded bank … Calculated daily … resulting in reduction of … and savings of …
	<u>Management and Administration</u>. Administered … Operated computer to verify … Researched monthly … Supervised 3 to 6 employees … Managed retail stores with merchandise worth $300,000, reducing … and saving …
EXPERIENCE	
1992-1997	Manager, Boomers Stores, Washington DC
1991-1992	Assistant Manager, Sun Savings, Wheaton, Maryland
1990-1991	Accountant, Midway Bank, Midway, Maryland
1987-1990	Assistant Contracting Officer, United States Air Force
1980-1990	United States Air Force
EDUCATION	BS, General Business Administration, University of Pennsylvania, 1992
	Professional Senior Management Course, University of Northern Virginia, 1993
	Middle management training, United States Air Force, 1990

Targeted résumé

KEVIN JONES
6953 Oakside Drive
Harlem, Georgia 30814-7606
(706) 277-5084

JOB OBJECTIVE   Senior credit analyst in an engineering department with potential for advancement within the corporation.

CAPABILITIES
- Analyzed credit data to ...
- Prepared reports of ...
- Studied, researched, reported ...
- Evaluated ... and prepared reports ...
- Consulted with ... on ...

ACHIEVEMENTS
- Responsible for ...
- Supervised a staff of ... responsible for $2 million inventory of ...
- Maintained ...
- Acted as ...
- Saved ... work hours and ... dollars ...

EXPERIENCE

1992-present   Senior Credit Analyst, Georgia South Corporation, Macon, Georgia

1991-1992   Credit and Collection Manager, General Electric, Clinton, New Jersey

1988-1991   Claims Examiner, Great Western, Billings, New Jersey

EDUCATION   MBA in Finance, Pace University, Monroe, Connecticut, 1991

BS in Accounting, Northeast College, Penham, Massachusetts, 1988 (Honors graduate)

Federal Government résumé

CALVIN SMITH
2616 Elberta Street
Northport, Alabama 35475-4924
Day (205) 333-6327
Evening (205) 333-8534
E-mail: csmith@aol.com

Citizenship: United States   **Federal Status:** Civil Engineer, GS-802-11
SSAN: 555-123-4487   **Veteran's Status:** USAF, 1976-1996, Retired

OBJECTIVE: Civil Engineer, GS-12; Announcement No. 97-113

PROFILE: Civil engineer with 10 years' experience with an emphasis ... projects ranging from $50K to $900K ... Report to project civil engineer concerning construction analysis, planning ... Projects include: parking lots, roads, bridges ...

SUMMARY OF QUALIFICATIONS

Management and Administration. Administered ... Operated computer to verify ... Researched monthly ... Supervised 3-6 employees ... Managed projects worth ... reducing ...and saving ...

EDUCATION AND TRAINING

BS   Civil Engineer   Princeton University, 1986

WORK EXPERIENCE

TUSCALOOSA ENGINEERING, Tuscaloosa, Alabama

Supervisor: Jack Ward   June 1994 to present
Starting Salary: $30,000   40 hours a week
Contact may be made

Project Manager
- Project manager for ... ranging from ...
- Manages ... simultaneously ... Customers include ... monthly ... Coordinates ...

Engineering Technician
- Plans ... Designs ... Manages ... Interacts with ... to fulfill project ...
- Reviews ...

OTHER QUALIFICATIONS

Security clearance   Computer skills

ENVELOPE

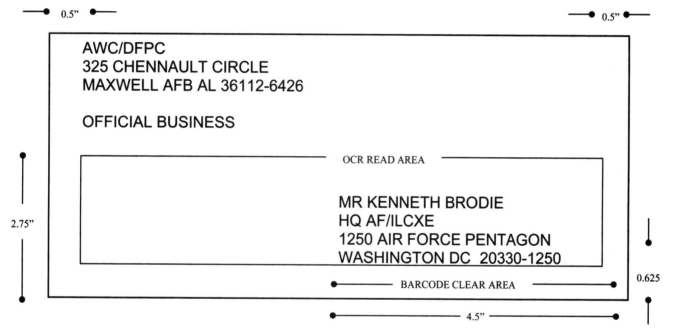

Using a standard envelope format helps the Postal Service identify the correct delivery address the first time that the mail piece is processed on postal equipment. Optical character readers (OCR) can read a combination of uppercase and lowercase characters in addresses but *prefers all uppercase characters*. Even though OCR enhancements now allow effective reading of punctuation in addresses, it is still suggested that punctuation be omitted, when possible. For faster, more accurate processing, include in the delivery address the street designators (for example, BLVD, DR); directional designators (for example, NE, SW); the apartment, suite, or room number; and a ZIP+4 code. Do not use bold, italic, script, artistic, or other unusual typefaces. Handwritten or rubber stamps should not be used.

ADDRESSING THE ENVELOPE

When preparing communications for dispatch, remember to use the appropriate size envelope for all correspondence. Use a rectangular envelope only slightly larger than the correspondence. Envelopes should be no smaller than 3.5 by 5 inches and no larger than 6.125 by 11.5 inches. When using window envelopes, be sure to adjust the MEMORANDUM FOR element to align the address with the envelope window. For consolidated mailing, the largest item that cannot be folded determines the envelope size. Do not use envelopes with clasps, staples, string buttons, or similar securing devices. Limit the thickness of any envelope to one-fourth inch or less when sealed. Write "nonmachineable" above the address on any envelope if it is more than one-fourth inch thick. There should be no printed or stamped markings of slogans or designs. This applies to labels, post cards and self-mailers.

⊟ **Envelope Format.** Use Courier New font 12 point (or similar simple sans serif font). Be sure characters are not too close together and do not touch or overlap. Leave margins at least 0.5-inch from the left and right edges of the envelope and at least 0.625-inch from the bottom of the envelope. The last line of the address should be no lower than 0.625-inch and no higher than 2.75 inches from the bottom of the envelope.

⊟ **Return Address Format.** Place the return address in the upper left corner of the envelope. Do not include names in return addresses. Type in uppercase with no punctuation except the hyphen in the zip code. Use the complete mailing address. Type "OFFICIAL BUSINESS" at least two line spaces below the return address. Rubber stamps may not be used in the return address portion of the envelope.

⊟ **Delivery Address Format.** Use block style with a left margin, parallel to the long edge of the envelope. Single-space the address block and type the entire address in uppercase. Use one or two spaces between words. Do not use punctuation in the last two lines of the address block except for the dash in the ZIP+4 code. Address area limited to five lines.

Optional Address Data Line. Use this line for any nonaddress data such as account numbers, presort codes, or mail stop codes.

Optional Attention (ATTN) Line. Use this line to direct mail to a specific person. It is the first line of the address. If the letter is sensitive information or for recipient eyes only, type "PERSONAL FOR MS JANE DOE" on this line.

Organization Abbreviation/Office Symbol Line. Use the organization abbreviation and office symbol separated by a virgule (/).

Delivery Address Line. Used for the street or post office (PO) box number, and the room or suite number. Use standard address abbreviations on page 313. Do not use punctuation on this line. You can spell out or abbreviate portions of the street address, e.g., AVE or AVENUE, STE or SUITE; both are acceptable. When addressing mail, do not use both the post office box and the street address—pick one.

City or Base, State, ZIP+4 Code Line. Use one or two spaces between words. Use the two-letter state abbreviations. Do not use punctuation except for the dash in the ZIP+4 code. With overseas addresses, do not use the APO or FPO number and geographical location together; this will cause the mail to enter the international mail channels. Do not type below the last line of the delivery address. Addresses are limited to five lines. Most printers today have the ability to print barcodes. This added function can help route your letter.

POUCH MAIL

Used to consolidate mail going to the same location. Contact your local official mail manager or base information transfer center for local pouch listing.

HOLEY JOE (STANDARD FORM 65-C)

Used to send mail within an organization or base. Address to the organizational designation and office symbol. If more than one person falls under the office symbol, consider using an ATTN line.

PARCELS

Labels should be placed on the top of the box or package. Any container used should be only slightly larger than the mail being sent. Seal boxes and package with paper tape.

To help you select the most cost-effective class of mail, review the definition listed on the next page.

DEFINITIONS

Official Mail. Official matter mailed as penalty mail or on which the postage of fees have been prepaid.

Official Matter. Official matter is any item belonging to or exclusively pertaining to the business of the US Government.

Official Mail Policies

The United States Postal Service (USPS) shall be used only when it is the least costly transportation method which will meet the required delivery data (RDD), security, and accountability requirements. When mailed, official matter shall move at the lowest postage and fees cost to meet the RDD, Security, and accountability requirements.

Official matter becomes official mail when it is postmarked by a distribution center or is placed under USPS control, whichever occurs first. Official matter ceases to be official mail when control passes from USPS or its representatives to someone else.

Classes of Mail

Express Mail
Fastest and most costly. Use only to prevent mission failure or financial loss.

First-Class
Any mailable item weighting 11 ozs. or less. Certain items must be mailed First-Class such as letters, handwritten or typewritten and post/postal cards.

Priority
Any mailable First-Class matter weighing over 11 ozs. but less than 70 lbs. Must be marked PRIORITY.

Second-Class
For magazines and other periodicals issued at regular, stated frequency of no less than four times per year.

Third-Class
For printed matter and parcels under 1 lb. Four ounces or less-same rate as first-class. Special bulk rates for larger mailings (at least 200 pieces or 50 lbs.)

Fourth-Class
"Parcel Post" For packages 1 to 70 lbs.

Military Ordinary Mail (MOM).
Goes by surface transportation within CONUS and by air transportation overseas. Add MOM to the second-, third-, or fourth-class endorsement on matter having a RDD not allowing sufficient time for surface transportation. Additional postage is not required.

Special Services

Registered
Provides added protection. Use only if required by law or a directive. Slow and expensive. For use only with First-Class and Priority Mail.

Certified
Provides a receipt to sender and a record of delivery at destination. For use with First-Class and Priority Mail.

Insured
Numbered insured service provides a method to obtain evidence of mailing and a record of delivery.

Certificate of Mailing (AF units must not use this service)
Provides evidence of mailing.

Special Handling (AF units must not use this service)
Provides preferential handling to the extent practical in dispatch and transportation, but does not provide special delivery. Applicable to third-and fourth-class mail.

Addressing Mail

Make sure mailing address is correct.

Use of office symbol reduces mail handling time.

Place city, state, and ZIP+4 in the last line of the address.

Return address is a must. Use your office symbol and ZIP+4.

Rules for Employees

Have personal mail sent to your home, not the office.

Use personal postage to mail job applications, retirement announcements, greeting cards, personal items, etc.

Tips for Cost Savings

Mailing 7 sheets or less of bond paper,-use letter size envelope. (Limit thickness to 1/4" or less when sealed.)

Manuals, pamphlets, etc., weighting over 4 ozs—mail third-class, special fourth-class, bulk rate, or bound printed matter rate.

Mailing several items to one address—cheaper to mail everything in one envelope. Check with your mailroom for activities serviced by consolidated mailings.

Check with your mailroom for activities/agencies within the local area that are serviced by activities/agencies courier—no postage required.

APPENDIX 1

THE MECHANICS OF WRITING

This appendix covers:

- A glossary of common grammatical terms.

- An alphabetical list of punctuation guidelines.

- Rules on capitalization of words and symbols.

- Guidelines on using abbreviations and writing numbers in text.

This is not an all-inclusive style manual. It's an Air Force quick-reference desktop guide to cure your most common trouble spots and to encourage standardization and consistency within the Air Force—especially during your professional military education. There are many style manuals and writers' guides available today and no two are exactly alike. Other commonly used style guides are *The Chicago Manual of Style, The Gregg Reference Manual, Air University Style Guide for Writers and Editors, US Government Printing Office Style Manual, Writer's Guide and Index to English*—not necessarily listed in order of preference. **If your organization or command has a "preferred" style of using capitals, abbreviations, numerals and compound words, use it.** If not, this guide is designed to serve that purpose.

GLOSSARY OF COMMON GRAMMATICAL AND WRITING TERMS

Grammar terminology is useful when we describe and correct problems with writing. Though we've tried to de-emphasize terminology and teach through examples throughout this book, sometimes you need a definition. We've tried to emphasize areas that are both commonly used and commonly misunderstood, such as the use of modal auxiliaries like can, could, shall, should, etc. Punctuation marks are not included in this list; they have a separate section in this Appendix.

"People who are experts in grammar don't always write well, and many people who write well no longer think consciously about grammar ... but when something goes wrong in a sentence, a knowledge of grammar helps in recognizing the problem and provides a language for discussing it."

— H. Ramsey Fowler

a/an	Use *a* before *consonant sounds* and *an* before *vowel sounds* <*a* historical event, *an* emergency.>
Active Voice	Shows the subject as the actor. <The ***girl*** *sang* a song> (pages 73-74).
Adjective	Describes or limits a noun or pronoun. It answers "Which one? What kind? or How many?" <*blue* box, *short* coat, *gregarious* man, *four* stools>
Adverbs	Modifies or limits a verb, adjective or another adverb and answers "When? Where? Why? How much? How far? To what degree?" <*quickly* run, *very* dull, *very* loudly>
	Conjunctive or Connective Adverb—transition words that often appears to connect clauses. <*however, therefore, etc.*>
Antecedent	Noun, phrase or clause to which a pronoun refers or replaces. (pages 99-100)
Appositive	Word, phrase or clause preceding or renaming a noun. <My dog Maggie.>
Article	Small set of words used with nouns to limit or give definiteness to the application. <*a, an, the*>
Bibliography	A list of books, articles and other works used in preparing a manuscript or other written product. (See "The Mechanics of Research," pages 345-347.)
Bullets	Any punctuation symbol used to emphasize specific items. (See "Display Dot" Punctuation Guidelines, pages 289-290.)
Case	Forms that nouns and pronouns take when they fit into different functions of the sentence. There are three:
	Nominative—for subjects, predicate nominatives and appositives. <*I*>
	Objective—for objects and their appositives. <*me*>
	Possessive—to show ownership, hence adjectival, functions. <*my*>
Clause	A group of related words containing a subject and a verb.
Conjunctions	Connects words, phrases, clauses, or sentences (*and, or, but, nor*).
Consonants	All letters of the alphabet except the ***vowels*** *a, e, i, o,* and *u.* In some words (*synergy),* the letter *y* acts as a vowel.
Glossary	An alphabetical list of unfamiliar terms and their definitions.
Interjection	Words used to express emotion or surprise (*ah, alas, great, hooray, help, etc.)* Strong interjections are punctuated with an exclamation point. (***Wow!*** *That's profound.)* Milder interjections are often set off by commas, usually at the beginning of a sentence. <***Oh,*** *I guess it wasn't.* ***Ouch,*** *that hurts.*>

Modifier	Words or groups of words that limit or describe other words. If improperly placed, modifiers can confuse the reader or suggest an illogical relationship (see **dangling** and **ambiguous** modifiers, page 97).
Modal Auxiliary	Verbs that are used with a principal verb that are characteristically used with a verb of predication and that in English differs formally from other verbs in lacking *-s* or *-ing* forms.
can **could**	Primarily expresses ability; *cannot* is used to deny permission. Sometimes the past tense of can. *<We could see the Big Dipper last night.>* Otherwise, *could* expresses possibility, doubt or something dependent on unreal conditions. *<We could see the Big Dipper if it weren't overcast.>*
may	Originally meant "have the power" (compare the noun *might*). Now it means "permission." Also, *may* is used to indicate possibility. *<You may leave if you are finished with your work.>* May is also used in wishes. *<May you recover soon.>*
might	Sometimes functions as simple past tense of may. *<He said he might have time to talk to us.>* Often it is used to express a more doubtful possibility than may does. *<He might returned before then.>* Might is also used after contrary-to-fact conditions. *<If I were off today, I might go fishing.>*
shall/ **should**	*Shall* expresses futurity in the first person; *should* does also, but it adds a slight coloring of doubt that the action will take place. Notice the difference in meaning in these sentences. *<I shall be happy to call the VA Medical Center for you. I should be happy to call the VA Medical Center for you.>* In indirect discourse *should* replaces the *shall* of direct discourse. *<I shall call at once. I said that I should call at once.>* Many speakers who use *shall* in the first person use *would* in preference to *should*. *<I said I would call at once.>* *Should* is used to express likelihood. *<Sue Sizemore should be able to finish on time.>* *Should* expresses obligation. *<We should file these orders more carefully.>*
will/ **would**	Will is the common future auxiliary used in the second and third persons. In addition it is used with special emphasis to express determination. *<You will finish by 4 p.m.>* *Would* still indicates past time in expressing determination. *<You thought you would finish by 4 p.m.>* *Would* expresses customary action in past time. *<Our last supervisor would bring us doughnuts every Friday morning.>* *Would* points to future time, but adding doubt or uncertainty. Notice the difference in meaning. *<I will if I can. I would if I could.>* *Would* replaces *will* in indirect discourse. *<He said that he would call.>*
must	Expresses necessity or obligation. It is somewhat stronger than *should*. *<You must call the director's office immediately.>* *Must* also expresses likelihood. *<It must have rained last night.>*

ought — Originally the past tense of *owe*, but now it points to a present or future time. *Ought* expresses necessity or obligation, but with less force. See the difference. *<We must go. We ought to go.>* *Ought* is nearly the equivalent of *should*.

dare — Originally a modal only, it is now used primarily in negatives or questions. *<He dare not submit the report in that form. Dare we submit the report like this?>*

need — Not originally a modal auxiliary, *need* is now used to mean *have to*. *<He need only fill out the top form. He need not get upset about the delay.>* In the meaning "lack," *need* is always a regular verb. *<He needs a little help with this project.>*

Equivalents of modals:

be able to — Used instead of *can* or *could* to indicate the ability as a fact rather than a mere potentiality. It is used also to avoid the ambiguity that may result from using *can* to express permission. *<He is able to support his mother.>*

be to — Indicates future events but hints at uncertainty. *<He is to have that report to us tomorrow.>*

have to — Commonly substitutes for *must*. It is a stronger expression of necessity. *<You have to have that done.>*

Other modals are used in speech, but they are inappropriate in writing.

had rather instead of *would rather*
had better instead of *should* or *ought*
(In speech, *had better* is emphatic in threats.)
have got to instead of *have to*

Modals are used with the infinitive of the perfect or progressive.

Can be going. Could have gone. Ought to be going. Ought to have gone.

Nouns — Names a person, place, thing, action or abstract idea. *<woman, office, pencil, game, Ohio, Maxwell AFB, democracy, freedom>*

Abstract Noun—nouns that name qualities rather than material things. *<love, danger>*

Collective Noun—nouns that are singular in form but plural in meaning; names a group of persons or things. *<audience, army, company, flock, committee, trio>*

Concrete Noun—nouns that can be seen or touched. *<table, book>*

Proper Noun—nouns that are capitalized and name specific persons, places, or things. *<Major Palmisano, Ohio, Air War College>*

Number — Shows the singular or plural of nouns, pronouns, or verbs.

Object
Noun or pronoun that is affected by the verb. *<The man read **the book**.>*

Parts of Speech
The basic building blocks of language: nouns, pronouns, verbs, adverbs, adjectives, prepositions, interjections, and conjunctions

Passive Voice
Shows the subject as receiver of the action. *<A song was sung by **her**.>*

Person
Pronouns that denote the speaker (first person; *I, we*), the person spoken to (second person, *you*), or the person spoken of (third person; *she, they*).

Phrases
Groups of words without a subject or predicate that function as a unit (adjective, adverbial, gerund, or infinitive phases).

Plagiarism
Using someone else's writing as if it were your own. This serious offense can lead to severe professional and legal consequences. If using another person's material, identify the borrowed passage and credit the author. (See "The Mechanics of Research" page 342.)

Predicate
Tells what the subject does or what is done to the subject, or the state of being the subject is in.

Preposition
Shows the relationship between a noun or pronoun to another word in the sentence. *<by, at, up, down, between, among, through>*

Pronouns
Substitutes for a noun. Here are three:

Definite—includes *I, you, he, she, it, we, they*, and all of their forms.

Indefinite—includes words like *someone, no one, each, anyone, and anybody*.

Relative—includes words like *who, whom, which, that*.

Sentence
Expresses one complete thought with one subject and one verb; either or both may be compound.

Complex Sentence—contains one main clause and at least one subordinate clause. *<When it rains, it pours.>*

Compound Sentence—contains two or more main clauses and no subordinate clauses. *<It rains, and it pours.>*

Subject
Tells what the sentence is about; the person, place or thing that performs the action or that has the state of being indicated by the verb.

Tense
Shows the time of the action, condition or state of being expressed. The three tenses—past, present, future—can be expressed in the simple, perfect, or progressive.

Verbals
Past and present participle forms of the verbs that act as nouns or adjectives. There are three:

Gerund—ends in *-ing* and functions as a noun. *<talking, singing>*

Infinitive—simple verb form used as a noun, adjective, or adverb and usually preceded by *to*. *<to go, to type>*

Participle—used as an adjective and acts as a modifier in present (-*ing*), past (*-ed, lost*), and perfect (*having lost*) forms.

Verbs

Expresses action or state of being of the sentence. There are six:

Transitive—transfers action from the subject to the object.

Intransitive—transfers no action and is followed by an adverb or nothing.

Linking—acts as an equal sign connecting the subject and the complement.

Auxiliary or Helping Verb—verb used with another verb to form voice or perfect and progressive tenses. <We *have eaten* there before.>

Principal Verb—last verb in a verb phrase.
Irregular Verb—verbs (see below) that form past tense and past participle differently:

PRESENT	PAST	PAST PARTICIPLE
become	became	become
begin	began	begun
bet	bet	bet
blow	blew	blown
break	broke	broken
bring	brought	brought
buy	bought	bought
catch	caught	caught
choose	chose	chosen
come	came	come
cut	cut	cut
draw	drew	drawn
drink	drank	drunk
drive	drove	driven
eat	ate	eaten
fly	flew	flown
forgive	forgave	forgiven
freeze	froze	frozen
give	gave	given
grow	grew	grown
keep	kept	kept
know	knew	known
ride	rode	ridden
ring	rang	rung
rise	rose	risen
set	set	set
shake	shook	shaken
sing	sang	sung
sink	sank	sunk
speak	spoke	spoken
spin	spun	spun

steal	stole	stolen
swear	swore	sworn
sweep	swept	swept
swim	swam	swum
swing	swung	swung
take	took	taken
tear	tore	torn
think	thought	thought
throw	threw	thrown
wear	wore	worn
weep	wept	wept

Vowel—The *a, e, i, o,* and *u.* In some words, the letter *y* acts as a vowel.

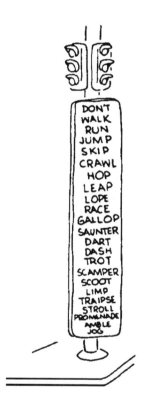

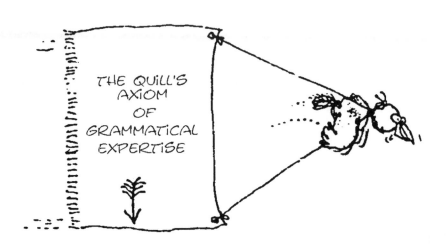

PUNCTUATION GUIDELINES: AN ALPHABETICAL LISTING

Punctuation marks are a writer's road signs they signal stops starts and pauses capitalization also helps writers communicate their meaning to their readers

Say what?! It's a jumbled mess, but with some effort you could grasp the writer's meaning. But look how easy it is when the proper punctuation is used.

Punctuation marks are a writer's road signs. They signal stops, starts, and pauses. Capitalization also helps writers communicate their meaning to their readers.

Punctuation marks are aids writers use to clearly communicate with others. Improper punctuation can confuse the reader or alter the meaning of a sentence. Excessive use of punctuation can decrease reading speed and make your meaning difficult to determine.

OPEN AND CLOSED PUNCTUATION

Open and closed punctuation: general philosophy.

Though many grammar rules are relatively clear-cut, there are some gray areas where the experts disagree. One area of debate is the issue of "open punctuation" versus "closed punctuation." Open punctuation advocates believe that writers should use only what's necessary to prevent misreading, while closed punctuation advocates are more apt to include punctuation whenever the grammatical structure of the material justifies it.

The following sentence illustrates how different writers might punctuate a particular sentence.

> If used incorrectly they may alter an intended meaning, and if used excessively they can decrease reading speed and make your meaning difficult to determine. **[Open punctuation—the meaning is clear without using all the punctuation that's needed by the grammatical structure.]**

> If used incorrectly, they may alter an intended meaning, and, if used excessively, they can decrease reading speed and make your meaning difficult to determine. **[Closed punctuation—using all required punctuation does not make meaning clearer and may slow reading speed.]**

In the Air Force, the general trend is to lean towards open punctuation in these types of cases. If you're confused about where to put commas, sometimes the best solution is to restructure the sentence to make the meaning clearer and eliminate the need for extra punctuation:

> If used incorrectly, they may alter an intended meaning; if used excessively, they can decrease reading speed and cause confusion. **[A slight change in sentence structure—fewer words to read and meaning is clear.]**

Open and closed punctuation: punctuating terms in a series.

So far, so good—much of this sounds reasonable to most people. Unfortunately, there IS one area where Air Force writers get conflicting guidance: the use of commas to separate three or more parallel words, phrases, or clauses in a series. Here's the rule and its two variants:

Use a comma to separate three or more parallel words, phrases, or clauses in a series.

> **In closed punctuation, include the comma before the final *and, or* or *nor*.**

>> Will you go by car, train, or plane?
>> You will not talk, nor do homework, nor sleep in my class.

> **In open punctuation, exclude the comma before the final *and, or*, or *nor*.**

>> Will you go by car, train or plane?
>> You will not talk, nor do homework nor sleep in my class.

Previous editions of *The Tongue and Quill* made a general recommendation to favor **open punctuation**. This **recommendation is unchanged**, but we'd like to acknowledge three reasons why closed punctuation guidelines might be used when punctuating three or more items in a series:

1. Closed punctuation of series is specified in most commercial grammar guides.

2. Closed punctuation of series is specified in some other Air Force references, including the *Air University Style Guide*.

3. The additional comma specified in closed punctuation may help clarify your meaning, especially when the items in the series are longer phrases and clauses.

Check to see which approach is preferred for the writing product you're working on. Award packages, performance appraisals, military evaluations, and other space-constrained formats typically use open punctuation. Research papers, academic publications, and books use closed punctuation. Technically speaking, either approach is acceptable, so consider your purpose and audience when deciding how to proceed.

Always remember that punctuation use is governed by its function: to help communicate the writer's meaning. Use the guidelines in the following section in the manner that best allows you to communicate your message to your readers.

"The only rule that doesn't have its exception is this one."

— The Quill

APOSTROPHE

USE AN APOSTROPHE...

- to create possessive forms of certain words
- to form contractions or to stand in for missing letters
- to form plurals for certain letters and abbreviations
- to mark a quote within a quote
- in technical writing to indicate units of measurement

1. Use an apostrophe to create possessive forms of nouns and abbreviations used as nouns.

   a. Add *'s* to singular or plural nouns that do not end with an *s*.

officer's rank	the oxen's tails
ROTC's building	the children's room

   b. Add *'s* to singular nouns that end with an *s*.

A business's contract	Mr. Jones's family tree
My boss's schedule	Marine Corps's Ball
	United States's policy

NOTE: This rule applies to most singular proper nouns, including names that end with an s: Burn's poems, Marx's theories, Jefferson Davis's home, etc. This rule does not apply to ancient proper names that end with an s, which take only an apostrophe: Jesus' teaching, Moses' law, Isis' temple, Aristophanes' play, etc.

   c. Add only the apostrophe to plural nouns that end in *s* or with an *s* sound, or to singular nouns ending with an *s* where adding an *'s* would cause difficulty in pronunciation.

The two businesses' contracts	for righteousness' sake
Our bosses' schedule	Officers' Wives Club; Officers' Club

   d. Add *'s* to the final word of compound nouns to show possession.

secretary-treasurer's report	mother-in-law's car; mothers-in-law's cars
attorney general's book	eyewitness' comment

   e. To show possession for indefinite pronouns (someone, no one, each, anyone, anybody, etc.), add *'s* to last component of the pronoun.

someone's car	somebody else's book

f. To show joint possession for two or more nouns, add the apostrophe or *'s* to the last noun. Add only the apostrophe to plural nouns ending in *s* and *'s* to singular nouns.

girls and boys' club
aunt and uncle's house

Diane and Wayne's daughters LaDonna,
Leah, Lynn, and Lori are ...

g. To show separate possession, place the possession indicators on each noun or pronoun identifying a possessor.

soldiers' and sailors' uniforms
king's and queen's jewels

Mrs. Williams's and Mr. Smith's classes
son's and daughter's toys

NOTE: Do not use an apostrophe when forming possessive pronouns (ours, theirs, its, his, hers, yours). One common mistake is using *it's* instead of *its*. Only use *it's* as a contraction of *it is*.

Its paw was caught in the trap.
Your savings account requires a minimum balance.

It's a bloody wound.
The reward was ours to keep.

NOTE: Don't confuse a possessive form with a descriptive form.

The Jones survey [a descriptive form: tells what survey you're talking about]
Jones' survey [a possessive form: shows to whom the survey belongs]

2. Use an apostrophe to mark omissions or form contractions.

can't (can not)
mustn't (must not)
don't (do not)
o'clock
I'm (I am)

the Roaring '20s
I've (I have)
won't (will not)
it's (it is)
wouldn't (would not)

jack-o'-lantern
you'll (you will)
let's (let us)
ne'er-do-well
rock 'n' roll

3. To form plurals of certain letters and abbreviations.
Make all individual lowercase letters plural by adding *'s* and make individual capital letters plural by adding *s* alone unless confusion would result. (For example, apostrophes are used with the plurals of A, I, and U because adding an s forms the words *As, Is,* and *Us*.) To plural most abbreviations (upper and lowercase), add a lowercase *s*. If the singular form contains an apostrophe, add *s* to form plural.

dotting the i's
OPRs, EPRs, TRs
1960s

S's, A's, I's, U's
bldgs (buildings)
Bs, 1s

the three Rs
B-52s
six the's

ain'ts
ma'ams
mustn'ts

4. Use apostrophes as single quotation marks for a quote within a quote.

"Let's adopt this slogan: 'Quality first.'"

5. Use an apostrophe in technical writing to indicate a unit of measurement (use the accent mark if the symbol is available).

a. As a length measure, use to specify the measurement of feet.

The room measures 16' by 29'.

b. As an angle measure, use to specify the measure of minutes (60 minutes = 1 degree).

NOTE: Angles identifying geographical latitude and longitude are specified in minutes and seconds (sixty seconds = 1 minute). When using the apostrophe or accent mark so specify minutes, use the quotation (") or double accent mark specify the measure of seconds.

The rendezvous coordinates are 35° 40' 30" N x 60° 20' 30" W.

ASTERISK

USE ASTERISKS ...

1. To refer a reader to footnotes placed at the bottom of a page.* Two asterisks identify a second footnote,** and three asterisks*** identify a third footnote. Number the footnotes if you have more than three, unless in a literary document number if more than one.

2. To replace words that are considered unprintable.

> If the camera was present when Smith called Schultz a *****, tonight's newscast would have had the longest bleep in TV history.

SPACING WHEN USING AN ASTERISK ...

- No space *before* following a word or punctuation mark within sentence or at the end of a sentence—unless replacing unprintable words, then one space before.
- One space *after* following a word or punctuation mark within a sentence.
- Two spaces *after* following a punctuation mark at the end of a sentence—unless manuscript format and using right justified, then one space *after*.
- No space *after* in a footnote.

NOTE: The *Tongue and Quill* favors two spaces after the end of a sentence. Rather to use one space or two is left up to the individual or organization. Either way is acceptable.

---

*Asterisk: A mark of punctuation to indicate a footnote. See "spacing."

**Use the asterisk with other punctuation as shown here.

***Number the footnotes if you have more than three—unless a literary document (see "spacing").

BRACKETS

USE BRACKETS ...

1. To clarify or correct material written by others.

> He arrived on the 1st [2d] of June.
> The statue [*sic*] was added to the book of statutes.

NOTE: The italicized word *sic* in brackets tells the reader something is wrong with the word immediately in front of the first bracket but the word is reproduced exactly as it appeared in the original.

2. To insert explanatory words, editorial remarks, or phrases independent of the sentence or quoted material.

> "Tell them [the students] to report to Wood Auditorium now."
> The tank-versus-tank battles of Villers-Brettoneux is the last significant event for the
>   tank in World War I. [Other accounts of this battle give different versions.]

3. To indicate you've added special emphasis (underline, bold type, all capitals, italics) to quoted material when the emphasis was not in the original work. The bracketed material may be placed immediately following the emphasized word(s) or at the end of the quotation.

> "She [emphasis added] seemed willing to compromise, but his obstinate attitude prevailed."
> "Tell them NOW to report to Wood Auditorium. [Emphasis added.]"

4. To enclose a parenthetical phrase that falls within a parenthetical phrase.

> (I believe everyone [including the men] will wear costumes.)
> I believe everyone (including the men) will wear costumes.

SPACING WHEN USING BRACKETS ...

—opening

One space *before* when parenthetic matter is within a sentence.

Two spaces *before* when parenthetic matter follows a sentence (when parenthetic matter starts with a capital and closes with its own sentence punctuation)—unless manuscript format and using right justified, then one space *before*.

No space *after*.

—closing

No space *before*.

One space *after* when parenthetic matter is within a sentence.

Two spaces *after* when parenthetic matter is itself a complete sentence and another sentence follows—unless manuscript format and using right justified, then one space *after*.

No space *after* if another punctuation mark immediately follows.

NOTE: The *Tongue and Quill* favors two spaces after the end of a sentence. Rather to use one space or two is left up to the individual or organization. Either way is acceptable.

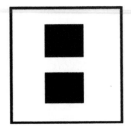

COLON

USE A COLON ...

1. To separate an introductory statement from explanatory or summarizing material that follows when there is no coordinating conjunction or transitional expression. (Capitalize the first word of the expression that follows the colon if it is the dominant element and is a complete sentence. For additional details, see the "Capitalization" section.)

> Living in base housing has many advantages: People can walk to work, shopping is convenient, and there are organized activities for the children.
> The board consists of three officials: a director, an executive director, and a recording secretary.

2. When a sentence contains an expression such as *following* **or** *as follows* **or is followed by a list or enumerated items.** [Notice the capitalization and punctuation.]

> The new directive achieved the following results: better morale and improved relations.
> Results were as follows: better morale, less work, and more pay.

> Consider these advantages when making your decision:
> 1. You won't have to be somewhere at 0800 every day.
> 2. You can get more involved in community activities.
> 3. You can pursue hobbies you haven't had time for in the last year.

3. To indicate a full stop before an enumerated or explanatory list.

> There are several possibilities: (1) the position could remain vacant, (2) it could be converted to a military position, or (3) another civilian within the organization could be temporarily detailed to the position.

4. With a quotation when the word *say* **or a substitute for** *say* **has been omitted, when the introductory expression is an independent clause, and when the quotation is typed in indented form on separate lines from the introductory clause.**

> The general turned [and said]: "Who gave that order?"
> The judge restated her ruling [independent clause]: "The defendant will remain in the custody of the sheriff until the trial begins."
> The speaker had this to say: "Please understand what I say here today represents my opinion alone. I am not here as a representative of the company for which I work."
> The speaker said:
>> The words you will hear from this stage today are the words and opinions of one man—me. I do not come as a representative of my company. I will not answer any question that is in any way related to the company for which I work.

5. To express periods of clock time in figures and to represent the word *to* **in proportions. Do not use a colon when expressing time on a 24-hour clock.**

> 8:30 a.m.
> 1:15 p.m.
>
> 1159 (24-hour-clock time)
> ratio of 2:1 or 3.5:1

6. When expressing library references to separate title and subtitle, volume and page number, city of publication and name of publisher in footnotes, and bibliographies.

Mail Fraud: What You Can Do About It
10:31-34 (Volume 10, pages 31 to 34)
New York: MacMillan Company

DO NOT USE A COLON ...

1. When the enumerated items complete the sentence that introduces them. [Notice punctuation.]

Liaison officers must
    a. become familiar with the situation,
    b. know the mission and
    c. arrange for communications.
[Not: Liaison officers must:]

2. When an explanatory series follows a preposition or a verb (except in rule 4 on page 280).

The editorial assistants in Publication Systems are Rebecca Bryant, Lisa McDay, and Yuna Braswell.
[Not: The editorial assistants are:]

3. To introduce an enumerated list that is a complement or the object of an element in the introductory statement.

Our goals are to (1) learn the basic dance steps, (2) exercise while having fun, and (3) meet new people.
[Not: Our goals are to:]

4. When the anticipatory expression is followed by another sentence.

The editorial assistants will bring the following items to eat. These food items will be heated and served at noon.
    Taco Bake
    tossed salad
    chips
    dip

SPACING WHEN USING COLONS ...

- No space *before.*
- Two spaces *after* within a sentence—unless manuscript format and using right justified, then one space *after.*
- No space *before* or *after* in expressions of time (8:20 p.m.) or proportions (2:1).

NOTE: The *Tongue and Quill* favors two spaces after the end of a sentence. Rather to use one space or two is left up to the individual or organization. Either way is acceptable.

See *Capitalization* section for more rules on capitalizing after a colon.

COMMA

USE A COMMA ...

1. With the coordinating conjunctions *and*, *but*, *or*, or *nor* when joining two or more independent clauses.

> **Right:** The art of war is constantly developing, but twentieth-century technology has so speeded up the change the military strategist now must run to keep pace.

> **Wrong:** The rapid expansion of the Air Force ensures a continuing need for qualified college graduates to fill existing vacancies, and also ensures ample opportunities for advancement. [This example contains only one independent clause with a compound verb; therefore, no comma is necessary.]

NOTE: No comma is needed if the sentence has one subject with a compound predicate connected with a coordinating conjunction because the second half of the sentence is *not* an independent clause.

> Martha Long received her master's degree December 2003 and is now pursuing her career.
> I am not only willing to go but also ready to stay a week.

2. To separate three or more parallel words, phrases, or clauses in a series.

In open punctuation, exclude the comma before the final *and*, *or* or *nor*.

> Will you go by car, train or plane? [open punctuation]

In closed punctuation, include the comma before the final *and, or,* or *nor.*

> You will not talk, nor do homework, nor sleep in my class. [closed punctuation]

NOTE: For longer phrases and clauses in a series, the additional comma specified in closed punctuation may help readability.

> Patients are classified as suitable for treatment at the installation, as requiring evacuation to the regional hospital, or as fit for duty.

NOTE: The use of *etc.* is discouraged in running text, but when used, it must be set off with commas. Do not use *etc.* when using *e.g., for example* or *such as*. These terms indicate you are only giving **some examples**; therefore, there is no need to imply there could be more.

> We will bake cookies, bread, cupcakes, etc., for the party.

3. With parallel adjectives that modify the same noun. If the adjectives are independent of each other, if the order can be reversed or if *and* can stand between the words, the adjectives are parallel and should be separated by a comma. However, if the first adjective modifies the idea expressed by the combination of the second adjective and the noun, do not use a comma.

> a hard, cold winter; a long, hot summer [the summer was long and hot]
> a heavy winter overcoat [winter modifies overcoat; heavy modifies winter overcoat]
> a traditional political institution [political modifies institution; traditional modifies political institution]

4. To separate two or more complementary phrases that refer to a single word that follows.

> The coldest, if not the most severe, winter Ohio has had was in 1996.

5. To set off nonessential or 'interrupting' words and phrases.

a. To set off nonessential words, clauses, or phrases not necessary for the meaning or the structural completeness of the sentence. You can tell whether an expression is nonessential or essential by trying to omit the expression. If you can omit the expression without affecting the meaning or the structural completeness of the sentence, the expression is nonessential and should be set off by commas. (For more examples, see page 284.)

> They want to hire Yuna Braswell, who has 10 years of experience, to run the new center. [The phrase "who has 10 years of experience" is nonessential information.]
> They want to hire someone who has at least 10 years of experience to run the center. [The phrase "who has at least 10 years of experience" is essential information.]
> There is, no doubt, a reasonable explanation. [This sentence would be complete without "no doubt."]
> There is no doubt about her integrity. [This sentence would be incomplete without "no doubt."]

NOTE: This rule includes interrupting words, phrases, or clauses that break the flow of the sentence.

> The faculty and staff, military, and civilian, are invited.
> She is a lieutenant colonel, not a major, and will be our new executive officer.
> The major, a recent promotee, is an experienced pilot.

b. With transitional words and phrases, such as *however, that is (i.e.), namely, therefore, for example (e.g.), moreover, consequently,* and *on the other hand,* when interrupting the flow of the sentence. A comma is normally used after these expressions, but the punctuation preceding is dictated by the magnitude of the break in continuity. However, when these words or phrases are used to emphasize meaning, do not set off with punctuation.

> It is important, therefore, we leave immediately.
> It is therefore vitally important we don't postpone the trip.
> A. Eaves is highly qualified for the job; i.e., he has 16 years of experience!
> Rebecca and Julie say they will attend—that is, if Robert and Lisa are attending.
> Planes from a number of bases (e.g., Andrews, Lackland, Tyndall) will participate in the flyover.

c. To set off a phrase introduced by *accompanied by, along with, and not, as well as, besides, except, in addition to, including, plus, not even, rather than, such as, together with,* or a similar expression when it falls between the subject and the verb.

> The faculty and staff, as well as the students, should be prepared to testify before the panel.
> The fifth and sixth graders, plus their parents, will be transported by bus.

NOTE: When the phrase occurs elsewhere in the sentence, commas may be omitted if the phrase is clearly related to the preceding words.

> We agree, Miss Johnson, our policy was badly processed as well as lost in the mail.

d. With the adverb too (meaning also) when it falls between the subject and verb. Omit the comma before too if it occurs at the end of a sentence or clause.

> You, too, can save money by shopping selectively.
> You should try to improve your typing too.
> If you want to bring the children too, we'll have room.

e. To set off nonessential appositives. An appositive is a word or phrase appearing next to a noun that identifies it and is equivalent to it. If the appositive is nonessential, set it off by commas. If essential or restrictive in nature, do not set it off by commas.

> Our cost analyst, Mrs. Sherri Thomas, will handle the details. [In this hypothetical example, we have only one cost analyst, so *Mrs. Thomas* is "nonessential." If we eliminate her name, the meaning of the sentence would not change.]
> The battleship *Pennsylvania* was taken out of mothballs today. [*Pennsylvania* is "essential" to the sentence because there is more than one battleship in mothballs.]
> Their daughter Julie won the contest. The other daughters were really annoyed. [Since they have more than one daughter her name is essential to the sentence.]
> Edward shares a house with his wife Esther in Prattville, Alabama. [Strictly speaking, **Esther** should be set off by commas because he can have only one wife and giving her name is nonessential information; however, because the words **wife** and **Esther** are so closely related and usually spoken as a unit, commas may be omitted.]

f. To set off the title, position, or organization after a person's name or name equivalent. (Some cases under this rule are appositives; other cases are not.)

> Lieutenant General Don Lamontagne, Commander of Air University, will speak at ACSC this Thursday.
> The Commander, 42d Air Base Wing, is responsible for ...

g. To set off long phrases denoting a residence or business connection immediately following a name.

> Lieutenant Colonel Fernando Ordoñez, of the Peruvian air force in Lima, Peru, will be here tomorrow.
> Lt Col Ordoñez of Lima, Peru, will be here tomorrow. [The comma is omitted before *of* to avoid too many breaks in a short phrase.]

6. To set off introductory elements.

a. With introductory elements that begin a sentence and come before the subject and verb of the main clause. The comma may be omitted if the introductory phrase is five words or less except when numbers occur together. If you choose to use a comma following a short introductory phrase, do so consistently throughout the document.

> In 1923, 834 cases of measles were reported in that city.
> In 1913 the concept of total war was unknown.
> Of all the desserts I love, my favorite is the fruit trifle.
> Since the school year had already begun, we delayed the curriculum change.

b. After introductory words such as *yes, no,* or *oh*.

> Yes, I'll do it.
> Oh, I see your point.

7. To set off explanatory dates, addresses and place names.

> The change of command, 1 October 1996, was the turning point.
> The British prime minister lives at 10 Downing Street, London, England.

NOTE: Use two commas to set off the name of a state, county, or country when it directly follows the name of a city **except** when using a ZIP code. When including the ZIP code following the name of the state, drop the comma between the two (see *Envelope* section), but use one after the ZIP code number if there is additional text.

> We shipped it to 2221 Edgewood Road, Millbrook AL 36054-3644, but it hasn't been received yet.

8. To set off statements such as *he said*, *she replied*, *they answered*, and *she announced*.

> She said, "Welcome to the Chamber of Commerce. May I help you?"
> She replied, "I have an appointment with Lt Col Rick Jenkins at 10 a.m."

NOTE: If a quotation functions as an integral part of a sentence, commas are unnecessary.

> They even considered "No guts, no glory!" as their slogan.

9. To set off names and titles used in direct address.

> No, sir, I didn't see her.
> Linda McBeth, you're not changing jobs, are you?
> And that, dear friends, is why you're all here.

10. With afterthoughts (words, phrases or clauses added to the end of a sentence).

> It isn't too late to get tickets, is it? Send them as soon as possible, please.

NOTE: The word *too* does not require a comma if located at the end of a sentence—see Rule 5d.

11. In the following miscellaneous constructions:

a. To indicate omission of words in repeating a construction.

> We had a tactical reserve; now, nothing. [The comma replaces *we have*.]

b. Before *for* used as a conjunction.

> She didn't go to the party, for she cannot stand smoke-filled rooms.

c. To separate repeated words.

> That was a long, long time ago.
> Well, well, look who's here.

d. With titles following personal names. (Jr. and Sr. are set off by commas; 2d, 3d, II, and III are not.)

> Lee B. Walker, Sr.
> Henry Ford II
> William Price, Esq
> James Stokes 3d
> > *In text:* Lee B. Walker, Sr., is …

NOTE: When you must show possession drop the comma following *Jr.* and *Sr.*
> Lee Walker, Sr.'s car is …

e. When names are reversed.

> Adams, Angie
> Baldwin, Sherwood, Jr.
> Brown, Carolyn
> Jones, Kevin

> Middleton, Mary
> Parks, James, III
> Price, William, Esq
> Walker, Lee B., Sr.

f. With academic degrees.

Scott H. Brown, PhD
James Parks III, MBA
> *In text:* Houston Markham, EdD, will …

g. To prevent confusion or misreading.

To John, Smith was an honorable man.
For each group of 20, 10 were rejected.
Soon after, the meeting was interrupted abruptly.

SPACING WHEN USING COMMAS ...

- No space *before*.
- One space *after*, unless a closing quotation mark immediately follows the comma.
- No space *after* within a number.

DASH

USE AN EM DASH (—) ...

1. To indicate a sudden break or abrupt change in thought.

> He is going—no, he's turning back.
> Our new building should be—will be—completed by June 2004.

2. To give special emphasis to the second independent clause in a compound sentence.

> Our new, but used, pickup truck is great—it's economical too!
> You'll double your money with this plan—and I'll prove it!

3. To emphasize single words.

> Girls—that's all he ever thinks about!
> They're interested in one thing only—profit—nothing else matters.

4. To emphasize or restate a previous thought.

> One day last week—Monday, I think—Congress finally voted on the amendment.

5. Before summarizing words such as *these*, *they*, and *all* when those words summarize a series of ideas or list of details.

> A tennis racket, swimsuit and shorts—these are all you'll need for the weekend.
> Faculty, staff and students—all are invited.

6. In place of commas to set off a nonessential element requiring special emphasis.

> There's an error in one paragraph—the second one.
> We will ensure all students—as well as faculty members—are informed of the
> Chief of Staff's visit.

7. To set off a nonessential element when the nonessential element contains internal commas.

> Certain subjects—American government, calculus and chemistry—are required courses.

8. Instead of parentheses when a nonessential item requires strong emphasis (dashes emphasize; parentheses de-emphasize).

> Call Lieutenant Colonels Kessler, Sims, and Forbes—the real experts—and get their opinion.

9. In place of a colon for a strong, but less formal, break in introducing explanatory words, phrases or clauses.

> Our arrangement with the Headquarters USAF is simple—we provide the camera-ready copy and they
> handle the printing and distribution.

10. With quotation marks. Place the dash outside the closing quotation mark when the sentence breaks off after the quotation and inside the closing quotation mark to indicate the speaker's words have broken off abruptly.

> If I hear one more person say, "See what I'm saying!?"—
> Thomas Hardy said, "When I get to 25 Barberry Street, I'll —"

11. With a question mark or an exclamation mark:

a. When a sentence contains a question or exclamation that is set off by dashes, put the appropriate punctuation mark before the closing dash.

> I'll attend Friday's meeting—is it being held at the same place?—but I'll have to leave early for another appointment.
> He's busy now, sir—wait, don't go in there!—I'll call you when he's free.

b. When a sentence abruptly breaks off before the end of a question or exclamation, put the end punctuation mark immediately following the dash.

> Shall I do it or —?
> Look out for the —!

USE AN EN DASH (–) ...

12. Before the source of a quotation or credit line in typed material (use an en dash in printed material).

> The ornaments of a home are the friends who frequent it.
> – Anonymous

13. To indicate inclusive numbers (dates, page numbers, time) when not introduced by the word *from* or *between*.

> Some instructions are on pages 15–30 of this article and from pages 3 to 10 in the attached brochure.
> My appointment is 0800–0900. I will be there between 0745 and 0800.
> She worked in the Pentagon from 1979 to 1996 and she said the 1990–1996 period went by quickly.

14. In a compound adjective when one element has two words or a hyphenated word.

New York–London flight   Air Force–wide changes   quasi-public–quasi-judicial
                                                                                            body

SPACING WHEN USING DASHES ...

- No space *before* or *after* an em dash (—) or en dash (–) within a sentence.
- Two spaces *after* the em dash at the end of a sentence that breaks off abruptly (rule 10)— unless manuscript format and using right justified, then one space *after*.
- No space *before*, *between* or *after* the em dash when inputting material with a typewriter. An em dash is made using two hyphens (--) when typed.

NOTE: The *Tongue and Quill* favors two spaces after the end of a sentence. Rather to use one space or two is left up to the individual or organization. Either way is acceptable.

> "All generalizations are false to a certain extent—including this one."
>
> – The Quill

DISPLAY DOT

USE A DISPLAY DOT OR BULLET*...

To emphasize specific items in either complete or incomplete sentences that are parallel in grammatical structure.

1. Use display dots when one item is not more important than the others, and the items do not show a sequence. (If the items show a sequence, a numbered list is recommended.)

2. Capitalize the first word of each item in the list when a complete sentence introduces them. (The complete sentence may end with either a period or a colon.)

> The prospect for growing drug abuse worldwide can be correlated with the prevalence of the following ingredients:
> - An awareness of drugs.
> - Access to them.
> - The motivation to use them.

> The Coast Guard is a multimission agency with broad, general mission areas in the maritime arena.
> - Safety.
> - Environmental protection.
> - Law enforcement.
> - Political-military.

3. Use a period (or other appropriate end punctuation) after each item in a vertical list when at least one of the items is a complete sentence.

> After listening intently to the defense attorney's closing remarks, the jury was convinced of three things:
> - Witnesses lied.
> - False evidence had been presented.
> - The defendant deserved a new trial.

> Two questions continually present themselves to commanders:
> - What is actually happening?
> - What (if anything) can I or should I do about it?

---

*A "bullet" is a generic term for any graphical symbol used to emphasize different items in a list. Display dots, squares, dashes, and arrows are the most common symbols used for this purpose, but today's software makes any number of designs possible. Regardless of your choice of bullet graphic, the above guidelines will help readability.

4. When the list completes a sentence begun in the introductory element, omit the final period unless the items are separated by other punctuation.

There is a tendency to speak of the commander, but there are, in fact, many interrelated commanders, and each commander uses a separate command and control process to
- make information decisions about the situation,
- make operational decisions about actions to be taken, and
- cause them to be executed within a structure established by prior organizational decisions.

5. A colon can be used to indicate a full stop before a list. A colon is often used with expressions such as *the following items* or *as follows*.

Consider the following advantages when making your decision:
1. You won't have to be somewhere at 0800 every day.
2. You can get more involved in community activities.
3. You can pursue hobbies you haven't had time for in the last year.

6. Do not use a colon when the listed items complete the sentence that introduces them.

Liaison officers must
- become familiar with the situation,
- know the mission, and
- arrange for communications.   [Not: Liaison officers must:]

The editorial assistants in Publication Systems are
- Rebecca Bryant,
- Lisa McDay, and
- Yuna Braswell.   [Not: The editorial assistants are:]

SPACING WHEN USING DISPLAY DOTS AND BULLETS ...

- No space *before*.
- Two spaces *after*—unless manuscript format and using right justified, then one space *after*.
- Hang indent all remaining lines.

NOTE: The *Tongue and Quill* favors two spaces after the end of a sentence. Rather to use one space or two is left up to the individual or organization. Either way is acceptable.

ELLIPSIS

USE AN ELLIPSIS ...

1. To indicate a pause or faltering speech within a quoted sentence or at the end of a sentence that is deliberately incomplete.

> "I ... I don't know ... I mean I don't know if I can go."
> Can you tell me what famous document begins with "Four score and seven ..."?

2. To indicate an omission of a portion of quoted material.

> "Four score and ... our ... brought forth...."

> **a. Use four periods (ending period plus ellipsis) to indicate an omission at the end of a sentence.**

> Work measurement is the volume of work....

NOTE: If quotation is intended to trail off, omit ending punctuation.

> He could have easily saved the situation by ... But why talk about it.

> **b. When a sentence ends with a question mark or exclamation point, use an ellipsis (three periods) and the ending punctuation mark.**

> What work measurement tool was used to determine...?

> **c. To indicate one or more sentences or paragraphs are omitted between other sentences, use the ellipsis immediately after the terminal punctuation of the preceding sentence.**

> In the last few years, we have witnessed a big change in the age groups of America's violence.... How far and wide these changes extend, we are afraid to say.

> **d. When a fragment of a sentence is quoted *within another sentence*, it isn't necessary to signify the omission of words before or after the fragment.**

> Technicians tell us it "requires a steady stream of accurate and reliable reports" to keep the system operating at peak performance.

SPACING WITH PUNCTUATION MARKS ...

- No space *between* the three periods within the ellipsis itself.
- One space *before* and *after* within a sentence.
- No space *before* when an opening quotation mark precedes the ellipsis.
- Two spaces *after* ellipsis with a period, question mark or exclamation point at the end of a sentence—unless manuscript format and using right justified, then one space *after*.

NOTE: The *Tongue and Quill* favors two spaces after the end of a sentence. Rather to use one space or two is left up to the individual or organization. Either way is acceptable.

EXCLAMATION MARK

USE AN EXCLAMATION MARK ...

1. At the end of a sentence or elliptical expression (condensed sentence, key words left out) to express strong emotion (surprise, disbelief, irony, dissent, urgency, amusement, enthusiasm).

> Congratulations on your new son!
> I suppose you consider that another "first"!
> Fantastic show!

2. In parentheses within a sentence to emphasize a particular word.

> He lost 67(!) pounds in 6 months.
> She said what(!)?

ALONG WITH OTHER PUNCTUATION ...

3. When an exclamation is set off by dashes within a sentence, use an exclamation mark before the closing dash.

> Our women's club—number 1 in the community!—will host a party for underprivileged children.

4. Use an exclamation mark inside a closing parenthesis of a parenthetical phrase when the phrase requires an exclamation mark and the sentence does not end with an exclamation mark.

> Jerry's new car (a 2004 Nissan Maxima!) was easily financed.
> The football game (Alabama versus Auburn) is always a super game!

5. An exclamation mark goes inside a closing quotation mark only when it applies to the quoted material.

> Lt Col Smith said, "Those rumors that I'm going to retire early simply must stop!"
> You're quite mistaken—Jane Palmisano clearly said, "Peachtree Grill at 1215"!
> Mark and Todd have both told him, "You had no right to say, 'Kimberly will be glad to teach Acquisition' without checking with her first!"

SPACING WITH PUNCTUATION MARKS ...

- Two spaces *after* the end of a sentence—unless manuscript format and using right justified, then one space *after.*
- No space *after* when another punctuation mark immediately follows (closing quotation mark, closing parenthesis, closing dash).

NOTE: The *Tongue and Quill* favors two spaces after the end of a sentence. Rather to use one space or two is left up to the individual or organization. Either way is acceptable.

HYPHEN

USE A HYPHEN ...

1. When dividing a word at the end of a line. When in doubt about the proper place to divide a word, consult a dictionary and apply the guidelines on page 297.

> Use a hyphen to indicate the continua-
> tion of a word divided at the end of a line.

2. To join unit modifiers. When you abbreviate the unit of measure, omit the hyphen.

> 4-hour sortie, 4 hr sortie
> rust-resistant cover
>
> long-term loan
> 24-gallon tank, 24 gal. tank

3. When expressing the numbers 21 through 99 in words and in adjective compounds with a numerical first element.

> Twenty-one people attended.
> Twenty-one people attended with at least 2 that failed to show up.
> Eighty-nine or ninety miles from here there's an outlet mall.
> I kept their 3-year-old child while they were away.
> There will be a 10-minute delay.

4. To join single capital letters to nouns or participles.

> U-boat
> T-shirt
>
> H-bomb
> T-bone
>
> X-height
> D-mark
>
> U-turn
> E-mail

5. To indicate two or more related compound words having a common base (suspended hyphen).

> It will be a 12- to 15-page document.
> The cruise line offers 2-, 3-, and 7-day cruises at special group rates.
> Long- and short-term money rates are available.

6. To join capital letter(s) and numbers in system designators and numerical identifiers.

> F-117
> KC-10
>
> B-1B
> Su-24TK
>
> F-16
> T-38

7. To form compound words and phrases. Some compound words are written as two words (post office, air brake, Mother Nature, fellow traveler), some as one (manpower, masterpiece, aircraft), some as a combination of words and joined by hyphens (father-in-law, great-uncle, secretary-treasurer, governor-general, men-of-war, grant-in-aid, mother-of-pearl), and some multiple-word compounds that include a preposition and a description (jack-of-all-trades, but flash in a pan and master of none). There's a growing trend to spell compound words as one word once widely accepted and used. However, sometimes the way you use a compound word or phrase will dictate how you write it—as one word, with a hyphen, or as two separate words. When in doubt, **consult an up-to-date dictionary** or treat as two words if the guidelines on the next pages don't fit.

   a. Use a hyphen with words and phrases that are combined to form a unit modifier immediately preceding the word modified (except with an adverb ending in *ly*). Do not hyphenate these phrases when they follow the noun.

> an up-to-date report; this report is up to date; a $500-a-week salary; a salary of $500 a week
> decision-making process; the process of decision making; red-faced man; the man with the red face
> X-rated movies; movies that are X rated; the X-ray equipment; the X-ray showed
> a well-known author; the author is well known
> a first-come, first-served basis; on the basis of first come, first served
> a highly organized group; a completely balanced meal

   b. Use a hyphen when two or more proper names are combined to form a one-thought modifier and when two adjectives are joined by the word *and* or *or*.

> Montgomery-Atlanta-Washington flight   life-and-death situation
> black-and-white terms   cause-and-effect hypothesis
> yes-or-no answer   go-no-go decision

   c. Use a hyphen when spelling the word solid creates a homonym.

> re-cover [cover again]; recover [to regain]   re-creation [create again]; recreation [play]
> re-count [count again]; recount [to detail]   pre-position [position again]; preposition
> re-create [create again]; recreate [refresh]   [word that forms a phrase]
> un-ionized [substance]; unionized [to organize]   re-mark [mark again]; remark [say]
> re-sign [sign again]; resign [quit]   multi-ply [as in fabric]; multiply [arithmetic function]
> re-start [start again]; restart [to start anew]   co-op [cooperative]; coop [to confine]
> re-treat [treat again]; retreat [withdraw]

   d. Use a hyphen to avoid doubling a vowel when the last letter of the prefix "anti," "multi," and "semi" is the same as the first letter of the word. Also, use a hyphen when the second element is a capitalized word or a number.

> anti-inflammatory; anti-Nazi; antiaircraft   semi-icing; semi-Americanized; semiofficial
> multi-industry; multielement; multimillion   pre-1914, post-World War II; ultra-German

   e. Use a hyphen to join duplicate prefixes.

> re-redirect   sub-subcommittee   super-superlative

DO NOT USE A HYPHEN ...

f. In compounds formed from unhyphenated proper nouns.

Methodist Episcopal Church   Southeast Asian country   Mobile Bay cruise

g. Between independent adjective preceding a noun.

hot water pipe   big gray cat   a fine old southern gentleman

h. In a compound adjective when the first element of a color term modifies the second.

sea green gown   grayish blue car

i. In a compound adjective formed with chemical names.

carbon dioxide formula   hydrochloric acid liquid

j. In a unit modifier with a letter or number as its second element.

Attachment 3 pages   Article 3 procedures

k. In a unit modifier enclosed in quotation mark unless it is normally a hyphenated term. Quotation marks are not to be used in lieu of a hyphen.

"blue sky" law   "tie-in" sale
"good neighbor" policy   right-to-work law

l. In a unit modifier to set off some prefixes and suffixes (ante, anti, bi, bio, co, counter, extra, infra, inter, intra, like, macro, meta, micro, mid, multi, neo, non, over, post, per, pre, pro, proto, pseudo, sub, re, semi, socio, super, supra, trans, ultra, un, under), BUT THERE ARE SOME EXCEPTIONS.

All words are hyphenated when used as an adjective compound.

all-inclusive background   all-out war   all-powerful leader

Best, better, full, high, ill, least, lesser, little, low, lower, middle, and *upper* compounds are hyphenated when used as an adjective before a noun; drop the hyphen when used following the noun.

ill-advised action; action is ill advised
lesser-regarded man; he was the lesser
   regarded
full-length dress; the dress is full length

upper-crust society; she is of the
   upper crust
high-level water; water is at the high level

best-loved book; the book was best loved
little-understood man; the man was
   little understood
least-desirable man; the man was least
   desirable
better-prepared man; the man was better
   prepared
middle-class house; he lives with the middle
   class

Cross and *half* words are hyphenated, but some aren't. Check your dictionary and, if not listed, hyphenate.

crosswalk   cross-pollination   cross section
halfback   half-dollar   half sister

Elect words are hyphenated, *except* when they consist of two or more words.

mayor-elect   county assessor elect   president-elect

Ex (meaning *former*) words are discouraged in formal writing; *former* is preferred. However, when you use *ex* in this context, use a hyphen.

ex-governor	ex-AU commander	ex-convict

Fold words are usually one word, *except* when used with numerals.

25-fold	tenfold	twofold

Like words are usually one word *except* when the first element is a proper name, words of three or more syllables, compound words, or to avoid tripling a consonant.

gridlike	lifelike	Grecian-like
mystery-like	squeeze-bottle-like	wall-like

Mid, Post, and *Pre* words are usually one word *except* when the second element begins with a capital letter or is a number.

midstream	post-Gothic	preeminent
mid-June	postgame	pre-Civil War
mid-1948	post-1900s	pre-1700s

Non words are usually one word *except* when the second element begins with a capital letter or consists of more than one word.

nonattribution	noncommissioned officer	nonsurgical
non-Latin-speaking people	non-civil-service position	non-European
non-line of sight		

Over and *under* words are usually one word *except* when the compound contains the word *the.*

over-the-counter drug	under-the-table kick	overbusy employee
overdone steak	underdone steak	understaffed office

Quasi words are always hyphenated.

quasi-judicial	quasi-public	quasi-legislative

Self as a prefix is joined to the root word by a hyphen. When *self* is the root word or is used as a suffix, do not use a hyphen.

self-made	selfish	herself
self-respect	selfless	itself
self-explanatory	selfsame	himself

Vice compounds are hyphenated *except* when used to show a single office or title.

a vice president; vice-presidential candidate	vice admiral; vice-admiralty; viceroy
the vice-consul; vice-consulate's office	vice-chancellor; vice-chancellorship

Well compounds are hyphenated when used as an adjective before a noun; drop the hyphen when used following the noun. *Well* used as a compound noun is always hyphenated.

well-made suit; suit was well made	the well-being of the family; consider her well-being
well-known author; author is well known	the well-bred dogs; the dogs were well bred

Wide words are usually one word *except* when long and cumbersome and when follows the noun.

worldwide	university-wide; the virus is university wide

SPACING WHEN USING A HYPHEN ...

- No space *before* or *after* to combine words, punctuation and/or numbers.
- One space *after* when dividing a word (see rule 1) or using suspended hyphen (rule 5).

DIVIDING WORDS AND PARAGRAPHS

WORDS

1. Never divide the last word on the first or last lines on a page; do not hyphenate the last words on two consecutive lines; avoid hyphenating more than five lines each page.

2. Never divide monosyllables (one-syllable words). [friend]

3. Never divide words at a vowel that forms a syllable in the middle of the word. [preju-/dice, **not** prej-/udice]

4. Never divide words at a final syllable whose only vowel sound is that of a syllabic "I." [prin-/cipal, **not** princi-/pals]

5. Never divide words of five or fewer letters even when they contain more than one syllable. [**not** i-/deal or ide-/a]

6. Never divide words by putting a single letter on a line. [**not** a-/round nor militar-/y]

7. Never further divide words that contain a hyphen—break these words at the built-in hyphen. [self-/control, **not** self-con-/trol]

8. Divide words containing double consonants between the consonants only when they do not end root words. [permit-/ted; spell-/ing]

9. When possible, divide words after the prefix or before the suffix rather than within the root word or within the prefix or suffix. [applic-/able, **not** applica-/ble; valu-/able, **not** val-/uable; pre-/requisite, **not** prereq-/uisite].

10. Never divide contractions. [**not** can'/t nor won/'t]

11. When necessary to divide a name, carry over only the surname (never separate a first name from a middle initial, an initial from a middle name or initials used in place of a first name). [Annette G./Walker; Ethel/Hall; R. A./Bowe]

12. Divide surnames, abbreviations and numbers only if they already contain a hyphen, and then divide only at the hyphen. [Johnson-/Roberts, **not** John-/son-Roberts; AFL-/CIO, **not** YM/CA; 249-/3513, **not** $55,-/000].

13. A person's rank or title should be on the same line with first name or initials, when possible. [Miss Duncan/Phillips; Dr. Louise/Miller-Knight; Major Larry/Lee]

14. When it's necessary to divide a date, separate the year from month—do not split the month from the day. [14 September/2004]

PARAGRAPHS

1. Never divide a paragraph of four or fewer lines.

2. When dividing a paragraph of five or more lines, never type less than two lines on either page.

 # ITALICS

USE ITALICS ...

1. In printed material to distinguish the titles of whole published works: books, pamphlets, bulletins, periodicals, newspapers, plays, movies, symphonies, poems, operas, essays, lectures, sermons, legal cases, and reports.

The Chicago Manual of Style	*The Phantom of the Opera*
AFM 33-326, *Preparing Official Communications*	*Star Trek*
The *Montgomery Advertiser*	*United States Government v. Bill Gates*

NOTE: When you use these titles in the plural, set the plural ending in Roman type.

There were five *Journal*s and two *Time*s on the shelf.

2. In place of the underscore to distinguish or givbe greater prominence to certain words, phrases, or sentences. Both the underscore and italics are acceptable, but not in the same document. Use sparingly.

Air Force *doctrine* has been the subject of much debate.
Air Force <u>doctrine</u> has been the subject of much debate.

3. In printed material to distinguish the names of ships, submarines, aircraft, and spacecraft. Italicize the name only, not initials or numbers preceding or following the name. (In typed material, the underscore is generally used for this purpose.) Do not italicize the class or make of ships, aircraft and spacecraft; and names of space programs.

USS *America*	*Nautilus*	B-1B *Lancer*	*Friendship* 7
frigate	U-boat	Concorde	*Columbia*
Spruance	KILO	Boeing 707	Gemini II

4. In typed material to distinguish foreign words *not* part of the English language. Once an expression has become part of the English language (in the dictionary), italics is unnecessary.

blitzkrieg	*Luftwaffe*	vis-à-vis	com d'étet
vakfiye	*poêle*	*le cheval*	Perestroika

5. When punctuation (except parentheses and brackets) marks immediately follow the italicized word, letter character, or symbol.

What is meant by *random selection?*	*Point:* one-twelfth of a pica
Luke 4:16*a;*	see *12b!*

"Few men are lacking in capacity, but they fail because they are lacking in application."
– Calvin Coolidge

PARENTHESES

USE PARENTHESES ...

1. To enclose explanatory material (a single word, a phrase or an entire sentence) that is independent of the main thought of the sentence.

> The ACSC students (542 of them) will begin classes the second week of June.
> The results (see figure 3) were surprising.

2. To set off nonessential elements when commas would be inappropriate or confusing and dashes would be too emphatic.

> Mr. Henry Anderson, Jr., is the general manager of the Montgomery (Alabama) branch. [Parentheses are clearer than commas when a city-state expression is used as an adjective.]
> All the classes will meet three days a week (Mondays, Tuesdays and Thursdays). [Parentheses are used in place of commas because the nonessential element contains commas.]
> I suggest you contact Edward Clinton (a true professional) for his recommendation. [Parentheses used in place of dashes to de-emphasize the nonessential element.]
> Contact Ms. Louise Robinson—the manager of the house in Tuscaloosa—and ask her if a room is still available. [Dashes are used in place of parentheses for emphasis.]

3. To enclose enumerating letters or numerals within a sentence.

> Our goals are to (1) reduce the number of curriculum hours, (2) eliminate the 90-minute lunch period, and (3) reduce the number of personnel needed to accomplish the mission.
> Also, include the following when you file your medical costs: (a) hotel charges, (b) meal costs (including gratuities), and (c) transportation costs.

4. To enclose numbers or letters identifying certain sections of an outline. In outlining, if you have a paragraph numbered *1,* you must have a paragraph numbered *2*; if you have a subparagraph *a*, you must have a *b* subparagraph.

> 1. xxxxx
>   a. xxxxx
>   (1) xxxxx
>   (a) xxxxx

5. To enclose a nickname or a descriptive expression when it falls between a person's first and last names. However, when it precedes or replaces a person's first name, simply capitalize it.

> George Herman (Babe) Ruth   Stonewall Jackson
> Major William F. (Clark) Kent   the Iron Duke

"And what he greatly thought, he nobly dared."
                                        – Homer

ALONG WITH OTHER PUNCTUATION ...

6. If an item in parentheses falls within a sentence, place comma, semicolon, colon, or dash outside (never before) the closing parenthesis.

> I'll see you later (probably Friday), but remember to collect your money.
> I'll attend the meeting (as I said I would); however, you'll have to go to the next one as I have another commitment.
> She's passionate about two important issues (and strives to support them): homeless children and a clean environment.

7. Use a period before a closing parenthesis only when the parenthetical sentence stands on its own or when the closing parenthesis is preceded by an abbreviation containing punctuation.

> The results were surprising. (See the analysis at atch 2.)
> Many heights of flowers (e.g., 6 in., 12 in., 36 in.) will be featured in the show.

8. Put a question mark or quotation mark before a closing parenthesis only when it applies to the parenthetical item and the sentence ends in a different punctuation.

> The Pentagon (you've been there, haven't you?) is a fascinating office building.
> Doris Williams said she would go. (In fact, her exact words were, "Go golfing? You bet! Every chance I get!")

9. When using an exclamation mark or question mark to emphasize or draw attention to a particular word within a sentence.

> You call this fresh(!) food.
> They said they will buy us four(?) machines.

SPACING WHEN USING PARENTHESES ...

—opening
- One space *before* when parenthetic matter is within a sentence.
- No space *before* when using exclamation or question marks to emphasize or draw attention to a particular word within a sentence.
- Two spaces *before* when parenthetic matter follows a sentence (when parenthetic matter starts with a capital and closes with its own sentence punctuation)—unless manuscript format and using right justified, then one space *before*.
- No space *after*.

—closing
- No space *before*.
- One space *after* when parenthetic matter is within a sentence.
- Two spaces *after* when parenthetic matter is itself a complete sentence and another sentence follows—unless manuscript format and using right justified, then one space *after*.
- No space *after* if another punctuation mark immediately follows.

NOTE: The *Tongue and Quill* favors two spaces after the end of a sentence. Rather to use one space or two is left up to the individual or organization. Either way is acceptable.

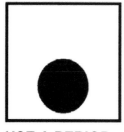

PERIOD

USE A PERIOD ...

1. To end declarative and imperative sentences.

> His work is minimally satisfactory.
> Don't be late.

2. To end an indirect question or a question intended as a suggestion or otherwise not requiring an answer.

> She wanted to know how to do it.
> He asked what the job would entail.
> Tell me how they did it.

3. With certain abbreviations. Most abbreviations today are written without punctuation (see *Abbreviations*, pages 309-315).

Ms.	Miss [not an abbreviation]	Sr.	no. [number; could be confused with the word *no*]
Mr.	Dr.	e.g.	in. [inch; could be confused with the word *in*]
Mrs.	Jr.	i.e.	etc.

4. To form ellipses (three periods that indicate a pause or faltering speech within a sentence, or an omission of a portion of quoted material). (See *Ellipsis* on page 291.)

5. In vertical lists and outlines.

   a. Use a period after each item in a vertical list when at least one of the items is a complete sentence. When the list completes a sentence begun in the introductory element, omit the final period unless the items are separated by other punctuation.

> After listening intently to the defense attorney's closing remarks, the jury was convinced of three things:
>   (1) Witnesses lied.
>   (2) False evidence had been presented.
>   (3) The defendant deserved a new trial.

> After listening to the defense attorney's closing remarks, the jury was convinced that
>   (1) several witnesses had perjured themselves,
>   (2) false evidence was presented and
>   (3) the defendant deserved a new trial.

> The following aircraft were lined up on the runway:
>   B-1B
>   T-38
>   F-16
>   F-117

b. Use periods after numbers and letters in an outline when the letters and figures are not enclosed in parentheses. If you have a numbered *1* paragraph, you must have a numbered *2;* if you have a subparagraph *a,* you must have a *b* subparagraph; and so on. For Air Force publications, follow guidance in AFI 33-360, Volume 1.

1.   ⇐outline sample
2.
   a.
   b.
     (1)
     (2)
       (a)
       (b)

ALONG WITH OTHER PUNCTUATION ...

6. With parenthetical phrases. Place a period inside the final parenthesis only when the item in the parentheses is a separate sentence or when the final word in the parenthetical phrase is an abbreviation that is followed by a period.

> I waited in line for 3 hours. (One other time I waited for over 5 hours.)
> One other committee member (namely, Dr. Glen Spivey, Sr.) plans to vote against the amendment.

7. With quotation marks with the period placed inside a closing quotation mark.

> She said, "I'll go with you."

8. With a dash only when used with an abbreviation that contains periods.

> Tony Lamar's desk is 48 in.—his is the only odd-sized desk.

SPACING WHEN USING A PERIOD...

- Two spaces *after* the end of a sentence—unless manuscript format and using right justified, then on space *after*.
- No space *before* unless an ellipsis (see *Ellipses*, page 291).
- One space *after* an abbreviation with a period within a sentence.
- No space *after* a decimal point or *before* within two numbers.
- No space *after* when another punctuation mark immediately follows (closing quotation mark, closing parenthesis, comma following an "abbreviation" period).
- Two spaces *after* a number or letter that indicates an enumeration (rule 5b).

NOTE: The *Tongue and Quill* favors two spaces after the end of a sentence. Rather to use one space or two is left up to the individual or organization. Either way is acceptable.

> "The pen is mightier than the sword."
> — Shakespeare
>
> "... unless you're on one."
> — The Quill

QUESTION MARK

USE A QUESTION MARK ...

1. To indicate the end of a direct question.

Did he go with you?
Will you be able to attend?

2. With elliptical (shortened) questions and to express more than one question within a sentence.

You rang? For what purpose?
Was the speaker interesting? Convincing? Well versed?
Who approved the sale? When? To whom? For what amount?

3. After an independent question within a larger sentence.

The question "Who will absorb the costs?" went unanswered.
When will the reorganization take place? will surely be asked.

4. To express doubt.

They plan to purchase three(?) new Pentium computers with individual scanners for us.
Jackie Baltzell and Gayle Magill have been associated with her since 1990(?).

ALONG WITH OTHER PUNCTUATION ...

5. Use a question mark before a closing parenthesis only when it applies solely to the parenthetical item and the sentence ends in a different punctuation mark.

At our next meeting (it's on the 16th, isn't it?), we'll elect a new president. As the gun opened fire (was it a .50-caliber gun?), all movement ceased. [Question marks were used within parentheses because sentences require a period at the end.]
Are tickets still available (and can I get two), or is it too late? [Question mark is omitted within parentheses because sentence ends with a question mark.]

6. A question mark is placed inside the closing quotation mark only when it applies to the quoted material or when the same punctuation is required for both the quotation and the sentence as a whole.

She asked, "Did you enjoy the trip?" [Question mark belongs with quoted material.]
Why did he ask, "When does it start?" [Question mark is same as ending punctuation.]
Did you say, "I'll help out"? [Quoted material is not a question; therefore, question mark applies to the sentence as a whole.]

7. When a question within a sentence is set off by dashes, place the question mark before the closing dash.

The new class—isn't it called Super Seminar?—begins tomorrow.

SPACING WHEN USING A QUESTION MARK...

- Two spaces *after* the end of a sentence—unless manuscript format and using right justified, then one space *after* (see page 302, spacing).
- No space *after* when another punctuation mark immediately follows (closing quotation mark, closing parenthesis, closing dash).

NOTE: The *Tongue and Quill* favors two spaces after the end of a sentence. Rather to use one space or two is left up to the individual or organization. Either way is acceptable.

English ... A Changing Language

Awful, terrible once meant "fear inspiring."

Barn once meant "barley-place."

Doom once meant "any legal judgment."

Girl once was used to refer to a child of either sex.

Hussy once meant "housewife."

Marshall once meant "stable boy" (one who looked after mares).

Meat once meant "food."

Nice once meant "ignorant."

Nimble once meant "good at taking things."

Shrewd once meant "wicked."

Silly once meant "fortunate."

Smart once meant "causing pain."

Starve once meant "die."

Villain once meant "farm worker."

–found in *Building Better English*

QUOTATION MARKS

USE QUOTATION MARKS ...

1. To enclose the exact words of a speaker or writer. With few exceptions, a quotation must be copied exactly as it appears in the original. If the quotation is woven into the flow of the sentence, do not use punctuation preceding the opening quotation mark. When words interrupt a quotation, close and reopen the quotation.

> Robert Frost said, "The brain is a wonderful organ; it starts working the moment you get up in the morning and doesn't stop until you get to the office."
> Why does she insist on saying "It just won't work"?
> "A pint of sweat" says General George S. Patton, "will save a gallon of blood."

NOTE: Do not set off indirect quotations.

> Why does she insist on saying that it just won't work?

2. To enclose slogans or mottoes, but not signs or notices.

> He had a "do or die" attitude.   He has a No Smoking sign in his car.
> "All's well that ends well" is a popular slogan.   There is a Gone Fishing notice on his door.

3. To enclose words or phrases used to indicate humor, slang, irony, or poor grammar.

> They serve "fresh" seafood all right—fresh from the freezer!
> For whatever reason, she just "ain't talkin'."

NOTE: When using quotation marks with other punctuation, the comma and period are always placed inside the closing quotation marks; the semicolon is always placed outside the closing quotation marks; the dash, exclamation mark, and question mark are placed according to the structure of the sentence (see guidelines on pages 288, 292, and 303).

4. With words and phrases that are introduced by such expressions as *cited as, classified, designated, entitled, labeled, marked, named, signed, the term, the word* when the exact message is quoted. Capitalize the first word when it begins a sentence, when it was capitalized in the original, when it represents a complete sentence, or when it is a proper noun.

> The card was signed "Your friend, Diane."
> The article was entitled "How to Write English That is Alive."
> "Fragile" was stamped on the outside of the package.
> The report is classified "secret" and can't be distributed.
> Our organization received an "Outstanding" Quality Air Force Assessment (QAFA) rating.

NOTE: Do not enclose these expressions: called, known as, so-called, etc.

> The flower was called an American Beauty rose.
> The boy whose name is "Bill Kent" was known as Clark Kent.
> The so-called secret report can now be distributed.

5. To enclose the title of any part (chapter, lesson, topic, section, article, heading) of a published work (book, play, speech, symphony, etc.). The title of the published work should be underlined in typed material and italicized in printed material.

> The Appendix 1 section in AFH 33-337 is "The Mechanics of Writing."
> When you read "Air Force Writing Products and Templates" section of *The Tongue and Quill*,
> keep in mind …

6. To enclose titles of complete but unpublished works such as manuscripts, dissertations and reports.

> We need to get a copy of the "The Evolution of a Revolt" document as soon as possible.
> The title of his dissertation is "Why Smoking Should be Banned from All Public Places."

7. To enclose the titles of songs and radio and television shows.

> They sang "The Star Spangled Banner" before the game began.
> "M.A.S.H." is still being shown on TV.

8. To denote inches.

> 6″ × 15″ [use inch (″) mark and multiplication (×) mark if using typewriter or computer that has
> these keys]

9. To enclose a nickname or descriptive expression when it falls between a person's first and last names. However, when it precedes or replaces a person's first name, simply capitalize it.

> George Herman "Babe" Ruth   Stonewall Jackson
> the Iron Duke   Major William F. "Clark" Kent

10. To enclose misnomers, slang expressions, nickname, coined words, or ordinary words used in an arbitrary way.

> His report was "bunk."   It was a "gentlemen's agreement."
> The "invisible government" is responsible.   *but* He voted for the lameduck amendment.

SPACING WHEN USING QUOTATION MARKS…

—opening
- Two spaces *before* when quoted matter starts a new sentence or follows a colon—unless manuscript format and using right justified, then one space *before*.
- No space *before* when a dash or an opening parenthesis precedes.
- One space *before* in all other cases.
- No space *after*.

—closing
- No space *before*.
- Two spaces *after* when quoted matter ends the sentence—unless manuscript format and using right justified, then one space *after* (see page 302, spacing).
- No space *after* when another punctuation mark immediately follows (semicolon, colon).
- One space *after* in all other cases.

NOTE: The *Tongue and Quill* favors two spaces after the end of a sentence. Rather to use one space or two is left up to the individual or organization. Either way is acceptable.

SEMICOLON

USE A SEMICOLON ...

1. To separate independent clauses not connected by a coordinating conjunction (*and*, *but*, *for*, *or*, *nor*, and *so*), and in statements too closely related in meaning to be written as separate sentences.

> The students were ready; it was time to go.
> It's true in peace; it's true in war.
> War is destructive; peace, constructive.

2. Before transitional words and phrases (*accordingly, as a result, besides, consequently, for example, furthermore, hence, however, moreover, namely, nevertheless, on the contrary, otherwise, that is, then, therefore, thus,* and *yet*) when connecting two complete but related thoughts and a coordinating conjunction is not used. Follow these words and phrases with a comma. Do not use a comma after *hence, then, thus, so* and *yet* unless a pause is needed.

> Our expenses have increased; however, we haven't raised our prices.
> Our expenses have increased, however, and we haven't raised our prices.
> The decision has been made; therefore, there's no point in discussing it further.
> The decision has been made so there's no point in discussing it.
> The general had heard the briefing before; thus, he chose not to attend.
> Let's wait until next month; then we can get better result figures.

3. To separate items in a series that contain commas (when confusion would otherwise result).

> If you want your writing to be worthwhile, organize it; if you want it to be easy to read, use simple words and phrases; and, if you want it to be interesting, vary your sentence and paragraph lengths.

> Those who attended the meeting were Colonels Jim Forsyth, Dean of Education; Michael Harris, Dean of Distance Learning; Mark Zimmerman, Chairman of Leadership and Communications Studies; and Phil Tripper, Chairman of Joint Warfare Studies.

4. To precede words or abbreviations that introduce a summary or explanation of what has gone before in the sentence.

> We visited several countries on that trip; i.e., England, Ireland, France, Germany, and Finland.
> There are many things you must arrange before leaving on vacation; for example, mail pickup, pet care, yard care.

SPACING WHEN USING A SEMICOLON...

No space before.
One space *after*.

FUNNY SIGNS

1. IN A LAUNDROMAT: Automatic washing machines. Please remove all your clothes when the light goes out.

2. IN A LONDON DEPARTMENT STORE: Bargain Basement Upstairs

3. IN AN OFFICE: Would the person who took the step ladder yesterday kindly bring it back or further steps will be taken.

4. IN ANOTHER OFFICE: After the tea break, staff should empty the teapot and stand upside down on the draining board.

5. ON A CHURCH DOOR: This is the gate of Heaven. Enter ye all by this door. (This door is kept locked because of the draft. Please use side entrance)

6. OUTSIDE A SECOND-HAND SHOP: We exchange anything—bicycles, washing machines etc. Why not bring your wife along and get a wonderful bargain.

7. QUICKSAND WARNING: Quicksand. Any person passing this point will be drowned. By order of the District Council.

8. NOTICE IN A DRY CLEANER'S WINDOW: Anyone leaving their garments here for more than 30 days will be disposed of.

9. IN A HEALTH FOOD SHOP WINDOW: Closed due to illness.

10. SPOTTED IN A SAFARI PARK: Elephants Please Stay In Your Car

11. SEEN DURING A CONFERENCE: For anyone who has children and doesn't know it, there is a day care on the first floor.

12. NOTICE IN A FIELD: The farmer allows walkers to cross the field for free, but the bull charges.

13. MESSAGE ON A LEAFLET: If you cannot read, this leaflet will tell you how to get lessons.

14. ON A REPAIR SHOP DOOR: We can repair anything (Please knock hard on the door—the bell doesn't work).

15. SPOTTED IN A TOILET IN A LONDON OFFICE BLOCK: Toilet out of order. Please use floor below.

ABBREVIATING
ABCs

ə-,bre-ve-'a-shən

A shortened form of a written word or phrase used in place of the whole.

– *Webster's Tenth New Collegiate Dictionary*

What's the appropriate abbreviation? Can I abbreviate in this document? How do I write it—all capital letters, all lowercase letters, or caps and lowercase letters? Can I use just the abbreviation or must I spell it out? How do I make it plural—add an *s*, or an *'s*? Where do I go for answers?

Though these questions are insignificant when compared with some Air Force problems, thousands of people confront these issues on a daily basis. To clear the smoke surrounding the use of abbreviations, we've listed some types of abbreviations used in Air Force writing and *some* general guidelines regarding their proper use.

ACRONYMS: Pronounceable words formed by combining initial letter(s) of the words that make up the complete form. Most acronyms are written in all caps without punctuation, but some are so commonly used they are now considered words in their own right.

AAFES (**A**rmy **a**nd **A**ir **F**orce **E**xchange **S**ervice)
NATO (**N**orth **A**tlantic **T**reaty **O**rganization)
POW (**p**risoner **o**f **w**ar)
laser (**l**ight **a**mplification by **s**imulated **e**mission of **r**adiation)
Modem (**mo**dular/**dem**odulator)
SALT (**s**trategic **a**rms **l**imitation **t**alks)
scuba (**s**elf-**c**ontained **u**nderwater **b**reathing **a**pparatus)
ZIP code (**Z**one **I**mprovement **P**lan code)

BREVITY CODES: Combinations of letters—*pronounced letter by letter*—designed to shorten a phrase, sentence or group of sentences.

CFC (**C**ombined **F**ederal **C**ampaign)
DDALV (**d**ays **d**elay en route **a**uthorized chargeable as **l**eave)
DOD (**D**epartment **o**f **D**efense)
PCS (**p**ermanent **c**hange of **s**tation)
TDY (**t**emporary **d**uty)
CNN (**C**able **N**etwork **N**ews)

***NOTE*:** When brevity codes begin with *b, c, d, g, j, k, p, q, t, u, v, w, y,* or *z*, the indefinite article **a** is used. With *a, e, f, h, i, l, m, n, o, r, s,* or *x*, use **an.**

CONTRACTIONS: Shortened forms of words in which an apostrophe indicates the deletion of letters.

can't (cannot)	let's (let us)
don't (do not)	mustn't (must not)
I'll (I will)	they're (they are)
I'm (I am)	we're (we are)
I've (I have)	won't (will not)
isn't (is not)	wouldn't (would not)
it's (it is)	you've (you have)

ABBREVIATIONS IN GENERAL:

- Use in *informal* documents, manuals, reference books, business and legal documents, scholarly footnotes, etc., when needed to save space. Avoid using in formal documents when style, elegance and formality are important.

- In formal writing "United States" is a noun (The United States and Canada were founding …); "US" is an adjective (US policy regarding …).

- Use sparingly, correctly and consistently.

- Spell out the word (or words) the first time used and enough times within the document to remind readers of its meaning.

- Use a figure to express the quantity in a unit of measure (without a hyphen in a unit modifier) when using an abbreviation. [3 mi, 55 mph, 50 lb, 33 mm film]

- Write abbreviations "first," "second," "third," "fourth," etc., as 1st, 2d, 3d, 4th, etc.

- Use when there's a choice between using an abbreviation and a contraction. [gov vs gov't]

- Use the shortest form that doesn't jeopardize clarity when there's more than one way to abbreviate a word or phrase. [con, cont, contd]

- Avoid beginning a sentence with an abbreviation (except Mr., Mrs., Ms., Dr.), acronyms and brevity codes.

- Avoid using in main headings.

- Avoid using words that are offensive, profane or repulsive when assigning acronyms, brevity codes and contractions.

- Write without punctuation unless confusion would result. [The abbreviation for *inch* (in.) might be confused with the word *in*, the abbreviation for *number* (no.) might be confused with the word *no,* etc.]

- Write abbreviations for single words in lowercase letters. [hospital - hosp; letter - ltr]

- Use the same abbreviation for singular and plural forms after spelling it out. [area of responsibility (AOR) - areas of responsibility (AOR)]

- When ambiguity could result, form the plural with a lowercase *s* and never use an apostrophe to form the plural. [letters - ltrs; travel requests - TRs; area of operations - AO; areas of operations - AOs]

- Do not cap the words just because the acronym or brevity code is capped. Check a source book, the library or the office of responsibility for the correct form. [OJT - on the job training; OPSEC - operations security; JIPC - joint imagery production complex; JCS - Joint Chiefs of Staff]

- Contact the office of primary responsibility for its proper use when writing articles, manuals, handouts, instructions, performance reports, award citations and narratives, and unit histories.

- Find out if your organization has a preference and use it. Otherwise, consult the latest dictionary or use the Joint Publication 1-02, *Department of Defense Dictionary of Military and Associated Terms*, for terms and definitions.

- Use the "/" (slash) when punctuating some abbreviations. [with – w/ or without – w/o; input/output – I/O]

- **SPELL IT OUT if there's still doubt!**

SOME ABBREVIATIONS USED BY AIR FORCE WRITERS ...

Air Force Ranks

Airman Basic	AB
Airman	Amn
Airman First Class	A1C
Senior Airman	SrA
Staff Sergeant	SSgt
Technical Sergeant	TSgt
Master Sergeant	MSgt
Senior Master Sergeant	SMSgt
Chief Master Sergeant	CMSgt
Chief Master Sergeant of the Air Force	CMSAF
Second Lieutenant	2d Lt
First Lieutenant	1st Lt
Captain	Capt
Major	Maj
Lieutenant Colonel	Lt Col
Colonel	Col
Brigadier General	Brig Gen
Major General	Maj Gen
Lieutenant General	Lt Gen
General	Gen

Days	Months		Years	
Sun	Jan	Jul	1999	99
Mon	Feb	Aug	2000	00
Tues	Mar	Sep	2001	01
Wed	Apr	Oct	2002	02
Thurs	May	Nov	2003	03
Fri	Jun	Dec	2004	04
Sat			etc.	

ZIP Code, State and Possession Abbreviations

Alabama	AL	Ala	Montana	MT	Mont
Alaska	AK		Nebraska	NE	Nebr
Arizona	AZ	Ariz	Nevada	NV	Nev
Arkansas	AR		New Hampshire	NH	
California	CA	Calif	New Jersey	NJ	
Colorado	CO	Colo	New Mexico	NM	NMex
Connecticut	CT	Conn	New York	NY	
Delaware	DE	Del	North Carolina	NC	
Florida	FL	Fla	North Dakota	ND	NDak
Georgia	GA	Ga	Ohio	OH	
Hawaii	HI		Oklahoma	OK	Okla
Idaho	ID		Oregon	OR	Oreg
Illinois	IL	Ill	Pennsylvania	PA	Pa
Indiana	IN	Ind	Rhode Island	RI	
Iowa	IA		South Carolina	SC	
Kansas	KS	Kans	South Dakota	SD	SDak
Kentucky	KY	Ky	Tennessee	TN	Tenn
Louisiana	LA	La	Texas	TX	Tex
Maine	ME		Utah	UT	
Maryland	MD	Md	Vermont	VT	Vt
Massachusetts	MA	Mass	Virginia	VA	Va
Michigan	MI	Mich	Washington	WA	Wash
Minnesota	MN	Minn	West Virginia	WV	Wva
Mississippi	MS	Miss	Wisconsin	WI	Wis
Missouri	MO	Mo	Wyoming	WY	Wyo
American Samoa		AS	Northern Mariana Islands	MP	
District of Columbia		DC	Palau	PW	
Federated States of Micronesia		FM	Puerto Rico	PR	
Guam		GU	Virgin Islands	VI	
Marshall Islands		MH			

Field Operating Agencies

AF Agency for Modeling and Simulation	AFAMS
AF Audit Agency	AFAA
AF Base Conversion Agency	AFBCA
AF Center for Environmental Excellence	AFCEE
AF Center for Quality and Management Innovation	AFCQMI
AF Civil Engineer Support Agency	AFCESA
AF Communications Agency	AFCA
AF Cost Analysis Agency	AFCAA
AF Flight Standards Agency	AFFSA
AF Historical Research Agency	AFHRA
AF History Support Office	AFHSO
AF Inspection Agency	AFIA
AF Legal Services Agency	AFLSA
AF Logistics Management Agency	AFLMA
AF Medical Operations Agency	AFMOA
AF Medical Support Agency	AFMSA
AF National Security, Emergency Preparedness Office	AFNSEPO
AF News Agency	AFNEWS
AF Office of Special Investigations	AFOSI
AF Operations Group	AFOG
AF Personnel Center	AFPC
AF Personnel Operations Agency	AFPOA
AF Program Executive Office	AFPEO
AF Real Estate Agency	AFREA
AF Review Boards Agency	AFRBA
AF Safety Center	AFSC
AF Service Agency	AFSVA
AF Studies and Analyses Agency	AFSAA
AF Technical Applications Center	AFTAC
AF Intelligence Agency	AIA
Air National Guard Readiness Center	ANGRC
Air Reserve Personnel Center	ARPC
Air Force Weather Agency	AFWA
Joint Services Survival, Evasion, Resistance and Escape Agency	JSSA

Phonetic Alphabet

A	Alfa	N	November
B	Bravo	O	Oscar
C	Charlie	P	Papa
D	Delta	Q	Quebec
E	Echo	R	Romeo
F	Foxtrot	S	Sierra
G	Golf	T	Tango
H	Hotel	U	Uniform
I	India	V	Victor
J	Juliett	W	Whiskey
K	Kilo	X	Xray
L	Lima	Y	Yankee
M	Mike	Z	Zulu

Academic Degrees

Bachelor of Arts	BA
Bachelor of Science	BS
Master of Arts	MA
Master of Science	MS
Doctor of Philosophy	PhD
Doctor of Law	LLD
Doctor of Medicine	MD
Doctor of Dentistry	DDS

Secondary Address Unit Indicators

Apartment	APT
Building	BLDG
Department	DEPT
Floor	FL
Room	RM
Suite	STE

Latin Abbreviations

A.M.	ante meridiem	before noon
c. or ca	circa	about, approximately
e.g.	exempli gratia	for example, for instance
et al.	et allii, et alia	and other people/things
etc.	et cetera	and so on, and other things
ib, ibid.	ibidem	in the same place
i.e.	id est	that is to say
loc. cit.	loco citato	in the place cited/mentioned
op. cit.	opere citato	in the work cited/mentioned before
P.M.	post meridiem	after noon
P.S.	post scriptum	after writing
Pro tem.	pro tempore	for the time, temporarily
Q.E.D.	quod erat demonstrandum	which was to be shown
Sc.	sic	thus used, spelt, etc.
v.,	versus	vs. against
v.v.	vice versa	the other way around

Direct Reporting Units

AF Doctrine Center	AFDC
AF Operational Test and Evaluation Center	AFOTEC
AF Security Forces Center	AFSFC
United States Air Force Academy	USAFA
11th Wing	11 WG

Units of Measure

gallon	gal
hertz	hz
kilogram	kg
miles per hour	mph
70 Degrees Celsius	70° C
revolutions per minute	rpm
inch	in.
foot	ft
mile	mi
kilometer	km
millimeter	mm
pounds per square inch	psi
nautical miles	NM

Major Commands

Air Combat Command	ACC
Air Education and Training Command	AETC
AF Materiel Command	AFMC
AF Reserve Command	AFRC
AF Space Command	AFSPC
AF Special Operations Command	AFSOC
Air Mobility Command	AMC
Pacific Air Forces	PACAF
United States Air Forces in Europe	USAFE

Researcher's Guide to Abbreviations

app	appendix	l (el)	line (plural: ll) [not recommended because the abbreviation in the singular might be mistaken for "one" and the plural for "eleven"]
art	article (plural: arts)	n	not, footnote (plural: nn)
b	born	nd	no date
bk	book (plural: bks)	no.	number (plural: nos)
c	copyright	np	no place; no publisher
ca	circa, about, approximately	NS	new series
cf	confer, compare [confer is Latin for "compare": *cf* must not be used as the abbreviation for the English "confer," nor should *cf* be use to mean "see")	op cit	opere citato in the work cited
ch	chapter (in legal references only)	OS	old series
chap	chapter (plural: chaps)	p	page (plural: pp) [it always precedes the numbers; when "p" follows a number, it can stand for "pence"]
col	column (plural: cols)	para	Paragraph (plural: pars)
comp	complier (plural: comps); complied by	passim	here and there
d	died	pt	part (plural: pts)
dept	department (plural: depts)	qv	*quod vide*, which see (for use with cross-references
div	division (plural: divs)	sc	scene
e.g.	*exempli gratia*, for example (use without etc.)	sec	section (plural: secs)
ed	edition, edited by editor (plural: eds)	[sic]	so, thus; show erroneous material intentionally kept in text
et al	*et alii*, and others	sup	supplement (plural: sups)
et seq	*et sequens*, and the following	supra	above
etc.	*et cetera*, and so forth	sv	*sub verbo*, *sub voce*, under the word (for use in references to listing in encyclopedias and dictionaries)
fig	figure (plural: figs)	trans	translator, translated by
fl	*flourit* flourished (for use when birth and death dates are not known)	v	verse (plural: vv)
i.e.	*id est*, that is (use with etc.)	viz	videlicet, namely
ibid	*ibidem*, in the same place	vol	volume (plural: vols)
id	idem, the same (refers to persons, except in law citations; not to be confused with ibid)	vs	versus, against (v in law references)
infra	below		

something to consider …

Build effective sentences with active voice, less garbage, positive tone (pages 23-24) and correct words ~ see pages 73-87

Drafting an effective paragraph with transitional devices ~ see pages 68-73

Writing your draft ~ see pages 64-90

Editing your work ~ see pages 91-103

CAPITALIZATION GUIDELINES

Air Force writers and reviewers spend an excessive amount of time trying to determine the appropriate use of capital letters (and abbreviations and numbers, as well). Everybody seems to want it a different way. Authoritative sources don't even agree! Put a half dozen style manuals in front of you and compare the rules—two out of the six might agree in some cases.

The reason for using capital letters is to give distinction or add importance to certain words or phrases. "But," you might say, "I thought it was important and should be capitalized, but the Command Section kicked it back to be changed to lowercase letters." It's unfortunate, but, if someone else is signing the document, that person has the last word. **The best advice we can give you is to find out what style your organization prefers and use it consistently.**

Although we can't possibly cover every situation, what follows is designed to provide some measure of consistency within the Air Force. You must ensure consistency within everything you write or type. *A word of caution:* When you're preparing Air Force publications, performance reports, forms/IMTs, awards, or other unique packages, consult the appropriate manuals or the office of primary responsibility to determine their unique requirements; e.g., Joint Publication 1-02, *Department of Defense Dictionary of Military and Associated Terms*.

FIRST WORDS

1. CAPITALIZE THE FIRST WORD...

a. of every sentence.

Twenty-one people attended the secret presentation given by the chief of staff.
Nonessential government employees were furloughed from 14 to 19 November 2002.

b. of every sentence fragment treated as a complete sentence.

Really? No! So much for that.
More discussion. No agreement. Another hour wasted.

c. of direct questions and quotations placed within a sentence even if quotation marks are not used.

The commander asked this question: How many of you are volunteers?
The order read "Attack at dawn."

d. of items shown in a list (using numbers, letters, or display dots) when a complete sentence introduces them.

The commander listed the following responsibilities of liaison officers:
a. Become familiar with the situation.
b. Know the mission.
c. Arrange for communications.

e. in the salutation and complimentary closing of a letter.

Dear Mr. McBride   Sincerely   Respectfully yours

f. after a hyphen when the hyphenated word is followed by a proper noun or adjective.

non-Latin speaking people

g. after a colon when the

(1) word is a proper noun or the pronoun I.

Two courses are required: English and Economics.

(2) word is the first word of a quoted sentence.

When asked by his teacher to explain the difference between a sofa and a love seat, the nursery school boy had this to say: "Don't reckon I know, ma'am, but you don't put your feet on either one."

(3) expression after the colon is a complete sentence that is the dominant or more general element.

A key principle: Nonessential elements are set off by commas; essential elements are not set off.

(4) material following the colon consists of two or more sentences.

There are several drawbacks to this: First, it ties up our capital for three years. Second, the likelihood of a great return on our investment is questionable.

(5) material following the colon starts on a new line.

They gave us two reasons:
    1. They received the order too late.
    2. It was Friday and nothing could be done until Monday.

(6) material preceding the colon is an introductory word (NOTE, CAUTION, WANTED, HINT, or REMEMBER).

WANTED: Three editorial assistants who know computers as well as editing and typesetting.

h. each line in a poem. (Always follow the style of the poem, however).

I used to write quite poorly.
My boss said it made him ill.
But now he's feeling better
'Cuz I use *The Tongue and Quill!*
                - TSgt Keyes

2. DO NOT CAPITALIZE...

a. the first word of a sentence enclosed in parentheses within another sentence unless the first word is a proper noun, the pronoun *I*, the first word of a quoted sentence, or begins a complete parenthetical sentence standing alone.

The company finally moved (they were to have vacated 2 months ago) to another location.
One of our secretaries (Carolyn Brown) will record the minutes of today's meeting.
This is the only tree in our yard that survived the ice storm. (It's a pecan tree.)

b. part of a quotation slogan or motto if it is not capitalized in the original quotation.

General MacArthur said that old soldiers "just fade away."

c. items shown in enumeration when completing the sentence that introduces them.

Liaison officers must
    a. become familiar with the situation,
    b. know the mission and
    c. arrange for communications.
[Notice punctuation]

d. the first word of an independent clause after a colon if the clause explains, illustrates or amplifies the thought expressed in the first part of the sentence.

Essential and nonessential elements require altogether different punctuation: the latter should be set off by commas, whereas the former should not.

e. after a colon if the material cannot stand alone as a sentence.

I must countersign all cash advances, with one exception: when the amount is less than $50.
Three subjects were discussed: fund raising, membership, and bylaws.

PROPER NOUNS AND COMMON NOUNS

1. Capitalize all proper names (the official name of a person, place or thing).

Porie and Tourcoing	Anglo-Saxon	Cliff Brow
Judy Phillips-McDonald	Rio Grande River	Stratford-on-Avon
the Capitol in DC	the capital of Maine is …	Mönchengladbach
US Constitution	the Constitution	the Alamo

2. Capitalize a common noun or adjective that forms an essential part of a proper name, but not a common noun used alone as a substitute for the name of a place or thing.

Statue of Liberty; the statue	Potomac River; the river
Air War College; the college	Berlin Wall; the wall
Washington Monument; the monument	Vietnam Veterans Memorial; the memorial

3. If a common noun or adjective forming an essential part of a name becomes removed from the rest of the name by an intervening common noun or adjective, the entire expression is no longer a proper noun and is not capitalized.

Union Station; union passenger station	Eastern States; eastern farming states

4. Capitalize names of exercises, military operations, military concepts, etc.

Exercise GLOBAL SHIELD	Operation ENDURING FREEDOM
Principles of War	New Vision; Global Reach, Global Power
Air and Space Superiority	Precision Engagement
Information Superiority	Agile Combat Support

TITLES OF LITERARY AND ARTISTIC WORKS AND HEADINGS

1. Capitalize all words with four or more letters in titles and artistic works and in displayed headings.

2. Capitalize words with fewer than four letters *except*…

    a. Articles: the, a, an

    b. Short conjunctions: and, as, but, if, or, nor

    c. Short prepositions: at, by, for, in, of, off, on, out, to, up

***NOTE*:** Capitalize short verb forms like *Is* and *Be*, but not *to* when part of an infinitive.

> How to Complete a Goal Without Really Trying
> "Reorganization of Boyd Academy Is Not Expected to Be Approved"

3. Capitalize all hyphenated words, except articles and short prepositions; coordinating conjunctions; second elements of prefixes (unless proper noun or proper adjective); and *flat*, *sharp* and *natural* after musical key symbols.

English-Speaking	Run-of-the-Mill	Non-Christians	Follow-Through
Large-Sized Mat	Post-Prezhnev	Self-explanatory	Ex-Governor
Over-the-Hill Sayings	Twenty-first Century	One-eighth	E-flat Concerto

4. Capitalize articles, short conjunctions and short prepositions when:

a. the first and last word of a title.

"A Son-in-Law to Be Proud Of"

b. the first word following a dash or colon in a title.

Richard Nixon—The Presidential Years   *Copyright Issues of the Air Force: A Reexamination*

c. short words like *in*, *out* and *up* in titles when they serve as adverbs rather than as prepositions. These words may occur as adverbs in verb phrases or in hyphenated compounds derived from verb phrases.

"IBM Chalks Up Record Earnings for the Year"
"Wilmington Is Runner-Up in the Election"
"Sailing up the Rhein"

d. short prepositions like *in* and *up* when used together with prepositions having four or more letters.

"Driving Up and Down the Interstate"
"Events In and Around Town"

NAMES OF GOVERNMENT BODIES, EMPLOYEES, NATIONAL AND INTERNATIONAL REGIONS, DOCUMENTS

1. Capitalize, *except* when used in a general sense...

a. full and shortened names of national and international organizations, movements, and alliances and members of political parties.

Republican Party	republicanism, communism
Republican platform; Republican	Bolshevik; Bolshevists
Eastern bloc; Communist bloc	Common Market; Holy Alliance
Democratic Party; Federalist Party	Federalist; Russian Federation; Supreme Soviet
Communist Party; Communist	

b. full and shortened names of US national governmental and military bodies.

US Government	the Federal Government, government workers
US Congress, Congress	US Air Force, Air Force
Department of Defense (DOD)	US Navy, Navy; Marine Corps, the corps
Defense Department	House of Representatives, the House
armed forces, armed services	Reserve Component, Active Component
National Command Authorities	Joint Chiefs of Staff, the joint chiefs
Department of the Air Force	executive branch
Air Force Reserve, reserve officer, reservist	Air National Guard, the Guard

NOTE: If *Army, Navy,* or *Air Force* can be used logically for *Marines*, use *M.* If the word *soldier* or *soldiers* logically fits it, use *m.*

Michael Johnson enlisted in the Marines.	three marines
a Marine landing	a company of Marines

c. titles of government employees.

US President, the President	commander in chief
Presidential campaign	Russian President
Congressman Everett; a congressman	US Senate; a senator; Senator Clinton
Navy officer; naval officer	Secretary of State
Service component command chaplain	service chiefs; chief of staff
the Bush Administration, the Administration	British Prime Minister

d. full titles of departments, directorates and similar organizations.

Department of Labor, the department	Directorate of Data Processing, the directorate
Center for Strategic Studies, the agency	Special Plans Division, the division
Air War College, the college	Squadron Officer School, the school

e. full titles of armies, navies, air forces, fleets, regiments, battalions, companies, corps, etc., but lowercase *army*, *navy*, *air force*, etc., when part of a general title for other countries.

Continental army; Union army	Fifth Army; the Eighth; the army
Royal Air Force; British air force	British navy; the navy
Russian government	US Air Force, the Air Force; Navy's air force
People's Liberation Army	Red China's army; the army

f. full names of judicial bodies.

Supreme Court, the Court	traffic court, judicial court
California Supreme Court	state supreme court
Circuit Court of Elmore County	county court; circuit court
Cabinet members	

NAMES OF STATE AND LOCAL GOVERNMENT BODIES

1. Capitalize the full names of state and local bodies and organizations, but not the shortened names unless mentioned with the name of the city, county, or state.

Virginia Assembly; the assembly
Montgomery County Board of Health; the Board of Health of Montgomery County; the board of
    health will …

2. Capitalize the word *state* only when it follows the name of a *state* or is part of an imaginative name.

New York State is called the Empire State.
The state of Alaska is the largest in the Union.
After an assignment overseas, we returned to the States.

3. Capitalize the word *city* only when it is part of the corporate name of the city or part of an imaginative name.

Kansas City; the city of Cleveland, Ohio, is …
Chicago is the Windy City; Philadelphia, the City of Brotherly Love

4. Capitalize *empire, state, country*, **etc., when they follow words that show political divisions of the world, a county, a state, a city, etc., if they form an accepted part of it; lowercase if it precedes the name or stands alone.**

11th Congressional District	his congressional district
Fifth Ward	the ward
Indiana Territory	the territory of Indiana
Roman Empire	the empire
Washington State	the state of Washington

ACTS, AMENDMENTS, BILLS, LAWS, PUBLICATIONS, TREATIES, WARS

Capitalize the titles of official acts, amendments, bills, laws, publications, treaties and wars, but not the common nouns or shortened forms that refer to them.

Social Security Act	Intermediate-Range Nuclear Forces Treaty
US Code, Vol 28, Sec 2201-2	Gulf War; Seven Years' War
Fifth Amendment	Tet offensive; Cuban Missile Crisis
the income tax amendment	Korean War; Korean conflict
GI bill; Bill of Rights, food stamp bill	antitrust law; the law
Sherman Antitrust Law	World War II, WWII; the two world wars
Air Force Manual 33-326; the manual	Battle of the Bulge; Berlin Airlift; the airlift
Treaty of Versailles, Jay Treaty; the treaty	

PROGRAMS, MOVEMENTS, CONCEPTS

1. Capitalize the names of programs, movements, or concepts when used as proper nouns, but not when used in a general sense or latter day designations.

Medicare Act; medicare payments	Civil Rights Act; a civil rights leader
Socialist Labor Party; socialism	Veterans Administration; veteran benefits
Warfare Studies Phase; the phase	Nation-States

NOTE: Also capitalize their *imaginative* names.

the New Deal	The New Frontier	Pacific Rim
the Great Society	the War on Poverty	Iron Curtain

2. Capitalize terms like *democrat, socialist* **and** *communist* **when they signify formal membership in a political party, but not when they merely signify belief in a certain philosophy.**

a lifelong democrat [person who believes in the principles of democracy]
a lifelong Democrat [person who consistently votes for the Democratic Party]

independent voters	leftists
the right wing	fascist tendencies

3. Do not capitalize nouns and adjectives showing political and economic systems of thought and their proponents, except when derived from a proper noun.

bolshevism	communism	communist
democracy	fascism	fascist
socialism	socialist	Marxism-Leninism

MILITARY RANK, MEDALS, AWARDS

1. Capitalize military rank when it is used with a proper name, but not when it stands alone.

> Colonel Larry D. Grant and his secretary, Linda Wilson; the colonel
> We have 30 majors and 26 lieutenant colonels.
> She's a staff sergeant in the Air Force.

NOTE: After initially identifying by full grade and name, use only the surname with the short grade title. Do not mix abbreviations with full words (Lt Col, not Lt Colonel).

> Brigadier General Richard S. Glenn, Brig Gen, General or Gen Glenn
> Master Sergeant Stephanie Reed, MSgt or Sgt Reed
> Chief Master Sergeant Susan Sharp, Chief Sharp

2. Capitalize specific names of medals and awards.

Medal of Honor	Distinguished Flying Cross	congressional medal
Nobel Prize	Pulitzer Prize	Oscars and Emmys
Purple Heart	Legion of Merit	*Croix de Guerre*

TITLES

1. Capitalize official titles of honor and respect when being used with a proper name or in place of a specific proper name.

> **a. national officials such as the President, Vice President, cabinet members, the heads of government agencies, and bureaus.**

President Bush; the President's speech	every President, Presidential campaigns
Vice President Cheney	Secretary of State
Attorney General	Director of FBI
Commissioner of …	Chief Justice

> **b. state officials.**

the Governor	the Lieutenant Governor
the attorney general	the senator

> **c. foreign dignitaries.**

the Queen (of England)	Prime Minister
The Chancellor of Germany ...	Prince of Wales

> **d. international figures.**

the Pope	the Secretary General of the United Nations.

2. Capitalize any title (even if not of high rank) when it is used in direct address, except *madam, miss,* or *sir* if it stands alone without a proper name following.

> Please tell me, Colonel, what risks are involved in this campaign.
> I need to take some leave today, sir.
> I asked the colonel what risks were involved in this campaign.

3. Also capitalize imaginative names used to refer to specific organizations.

Big Blue [IBM]	the Big Board [the NY Stock Exchange]
Ma Bell [AT&T]	the Baby Bells [the US regional phone companies]

4. Do not capitalize:

a. organization officials.

The commander will visit …
The secretary's minutes were read and approved. [formal minutes]

b. job titles when they stand alone.

Marion Conroy has been promoted to the position of senior accountant.

c. general terms of classification.

The Commandant of ACSC; an intermediate service school commandant
Have your director of research call me.
Squadron Leader David Bye of the Royal Air Force
Samuel A. South, USAF, Retired, went …; Samuel A. South retired from …
United States senator
a state governor
every king
any ambassador
The 2d Security Forces Squadron Commander; the squadron commander

d. *former*, *late*, *ex-*, or *-elect* when used with titles.

the late President Truman   ex-President Bush   Mayor-elect Bawley

e. family titles when preceded by *my, your, his, her, our*, and *their* and describe a family relationship.

Let me ask my mother and dad if that date is open for them.
Do you think your brother Bobby would like to meet my sister Fern?
Frank wants us to meet his Uncle John. (Here Uncle John is a unit.)
Frank wants us to meet his uncle, John Cunningham.

f. *the* at the beginning of a title, except when actually part of the title or when used as part of an official name or title at the Secretariat or Air Staff level.

For extensive details check the *Encarta '99*.
Major Gregg's article was published in *The New York Times*.
The Adjutant General   The Inspector General
The Judge Advocate General   The Surgeon General

5. Do not capitalize when titles follow a personal name or used in place of a personal name.

a. departments within an organization.

Some civilians in Air Command and Staff College that will help you are: Linda Wilson, commandant's secretary; Shirley Keil, protocol officer; Glen Spivey, educational advisor; and Lisa McDay, Yuna Braswell, and Rebecca Bryant, typesetters.
I'm applying for a job in your Directorate of Education and Curriculum.
The vacancy in our directorate has been filled.

b. local governmental officials and those of lesser federal and state, except in writing intended for a limited readership where the intended reader would consider the official to be of high rank.

> Francis Fahey, mayor of Coventry, Rhode Island, appeared before a House committee today. The mayor spoke forcefully about the... [national news service release.]
> The Mayor promised only last fall to hold the city sales tax at its present level. [editorial in a local newspaper.]
> I have written for an appointment with the attorney general and expect to hear from his office soon.
> I would like to request an appointment with the Attorney General. [memo to the state attorney general's office.]

COLLEGES, UNIVERSITIES, ORGANIZATIONS, COMMITTEES, AGENCIES

Capitalize the proper names of colleges, universities, organizations, committees and agencies, but not the common nouns that refer to them.

> University of Alabama; the university
> National Labor Relations Board; the board
> Veterans Administration; the administration
> 42d Air Base Wing; the wing

> Air Command and Staff College; the college
> Organization of American States; the organization
> Committee on Foreign Affairs; the committee
> the National Security Agency; the agency

NOTE: When using the abbreviated form of a numbered organization (e.g., ABW versus Air Base Wing), do not use *th*, *st*, or *d* with the number. When writing it out in its entirety (Supply Squadron versus SUPS), add the *th*, *d*, or *st* to the number.

> 42d Air Base Wing or 42 ABW
> 42d Supply Squadron or 42 SUPS
> 101st Air Refueling Wing or 101 ARW

NOTE: The *preferred style* is to use the long method in written text and the shortened method in address elements, charts, graphs, notes, and bibliography.

> Maxwell Air Force Base, Alabama, has … or Maxwell AFB, Alabama, is …
> Maxwell AFB AL 36112-3648 or HQ USAF CO 80840-6254 [address use only]
> Maxwell AFB, Ala [notes, bibliography]

ACADEMIC DEGREES AND COURSE TITLES AND SUBJECTS

1. Capitalize the names of specific course titles, but not areas of study.

> American History 201 meets on Tuesdays and Thursdays.
> Esther is teaching kindergarten at Daniel Prattville Elementary and is taking EDL 609, Personnel Admin.
> Psychology of Career Adjustment will be offered next quarter.
> The Leadership and Command course employs an approach to further …

2. Capitalize academic degrees following a person's name and when the complete title of the degree is given, but not when they are used as general terms of classification.

> H. A. Schwartz, Doctor of Philosophy
> master's degree; bachelor's degree
> BA, MA, PhD, LLD, MD, DDS, EdS
> Master of Arts

> Bachelor of Arts Degree in Computer and Information Sciences
> bachelor of arts degree

NOUNS WITH NUMBERS AND LETTERS

Do not capitalize nouns followed by numbers or letters unless using full titles and then the first word and all-important words are capitalized.

annex A	chart 10	page 269	tab 2
appendix D	DD Form 282	paragraph 3	table 10
article 2	exhibit A	part II	task 3.1
attachment 2	figure 7	room 154	verse 3
book XI	line 4	rule 3	volume 1
building 1402	map 1	size 8	Annex A, Components
chapter 5	note 1	subtask 3.1.1	Tab 2, Directory of Terms

COMPASS DIRECTIONS

1. Capitalize compass directions when referring to specific regions or when the direction is part of a specific name, but not when merely indicating a general direction or location:

a. general direction/location.

travel north on I-65

southeastern states

the west side of town

East Side; Twin Cities

b. specific regions or a part of the world.

vacation in the Far East

brought up in the Deep South; but there are clouds forming in the south

visit Northern Ireland and New England
Sun Belt; West Coast; North Pole

Central Europe; the Continent [Europe]

c. part of a specific name.

Southland Dairy Company

Northeast Manufacturing Corporation

2. Capitalize words such as *northern, southern, eastern,* and *western* when referring to people in a region and to their political, social or cultural activities, but not when merely indicating a general location or region.

Southern hospitality
Eastern bankers
the South
southern California

Midwesterner
Western Hemisphere
the Northern vote
northern Maine

CELESTIAL BODIES

Capitalize the names of planets (*Jupiter, Mars*), stars (*Polaris, the North Star*), and constellations (*the Big Dipper, the Milky Way*). However, do not capitalize the words *sun, moon,* and *earth* unless they are used with the capitalized names of other planets or stars.

With this weather, we won't see the sun for a while.
We have gone to the ends of the earth to reorganize this unit.
Compare Mars, Venus and Earth.

DAYS OF THE WEEK, MONTHS, HOLIDAYS, EVENTS, PERIODS, SEASONS

Capitalize the days of the week, months, holidays, historic events and periods. Do not capitalize seasons or latter-day designations.

Sunday; Monday
January; February
Veterans Day; New Year's Day
Battle of the Bulge; World War II

Roaring Twenties; Gay Nineties; Roaring 20s
Dark Ages; Middle Ages; Ice Age; the Restoration
spring, summer, fall, winter
age of steam; nuclear age; space age; rocket age

NOTE: A numerical designation of an era is lowercased if it's not part of a proper noun; i.e., twenty-first century, the nineteen hundreds.

RACES, PEOPLES, LANGUAGES

Capitalize races, peoples, and languages.

the Sioux; Mandarin Chinese
English; French; Finnish; German

African-American; black; Caucasian; white
Hispanic; Latin American; Mexican

COMMERCIAL PRODUCTS

Capitalize trade names, variety names and names of market grades and brands, but not the common nouns following such names.

Elmer's glue; Krazy Glue
Microsoft Windows software
Macintosh computers; McIntosh apples
Xerox; Photostat; photocopy; fax
American Beauty rose

Choice lamb chops; White oats
Kleenex tissue; 501 Levi jeans
Band-Aid; Ace bandage; Ping-Pong, table tennis
Scotch tape; Post-It notes; Magic Maker; White-Out
Ivory soap; Coca-Cola; Coke; cola drink

RELIGIOUS REFERENCES

1. Capitalize all references to a supreme being.

God
the Lord
the Supreme Being
the Messiah

the Almighty
the Holy Spirit
Allah
Yahweh

2. Capitalize personal pronouns referring to a supreme being when they stand alone, without an antecedent nearby.

Give praise unto Him.
His loving care
Thy mercy

Seek the Lord for His blessing.
My Father
Our Father

3. Capitalize references to persons revered as divine.

the Apostles
John the Baptist
the Prophet

Buddha
the Blessed Virgin
Saint Peter

4. Capitalize the names of religions, their members, and their buildings.

Reform Judaism	Mormon	Saint Mark's Episcopal Church
Zen Buddhism	Methodists	Temple Beth Shalom

the Roman Catholic Church [the entire institution]

the Roman Catholic church on Bell Road [indefinite reference to a specific building]

5. Capitalize references to religious events.

the Creation	the Exodus	the Crucifixion
the Flood	the Second Coming	the Resurrection

6. Capitalize names of religious holidays.

Passover	Christmas	Hanukkah

7. In general, do not capitalize references to specific religious observances and services.

bar mitzvah	baptism	the Eucharist
seder	christening	the Mass

8. Capitalize (do not quote or underscore) references to works regarded as sacred.

the New International Bible	the Koran	the Ten Commandments
biblical sources	the Talmud	the Sermon on the Mount
the Revised Standard Version	the Torah	Psalms 23 and 25; Psalms 23-24
the Old Testament	the Our Father	Kaddish
the Book of Genesis	the Lord's Prayer	Hail Mary
Philippians 1:3	the Apostle's Creed	Psalms 23 and Joshua 9: 1-2, 5

"A man stopped in at a truck stop for a cup of coffee. When the waitress set it in front of him, he decided to strike up a conversation. 'Looks like rain,' he said. The waitress snapped back, 'It tastes like coffee, doesn't it?'"

– Anonymous

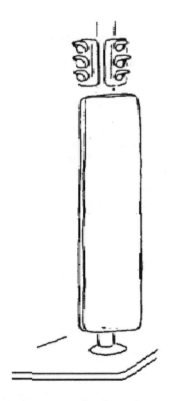

STOP, LOOK AND LISTEN

STOP function fluctuation and capitalization frustration

LOOK at pages 273-307 — the mechanics of writing

LISTEN for more to come... on numbers

STOP, LOOK AND LISTEN to the rule books on punctuation and capitalization

NUMBERS
NUMBERS
NUMBERS

Should we let the numbers speak for themselves?

It is impossible to establish an entirely consistent set of rules governing the use of numbers—we've tried! When expressing numbers, keep in mind the significant difference in the appearance of numbers. Figures will grab your attention immediately because they stand out more clearly from the surrounding words, while numbers expressed in words are unemphatic and look like the rest of the words in the sentence. **Figures emphasize; words de-emphasize.**

The following guidelines cover the *preferred* Air Force style of expressing numbers. Remember, however, that personal and organizational preference, and appearance may override these guidelines. If your organization has a preferred style—use it. If not, read on....

- In **general,** numbers 10 and above should be expressed in figures, and numbers one through nine should be expressed in words.

- In **scientific** and **statistical** material, all numbers are expressed in figures.

- In high-level **executive correspondence** and **nontechnical, formal,** or **literary manuscripts, citation, decoration, memo to the general, textbook, and articles,** spell out all numbers through one hundred and all round numbers that can be expressed in two words (one hundred, five thousand, forty-five hundred). All other numbers are written in figures (514). Turn to the next few pages and research the ones with checks (√) to know which to spell out in this style. It is appropriate, though, to use numbers in tables, charts, and statistical material.

FIGURE STYLE

1. The following categories are almost always expressed in figures, unless high-level executive correspondence and nontechnical, formal, or literary manuscripts. Those with checkmarks (√) are to be spelled out if in high-level executive correspondence and nontechnical, formal, or literary manuscripts. Also when you abbreviate a unit of measure in a unit modifier, do not use a hyphen.

TIME √

payable in 30 days	a note due in 6 months
waiting 3 hours	15 minutes later

AGE √

a 3-year-old filly	a boy 6 years old
52 years 10 months 5 days old	a 17-year-old German girl

CLOCK TIME

at 9:30 a.m. Eastern Standard Time; after 3:15 p.m. Greenwich Mean Time, after 1515 Z
6 o'clock [do not use a.m. or p.m. with o'clock]
0800 [do not use the word *hours* when expressing military time]

MONEY

a $20 bill	it costs 75 cents [if sentence contains other monetary amounts requiring the dollar sign, use $.75]
$5,000 to $10,000 worth; $2 million	
$3 a pound	a check for $125 [if sentence contains other monetary amounts requiring the cents, use $125.00]
$9.00 and $10.54 purchases	
US $10,000	10,000 US dollars
Can $10,000	10,000 Canadian dollars
Mex $10,000	10,000 Mexican dollars
DM 10,000	10,000 West German deutsche marks
£10,000	10,000 British pounds
¥10,000	10,0000 Japanese yen

NOTE: To form the British pound on the typewriter, type a capital *L* over a lowercase *f*. To form a Japanese yen, type a capital *Y* over an equal (=) sign.

MEASUREMENTS √

110 meters long	2 feet by 1 foot 8 inches
5,280 feet	8 1/2- by 11-inch paper
about 8 yards wide	200 horsepower
23 nautical miles	15,000 miles

DATES

5 June 2004 or 5 Jun 04 [when abbreviating the month, also abbreviate the year]	21st of July
	Fiscal Year 2004, FY04, the fiscal year
	Academic Year 2003, AY03, the academic year
from 4 April to 20 June 2005	Class of 2004 or Class of '04
July, August and September 2004	on the 13th send it to

DIMENSIONS, SIZES, TEMPERATURES

a room 4 by 5 meters

a 15- by 30-foot room

size 6 tennis shoes

thermometer reads 16 degrees

PERCENTAGES, RATIOS, PROPORTIONS, SCORES, VOTING RESULTS

a 6 percent discount [use % in technical writing, graphs, charts]

a 50-50 chance

an evaluation of 85

Alabama 14, Auburn 17

a vote of 17 to 6

a proportion of 5 to 1; a 5-to-1 ratio

20/20 or twenty-twenty vision

LATITUDE AND LONGITUDE

a. In nontechnical text:

the polar latitudes

from 10°20′ north latitude to 10°20′ south latitude

longitude 50° west

b. In technical work and tables:

lat 32°25′20″N

long 85°27′60″W

The map showed the eye of the hurricane to be at 32°25′60″N, 85°27′60″W.

NUMBERS REFERRED TO AS NUMBERS AND MATHEMATICAL EXPRESSIONS

pick a number from 1 to 10

multiply by 1/4

number 7 is considered lucky

No. 1—You're No. 1 in my book.

ABBREVIATIONS, SYMBOLS, SERIAL NUMBERS, DOCUMENT IDENTIFIERS

$25

paragraph 3

serial number 0958760

lines 5 and 13

46-48 AD

attachment 2

Proverbs 3:5-7

pages 273-278

UNIT MODIFIERS AND HYPHENATIONS √

5-day week

110-metric-ton engine

10-foot pole

8-year-old car

1 1/2-inch pipe; 1½-inch pipe; not 1-1/2

2. When a sentence contains numbers used in a related series and any number in the series is 10 or more, express all numbers in the series in figures (except the first word of the sentence if it is a number). √ When a number is always a figure, it doesn't change the other numbers to figures in the same sentence.

Six children ate 9 hamburgers, 14 hot dogs, and 6 Popsicles.

Our office has five officers, two sergeants and six civilians.

Our tiny office, which is only 200 square feet, contains five desks, two bookcases and five people.

3. Numerical designations of military units are written as follows:

a. Air Force units. Use figures to designate units up to and including wings. Use figures for numbered air forces only if using the abbreviation AF.

19th Logistics Group; 19 LG

Ninth Air Force, 9 AF

but 19TH LOGISTIC GROUP (address label)

347th Wing; 347 WG

42d Mission Support, 42 MSS

NOTE: Refer to AFMAN 33-326 for proper address elements, and keep in mind when you abbreviate the organizational name (CSG, TFW, AD, AF, etc.) do not use ***st, d,*** or ***th*** with the number.

b. Army units. Use figures to designate all army units except corps and numbered armies. Use Roman numerals for corps and spell out numbered armies.

2d Army Group	First Army
III Corps	2d Infantry Division
7th AAA Brigade	92d Infantry Regiment

c. Marine Corps units. Apply same rules as army units.

d. Navy units. Use figures to designate all navy units except fleet.

Seventh Fleet Carrier Group 8	VF31

4. Numbers expressed in figures are made plural by adding *s* alone.

in the 1990s	four 10s in the deck
temperature in the 80s	two F-16s at the base

NOTE: To plural a number that is used as part of a noun, place the *s* on the noun and not the number: DD Forms 282; but "file the 282s."

WORD STYLE

5. Spell out numbers from 1 through 9; use figures for numbers 10 and above in ordinary correspondence.

> I need nine copies of this article.
> At the conference, we got over 11 comments to start a new ...

6. Spell out numbers that introduce sentences. A spelled out number should not be repeated in figures (except in legal documents).

> Twelve people volunteered for the job; not twelve (12) people ...
> Eight children participated in the relay race.

7. Related numbers appearing at the beginning of a sentence, separated by no more than three words, are treated alike.

> Fifty or sixty miles away is Auburn University.
> Five to ten people will probably respond.

NOTE: Related numbers in the same set are also treated alike.

> The $12,000,000 building had a $500,000 tower. [Not written as *$12 million* because of its relation to *$500,000.*]
> We mailed 50 invitations and only received 5 RSVPs.

8. Spell out numbers in formal writing and numbers used in proper names and titles along with serious and dignified subjects such as executive orders and legal proclamations.

the Thirteen Colonies	the first Ten Amendments
The Seventy-eighth Congress	threescore years and ten

9. Spell out fractions that stand alone except with unit modifier.

> one-half of the vote; but 1/2-inch pipe (unit modifier) or ½-inch pipe
> six-tenths of a mile

NOTE: A mixed number (a whole number plus a fraction) is written in figures except at the beginning of a sentence.

> 1 1/2 miles; 1½ miles; not 1-1/2 miles   One and a half miles

10. Spell out compound modifiers and numbers of 100 or less that precede hyphenated numbers.

> three 10-foot poles   120 1-gallon cans
> one hundred 1-gallon cans   twenty 5-year-old children
> three 1 1/2-inch pipes; three 1½-inch pipes   two 4-hour sorties

11. Spell out rounded and indefinite numbers.

> the early nineties; but the early 1990s   the twentieth century
> hundreds of customers   nineteenth-century business customs
> a woman in her fifties   approximately six thousand soldiers

12. For typographic appearance and easy grasp of large numbers beginning with million, use words to indicate the amount rather than 0s (unless used with a related number).

> $12 million   less than $1 million
> $6,000,000 and later 300,000 …   2 1/2 billion or 2½ billion
> $2.7 trillion $300,000 (not $300 thousand)

13. Form the plurals of spelled-out numbers as you would the plurals of other nouns—by adding *s*, *es* or changing the *y* to *i* and adding *es*.

> ones   twenties
> twos   fifties
> sixes   nineties

STRIKE A BLOW FOR FREEDOM!!

Are all these rules making you numb? Why don't we put some sense in this silly nonsense and take it upon ourselves as rational men and women to make our *own* rule that will let us win at this numbers game. How about …

> "Always express numbers as figures unless the number starts the sentence,
> or unless the use of figures would confuse the reader … or would look weird."
>
> —*The Quill's* Law of Numerical Bingo

The simplicity of it is downright ingenious. Think how many pages out of the inconsistent grammar books we could eliminate. Save a forest! Be a leader!

ROMAN NUMERALS

Roman numerals are used most frequently to identify the major sections of an outline. They're also used (in lowercase form and in italics—*i*, *ii*, *iii*) to number pages in the front sections of books. The following table shows Roman numerals for some Arabic figures.

A dash above a letter tells you to multiply by 1,000.

ROMAN NUMERALS

I 1	XXIX..........29	LXXV 75	DC 600
II...............2	XXX 30	LXXIX 79	DCC.............700
III 3	XXXV........35	LXXX 80	DCCC 800
IV 4	XXXIX.......39	LXXXV...........85	CM.................900
V................5	XL 40	LXXXIX 89	M 1,000
VI 6	XLV 45	XC....................90	MD............1,500
VII...........7	XLIX..........49	XCV 95	MM...........2,000
VIII........8	L 50	XCIX................99	M̄V............4,000
IX 9	LV 55	C......................100	V̄ 5,000
X.............10	LIX............59	CL 150	X̄ 10,000
XV..........15	LX 60	CC 200	L̄............50,000
XIX 19	LXV 65	CCC................300	C̄ 100,000
XX.........20	LXIX..........69	CD..................400	D̄ 500,000
XXV.......25	LXX 70	D.....................500	M̄ 1,000,000

DATES

1600 - D̄C	1997 - M̄MXCVII
1900 - MCM	2000 - MM

Other combinations of Roman numerals are derived by prefixing or annexing letters. Prefixing a letter is equivalent to subtracting the value of that letter, while annexing is equivalent to adding the value.

49 is L minus X plus IX: XLIX

64 is L plus X plus IV: LXIV

APPENDIX 2

THE MECHANICS OF RESEARCH

This appendix covers:

- Getting started: planning your research schedule

- Defining the research problem and stating the research question.

- Reviewing the related literature.

- Research citations, quotations, and paraphrases

- Research methodology: qualitative and quantitative methods

Research is a structured and systematic way to create knowledge. Research searches for facts with a purpose in mind. Kuhn said research is like solving a puzzle: you have to follow specific rules, and the challenge of putting the puzzle together to get a complete picture is what usually motivates the researcher.[1] This appendix provides a summary of research issues Air Force members may encounter either during their assignments or during their professional military education. It also provides samples of citation formats used in research papers, including endnotes, footnotes, and textnotes.

GETTING STARTED: PLANNING YOUR RESEARCH SCHEDULE

"Don't put off for tomorrow what you can do today, because if you enjoy it today you can do it again tomorrow."

– James A. Michener

At first research may seem like a tedious and complex process. Actually though, some researchers find that once they begin to pursue a research question they are interested in, they really begin to enjoy the process. Do not let the idea of writing a huge paper keep you from getting started. As Henry Ford said "Nothing is particularly hard if you divide it into small jobs."

Here are some tips on getting yourself going:

• Don't wait to get started until you can set aside two or three entire days to do your research. This is one task you want to do in a few sittings and not all at once. If you keep waiting for the "perfect" time to get started, you may never get anything done at all.

• Create an overall plan for what you want to accomplish and when you want to have each step done. For example, here is a proposed research process broken down into manageable chunks:

Task	Desired Date	Actual Date
Topic selected		
Research question written		
Methodology selected		
Sources and data gathered		
Sources read and data analyzed		
Outline created		
Expand outline to draft all topic sentences		
First draft written		
Second draft written		
Final draft written to include front matter (abstract, preface) and back matter (endnotes, bibliography)		
Seek publication if appropriate		

Back up your milestones from whatever due dates you are given to allow for a little "slop" time and then reward yourself for any deadlines you meet. (Bribery does work—even if you are bribing yourself.)

• Not all tasks associated with research need to be done in the quiet of a library with an 8-hour stretch of time. Break your research-related tasks down into chunks and work them in when and where you can. For example, you don't necessarily need to devote an entire day to reading articles. Instead you can carry a few with you to read when you get stuck waiting somewhere.

- If your research involves a survey or an experiment that requires the collection of data involving people, be sure that you get proper approval before you get started. This can take up a bit of time, so the earlier you do this the better.

"If you want to make an easy job seem mighty hard, just keep putting off doing it."

– Olin Miller

SELECTING A RESEARCHABLE PROBLEM

Research starts with the selection of the problem. This can be harder than it sounds because not every problem may be researchable. For some problems it can be impossible to collect the data to support it. On the other hand, it's easy to become lost in data that is easy to collect and lose sight of the original question. Carr listed three laws to help define researchable problems:

1. The problem should be clearly formulated in a single sentence of 25 words or less. (Otherwise you could find yourself working with no direction and go off onto irrelevant tangents.)

2. You should be able to collect useful empirical data with observable and accessible criteria. Wherever possible that data should be numerical.

3. You should be able to directly, or indirectly, observe the events you plan to collect data about.[2]

These three simple rules should be followed as closely as possible whenever a problem is selected.

Another challenge in defining a problem is making it "not too big and not too small." Barzun and Graff discussed how your subject should, "when clearly presented in a prescribed amount of space, leave no questions unanswered within the presentation, even though many questions could be asked outside it."[3] Though this can be difficult, following these rules will prevent serious problems later on in the research process. Ideally, the problem you select to research will be one that you are interested in. The longer the research paper you will be writing, the more important this is. If you are writing a 5-page paper you can usually slog through almost any topic, but if this is your doctorate dissertation it better be something you feel passionate enough about to attack daily for a year or more.

> **What is NOT an appropriate research question?**
>
> 1. A ruse for achieving self-enlightenment.
>
> 2. Problems where the sole purpose is comparing two sets of data.
>
> 3. Problems seeking correlation between two sets of data merely to show a relationship between them.
>
> 4. Problems with a "yes" or "no" answer.
>
> **SOURCE:** Leedy's *Practical Research*, pp 47-48.

STATING THE RESEARCH QUESTION

Once the problem has been decided, the next step is to articulate it concisely and clearly in a research question.

1. **Name your topic:** I am studying …

2. **Imply your question:** Because I want to find out/show you who/how/why …

3. **State the rationale** for the question and the project: In order to understand/explain how/why/what …

Once you have your focus narrowed, and you know just what you want to study, it is time to begin the hunt for what others have written on the same, or similar, subjects.

REVIEWING AND CITING THE RELATED LITERATURE

"Read the best books first, or you may not have a chance to read them at all."

– Henry David Thoreau

READ ALL ABOUT IT!

Quotations
Paraphrases
Copyright
Footnote
Endnote
Bibliography
Formatting

Once you have your research question solidified, start your literature review as soon as possible. One reason to search the related literature right away is to make sure that someone else hasn't already researched the same topic. Keep in mind that if someone has already done the study you would like to do, you can still check their conclusions to see if they have recommended an area of further research. Also, check the date of their study. If it was quite some time ago, replicating their study with a few new twists just might expose some interesting conclusions.

There are several other reasons to conduct a thorough literature review:

- It will increase your confidence in your topic….
- It can provide you with new ideas and approaches….
- It can inform you about other researchers whom you may wish to contact….
- It can show you how others have handled methodological and design issues….
- It can reveal sources of data….
- It can introduce you to measurement tools other researchers have developed….
- It can reveal methods of dealing with problem situations….
- It can help you interpret and make sense of your findings….[4]

One major way to save time is to **record all your source data properly from the start.** There are several ways to gather your data as you review the related literature. You can use old-fashioned pen, ink, and note cards. You could take a laptop along and type your notes right in. Or you can purchase or photocopy everything. Note cards are convenient in that you can arrange your thoughts in order as you get ready to write. You can also do the same thing with computer notes by printing out your notes, cutting them into strips and then arranging them as you see fit. Just make sure that each line, or paragraph, has a source and page number before you start to cut up your notes pages. One big benefit of typing your notes on a computer is that you can copy and paste quotes right from your notes to your paper. If you can afford it, purchasing or photocopying your sources can be useful. This is especially true if you plan to do further research on the same topic. Being able to refer back to the original source for more information can be very helpful.

No matter how you capture your source data, be sure to include the details of where you got them. This is especially important if you are using sources from several different libraries and need to track one back down again. Another important thing to remember if you are typing or writing notes is to distinguish clearly between what is a direct quote, what is a paraphrase, and what are your own words and thoughts. If you don't indicate the difference now, while you initially type them in, you will forget and then run the risk of plagiarism.[5]

> **Read The Original**
>
> Wherever possible, you should try to read original works instead of someone's interpretation of another work. One rule is that if three others have cited the same source, you should probably hunt it down and read the original work yourself.
>
> **SOURCE:** Paul D. Leedy & Jeanne Ellis Ormrod, *Practical Research; Planning and Design* (Upper Saddle River, NJ: Merrill Prentice Hall, 2001), 76.

"Our two greatest problems are gravity and paperwork. We can lick gravity, but sometimes the paperwork is overwhelming."

– Dr. Wernher von Braun

CITATIONS

In the staff environment, we frequently reuse previously prepared data to save time and avoid "reinventing the wheel," but we rarely need to cite the source of such data. In the academic world, however, reusing another's work without giving that person credit and deliberately trying to pass it off as your own, is *plagiarism* and can get you into a heap of trouble.

When and where do you document sources? The rule is simple: **If the ideas and information in what you've written are not "common knowledge" or do not represent your own work, you must document where and from whom the "borrowed" ideas and information came.** As a writer, when you quote an authority word for word, paraphrase someone's thoughts or use someone's ideas, model, diagram, research results, etc., you need to do so at that point in the text. This is referred to as *citation*. Citation refers to one of several types of systems writers use to document their sources. The signals for citation may be footnotes, in-line notes or endnotes. Whatever the system, the purpose is to flag material for which the writer is indebted and to identify the source. The sum of all citations in a paper, together with the bibliography, is the documentation system of the paper. Citation, if done properly, fulfills a writer's responsibility for maintaining academic integrity. So, to keep yourself out of a literary (and perhaps legal) jam—give credit where credit is due and cite those sources!

Each community has its own standards for citations. This section relies very heavily on the *Air University Style Guide for Writers and Editors,* which can be accessed at http://www.maxwell.af.mil/au/aul/aupress/.

QUOTATIONS

A quotation (also called a direct quotation) occurs when a writer is indebted to a source not only for the source's ideas or facts but also for the wording of those ideas. When you are using a portion of a source word-for-word you must indicate so by using either quotation marks or a block quote. For shorter quotations, keep them in the text and simply enclose the words you are using from another source in double quotation marks (see the example below). Different style-guides have different criteria for how long a quote needs to be before you pull it out of the text and create a "block quotation." According to the *Air University Style Guide for Writers and Editors,* you should use a block quotation, "for passages easily set apart from the text, 10 or more typed lines, or exceeding one paragraph. Indent from both sides and single-space. Do not use quotation marks to enclose the block quotation, and do not indent its paragraphs. Use double quotation marks to enclose a direct quotation within a block quotation. Skip a line between paragraphs. The block quotation should reflect the paragraphing of the original."[6] (That was also an example of an in-text quotation.) If you are using quite a bit from a copyrighted work, you need to get written permission from the copyright holder. The *Air University Style Guide* also offers the following advice on direct quotations: (note that the quote below is an example of the block quote format)

> **Plagiarism—HOW TO AVOID IT**
> - Be aware of where your eyes are when you type and/or write: source or your page?
> - Realize when you rely heavily on a source: re-writing what you see?
> - Compare your work with sources: same words/phrases as in source?
> - Take good notes, note page references: check your work later
>
> **SOURCE**: Booth, Colomb & Williams, *The Craft of Research,* 170.

1. You may change single quotation marks to double quotation marks and vice versa, if necessary.
2. You may change the initial letter to a capital or lowercase letter.
3. You may omit the final period or change it to a comma, and you may omit punctuation marks where you insert ellipsis points.
4. You should usually omit original note-reference marks in a short quotation from a scholarly work. You may insert note references of your own within quotations.
5. You may correct an obvious typographical error in a passage quoted from a modern source, but you should usually preserve idiosyncratic spellings in a passage from an older work or manuscript source unless doing so would impair clarity. You should inform the reader of any such alterations, usually in a note.[7]

It is good to set the stage for every quote you use and then provide closure after each quote by showing how it relates back to the main point of your paper. Quotes just hanging there, without any stage setting or closure, can make your paper sound choppy and disjointed. What you want is a product that flows seamlessly between what the experts say (quotes) and the conclusions and creative recommendations you can draw from what they say.

PARAPHRASE

Paraphrase is a restatement of a text, passage, or work, giving the meaning in another form. It is not simply changing a couple of words or putting them in a different order. A paraphrase falls into a gray area between summary and quotation. Where a summary uses only the source's content, but not its words, a paraphrase uses the source's content stated in words and sentence structure that are similar to—but not exactly like—the source's. If you do paraphrase, always cite the source (and the appropriate page numbers).

There is no simple answer to the problem of deciding how many words you may use from a source before you are required to show you are quoting. A complete sentence taken from the source would certainly have to be treated as a quotation. But even a single word might have to be quoted, especially if it is a new technical word introduced or developed by the source. The *Air University Style Guide* says, "Ideally, you should introduce your paraphrase so that the reader has no question at all about where your own commentary ends and where your paraphrase begins."[8]

All in all, you need to make sure your research report consists of much more than just a string of quotes and paraphrases from other sources. It would be very inappropriate for example to put a citation mark next to a chapter heading to indicate that everything in an entire chapter of your research paper came from another source. (Yes, someone has tried to do that!) If you have that much from a single source, and if it really is key to understanding your research, then you can put it word-for-word in an appendix and cite the source there. Remember that research is not just a compilation and regurgitation of others' thoughts—your own thoughts need to be evident too.

COPYRIGHTS!

What is a copyright? A copyright is the exclusive legal right granted under Title 17, US Code, to the author of an original published or unpublished work (literary, dramatic, musical, artistic, and certain other intellectual works) to copy and send copies (paper or electronic), to make derivative works, and to perform or display certain types of works publicly.

> Research papers, or any other written material produced as part of your official government duties, are not subject to copyright protection, and are the property of the United States Government.

What are your rights and limitations? Ownership of the copyright is distinct from ownership of the material object (book, periodical, photograph, record, video or audio recording, music, etc.) in which the work is included. The owner is the boss—the head honcho who allows (or not) the work to be performed or displayed publicly. Be careful to not trespass on someone else's property or step on anyone's toes. However, there are exceptions that allow the use of the owner's work without requesting permission or obtaining a license. Find your organization's expert to keep you out of hot water ... or jail!

Can you make changes? You will not get your hand slapped for making *minor* style changes. But the changes, individually or cumulatively, should not significantly change the context or its meaning. Minor changes are only allowed to let you fit the work smoothly into the syntax and typography style of the product.

HOW TO CITE YOUR SOURCES

Footnotes, textnotes and endnotes are three common citation methods used to indicate where you got your information. For more extensive coverage of these and other methods, consult the style manuals such as *The Air University Style Guide, The Chicago Manual of Style*, or *The American Psychological Association Style Guide.* There are certainly plenty of style guides to choose from, but the *Air University Style Guide for Writers and Editors* is not only free from Air University Press, it is also available on-line.[1] Whatever style you choose within a particular document, be consistent.

1 FOOTNOTES

With today's computers, using footnotes is much easier than it used to be with a typewriter. If you are using footnotes, each footnote should appear on the same page you refer to it. Sometimes you might want to include more than just source data in the footnote. Keep in mind that a page containing more footnote material than text not only is unpleasant to the eye, but also may discourage all but the most determined reader.

Most computer programs have an option where you can choose to insert a footnote and it takes care of all the formatting for you, to include keeping the numbering straight—even if you move text around while editing. In Word, you can jump back and forth between the text and the citation by double-clicking on the footnote number. You simply can't beat having the computer do all the work for you, but if you do have to put footnotes in manually, they should be underneath a flush left, five-eighth inch line, that has at least one space above. Also, use an 8-point font size.   Single-space within each footnote and also between each footnote. Footnotes look just like endnotes (described next), except they are at the bottom of each page—just like the one below.

2 ENDNOTES

For an example of endnotes, check out the end of this chapter. The *Air University Style Guide for Writers and Editors* has other examples for a wide variety of sources.

Endnotes can be placed at the end of each chapter, or at the very end of the manuscript. Endnote numbers may run consecutively from beginning to end of the manuscript, or they may begin again with each new chapter. The typed format for endnotes included at the very end of the manuscript differs slightly in that chapter numbers could be included too.

The computer can set up your endnote formatting in the same way it does footnotes. In Word, when you select "Insert" and "Footnote," a menu pops up where you can choose between footnotes and endnotes. You can also specify where you want the endnotes to go, at the end of the section or the entire document. If your endnotes aren't going where you want them to go,

---

[1] Marvin Bassett, ed., *Air University Style Guide* (Maxwell AFB AL: Air University Press, 2001), 11. Web site http://www.au.af.mil/au/awc/awcgate/style/styleguide.pdf

you may need to insert (or remove) a section break. For example, sometimes they will appear after your bibliography, which is incorrect. In this case, put a section break before the bibliography. If you are creating endnotes manually, start on the page where you want your endnotes to appear and center the word "**Notes**" 1 inch from the top of the page or double-spaced below any text on the same page. Triple-space between the heading and the first text entry. Number each entry and arrange in numbered sequence as superscript numbers in text. Single-space within each entry as well as between entries. Indent each numbered entry one-half inch from left margin with subsequent lines flush left. When including chapter numbers, center "Chapter #" on the third line below "**Notes**" and begin the list two line spaces underneath.

3 TEXTNOTES

Textnoting is a means of identifying a source parenthetically at the appropriate point within the text. This method can be accomplished two ways. One way is to show the source within the text. For example:

> Samuel Huntington, in his book *The Soldier and the State*, observed, "The outstanding aspect of civil-military relations in the decade after World War II was the heightened and persistent peacetime tension between military imperatives and American liberal society."

Another way is to provide only the source name and page number in the text and prepare a bibliography page containing the complete source identification. When the bibliography contains more than one entry by the same author, include the date in your textnote to ensure the reader refers to the proper source.

> The definition of analysis (Motes: 23-31) used is …
>
> The definition of analysis (Motes, 1997: 23-31) …

Remember your audience when you are deciding which citation format to use. If you are writing for a school assignment, your school probably has a specific way they want you to cite your sources. If you are writing to seek publication, check with your target publications for their preferred style and format before you get started. If you are really serious about publishing widely, there are computer programs available which will convert your document from one style guide to another which may be worth the investment.

BIBLIOGRAPHY

A bibliography is an accurate list of all sources used to prepare your research manuscript. This means there could be sources in your bibliography that do not appear in your endnotes, footnotes, or textnotes. On the other hand, everything you have cited in your notes should definitely be in the bibliography. One way to prepare your bibliography is to copy all your endnotes and/or footnotes into the bibliography section once you are done writing your paper, and then put them in alphabetical order. However, you still aren't done. Take a close look at the differences in some style guides between the endnote and/or footnote format and the bibliography format. For example, these are the differences for citing the same source according to the *AU Style Guide*:

Bibliography:

> Leedy, Paul D. *Practical Research*. New York, N.Y.: Macmillan Publishing Co., 1989.

Footnote and/or Endnote:

> Paul D. Leedy, *Practical Research* (New York, N.Y.: Macmillan Publishing Co., 1989), 112.

That is just enough difference to drive you crazy, huh? One time-saver is to cite your sources in the correct format from the very start of your literature review. The last thing you want to do after spending lots of time on the text of your paper, is to have to go back to hunt down information that would have been really easy to collect the first time you looked at the data. This means you shouldn't just photocopy articles and stash them away hoping all of the information you want will be on the photocopied page. Make sure the volume number, journal title, page numbers, author, etc., are all captured somehow. Usually you need to look at the very front of some journals to find the volume number. This can be very easy to overlook, so if you have a comprehensive and structured process to capture all the pertinent details on each and every source you use, you will save yourself lots of trouble as you tie up the loose ends of your research.

GUIDE TO TYPING A BIBLIOGRAPHY

An example of the bibliography format is at the end of this book. Hopefully you will be building your entire research report off of a research paper template that lays everything out for you. In the event you are building all of this manually, here are some ideas to get you started.

- Use bond paper and center "***Bibliography***" 1 inch from top edge.

- Triple space between the heading and your first entry; single-space within each entry and double space between entries.

- Begin each entry at the left margin and indent subsequent lines five spaces.

- Arrange entries alphabetically, listing author's name in reverse order (last name, first name, middle initial). When no author is named and a title is used, omit initial articles—*a, an* and *the*—use the first major word of the title to alphabetize. If the title begins with a number, alphabetize as though the number is spelled out (76 Trombones would be alphabetized by the letter *s*).

- When you list two or more works by the same author, do not repeat the author's name. Use a line three dashes long for all entries after the first. List the works for the same author alphabetically by title.

- Items that involve a coauthor follow the works written by the first author alone.

- Set off titles of magazine articles in quotes and italicize the name of the periodical in which the article appeared. Follow with a volume number (may be the month and year) and date of issue in parentheses with page(s).

- Italicize a published report title just as a book title. Also, if there is no author, show the agency responsible for the report. Continue as you would for a book. If an organization is the author, alphabetize by organization name.

- While the *AU Style Guide* says it is not necessary to include interviews in a bibliography,[9] it can add to the credibility of your paper. You should make sure you have the interviewee's signed consent and they have checked over their parts of your text. List an interview with the names of the interviewee and the interviewer ("author" is used for the name of the author of the book or article in which the interview is listed), the place and date of the interview and, if possible, where it is stored.

♦ Even e-mails can and should be cited appropriately.[10]

The *Air University Style Guide* has page after page of examples, so if you are in doubt about how to cite something, be sure to check there or with another style guide.

WRITING STYLE

For some reason some people think that research papers must be written at a level that is above the complexity of day-to-day writing. This is not so. Here is an example of an abstract with clear writing:

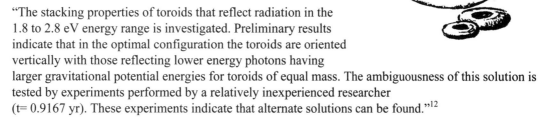

> "We had some fun with a stacking rings toy and learned something about how the perceptions of adults are different from those of babies."[11]

Compare that abstract with this one, which has been a bit "beefed-up."

> "The stacking properties of toroids that reflect radiation in the 1.8 to 2.8 eV energy range is investigated. Preliminary results indicate that in the optimal configuration the toroids are oriented vertically with those reflecting lower energy photons having larger gravitational potential energies for toroids of equal mass. The ambiguousness of this solution is tested by experiments performed by a relatively inexperienced researcher (t= 0.9167 yr). These experiments indicate that alternate solutions can be found."[12]

They both say the same thing; one just says it much more clearly than the other. Please don't think of your research report as an opportunity to throw together as many multisyllabic words as you can possibly think of. Remember, you will have to read this over a few times yourself.

FORMATTING ISSUES

One way to help your paper appear organized is through effective use of headings. In order to make sure you are consistent with your main points, the size and style of your headings and should follow what the introduction says the paper is going to cover and clearly indicate which of the issues are the most important. One of the ways to do this is to match up your outline with the appropriate heading levels.

Level One – Centered, bold, 16-font

Level Two – Centered, bold, 14-font

Level Three – Flush Left, Bold, 12-font

Level Four - Flush Left, Bold, 12-font. Followed immediately by text.

If possible, do not "stack headings" with one immediately following another, instead make sure some text separates your headings.[13]

Now that you know how to capture your sources, ideas, and thoughts—one last consideration is what methodology you use to collect, process and analyze your data.

INTRODUCTION TO RESEARCH METHODOLOGY

What is methodology? One good way to start is by describing what methodology is *not*. Methodology is not a running dialogue of how you did your research!! Methodology simply provides a structured way of gathering and analyzing your data.

Methods

I. Qualitative Research Methods
  A. The Historical Method
  B. General Qualitative Methods
    1. Case Study
    2. Ethnography
    3. Phenomenological Study
    4. Grounded Theory Study
    5. Content Analysis
II. Quantitative Research Methods
  A. Nonexperimental Methods
    1. Descriptive Survey
    2 Analytical Survey
  B. Experimental Method

There are as many different ways to group the research methods together as there are research methods themselves, but most of them fall into two main categories: qualitative and quantitative. Some research will even use a mix of the two so don't confine yourself to just one category. One of the major differences between these different research methods is how much control the researcher has over the situation.   This section will briefly survey a few of these methods, beginning with the ones where the researcher has little control and working up to experiments where the researcher has a great deal of control.

QUALITATIVE RESEARCH METHODS

Qualitative research tends to focus on studying things that occur in natural settings.[14] Qualitative methods fall into two main categories: one is the historical method and the other is a general category which includes case studies, ethnography, phenomenological study, grounded theory study and content analysis.

THE HISTORICAL METHOD

The historical method is one of the most commonly used research methods, and it is the only one available for studying the past. It "aims to assess the meaning and to read the message of the happenings in which men and women, the events of their lives, and the life of the world around them relate meaningfully to each other."[15] Historical research is *much much* more than listing events in chronological order, or restating historical data in a new and different format. This concept can be lost on students who mistakenly view research as nothing more than a beefed-up book report. The key element in the historical method's search for meaning requires the researcher to interpret what may have appeared to be simply chance. This method pulls together both things that are commonly known by the well educated, and also any special information that may be relevant to the historical question being studied. In order to help provide stronger conclusions, primary sources should be used as much as possible to increase the validity of the data. Unfortunately, in the historical research design the researcher has absolutely no control over the collection of the data, the subjects, or even anything that happened at the time the events took place. This can make it hard to be sure all of the needed data has been found, if the data on hand is accurate, and if or how much of the data has been distorted or destroyed.

The researcher using the historical method does not need to look at history only from a time dimension, but can also look at it from the dimension of where things happened.[16] By arranging historical data in different ways, such as on timelines, charts, or a map, new insights can be found to help find new meaning in the data.

Also, historical research does not restrict itself to just the study of events and people from the past. This methodology is also useful for exploring the origin, development and influence of ideas and concepts.[17] This is where the power of the historical research methodology lies. The ideas and concepts explored through this methodology could have as strong an influence on their ages as the rise or fall of a nation or civilization and would be powerful lessons to pull from history to help carry us into the future.[18]

GENERAL QUALITATIVE METHODS

In addition to the historical method, there are several other qualitative methods.

CASE STUDY

Case studies seek to understand a person or a situation in depth.[19] For example, someone could study in great detail the transition of a unit from one aircraft to another. You could do this by focusing on just one case, or make an even stronger analysis by looking at multiple cases and making comparisons between them. This is a good method to use if little is known about a situation or if you want to look at how things change over time.[20] Unfortunately, this method is not strong when it comes to being able to generalize the results.

ETHNOGRAPHY

Ethnographies are broader than case studies since they study entire groups in depth, particularly groups that share a common culture.[21] For example, instead of looking at one unit transitioning from one aircraft to another, a researcher could use this method to study the entire fighter pilot community. In this method, the researcher studies the group in their natural setting over a period of months, possibly even years, "with an intent to identify cultural norms, beliefs, social structures, and other cultural patterns."[22] This method was first used in cultural anthropology, but is just as applicable in today's organizational cultures.

PHENOMENOLOGICAL STUDY

A phenomenological study attempts to "understand people's perceptions, perspectives, and understandings of a particular situation" by looking at several different views of the same situation to make generalizations about what that situation is like.[23] This research method depends heavily on interviews. For example, the individuals who transitioned from one aircraft to another could be interviewed to see what the change meant for them.

GROUNDED THEORY STUDY

This method is interesting in that it is the reverse of most research methods, which normally begin with a theory and test it with data. A grounded theory study begins with the data and uses the data to develop a theory, and is "typically used to examine people's actions and interactions."[24]

CONTENT ANALYSIS

Content analysis takes the contents of a body of material such as books, films, transcripts, and searches for themes and patterns.[25] This method could be used to determine how much violence appears on TV in any given day, and describe what type of violent acts appear most frequently. This challenging task of sorting through mounds of data is now easier with computer programs that can help do this quickly. This method is being used more and more and in combination with other methods.

QUANTITATIVE RESEARCH METHODS

In quantitative research, numbers, and statistical analysis play a larger role, and the ability to generalize results is somewhat stronger than most qualitative methods. Quantitative research methods can be divided into two major categories: nonexperimental and experimental.

NONEXPERIMENTAL RESEARCH METHODS

In nonexperimental research methods, the researcher only controls the measurement of the item under study. Both the descriptive and analytical survey methods are examples of this kind of research.

THE DESCRIPTIVE SURVEY METHOD

> "Survey" is used here to mean, "to look, or see over or beyond."

The descriptive survey method uses data obtained through observation. While historical data is used to explore events of the past, survey research *looks at things as they are happening*. A descriptive survey investigates things as they are without interfering with them.

There are three basic forms of surveys: retrospective, current, and prospective. A retrospective survey reaches back into the past to find out how things used to be. A current survey looks at things the way they are now. And a prospective survey selects a population and follows it for a period of time into the future to see what changes take place. Retrospective surveys have some problems in that the data may be incomplete or even missing. Current surveys, which are used most often, are used to determine things like current attitudes towards specific issues. Prospective surveys do the same thing, but are conducted on the population more than once. This type of long-term study is not done as often as other studies are, mainly due to time and cost constraints.

> The questionnaire should be a "totally impersonal probe" which means the researcher must take several precautions when using this tool including:
>
> 1. The language must be unmistakably clear.
>
> 2. Questionnaires should be designed to fulfill a specific research objective.
>
> 3. Questionnaires succeed as their success is planned.
>
> 4. The initial letter is all-important.
>
> **SOURCE:** Leedy's *Practical Research*, 1989, pp 142-146.

There are several different tools for the actual collection of the data, including questionnaires, interviews, and the differential sliding scale checklist or inventory.

When using any survey method, setting the proper population and using the appropriate, deliberate sampling procedures are essential to produce trustworthy results. Accidental sampling, such as asking questions to the first five folks who walk in each day, makes absolutely no pretense of being representative of a population and makes no attempt to control for personal bias. Quota sampling selects respondents in the same ratios as they are found in the overall population being researched. What is essential is the process of randomization that is used to choose from the overall population. This concept of randomization and the elimination of bias are the two most important elements in successful survey research.

THE ANALYTICAL SURVEY METHOD

The analytical survey method is different because it relies more on quantitative data as opposed to qualitative data. This means where the descriptive survey method relies on describing a situation through words, the analytical survey method interprets situations through numerical data. Even though many descriptive survey methods will use some numbers, the calculations are not the major form in which data exists in analytical survey research.[26] In the analytical survey method, the researcher is analyzing a set of data to *test a hypothesis* and to look for meanings that may be hidden within the data. Once the data is gathered, statistics play a very large role in their interpretation.

Quick Reference Sampling Table	
Total Number in Sub-Population	Number of people to be surveyed
1-5	All
6-7	5
8-9	6
10-12	7
13-16	8
17-20	9
21-30	10
31-49	11
50-99	12
100	14
200	21
500	40
1,000	75
2,000	78
5,000	80
10,000	81

SOURCE: Wilkerson & Kellogg, 1994, 14.

How do statistics help interpret data?

1. Indicate the central point around which the data revolve.

2. Indicate how broadly the data are spread.

3. Show the relationship of one kind of data to another kind of data.

4. Provide certain techniques to test the degree to which the data conform to or depart from the expected operations of the law of chance or approximate an anticipated standard.

SOURCE: Leedy's *Practical Research*, p 186.

The biggest difference between the nonexperimental and experimental methods of research is survey research does not include the same highly controlled aspects of experimental research.

THE EXPERIMENTAL METHOD

There are many different subsets of the experimental method, but a common characteristic is the level of control the researcher has over both the event and the ability to collect data on the event. An experiment is a contrived event where the effect of a deliberate act is observed. One of the main purposes of experimental research is to determine cause-and-effect relationships. Unfortunately, some study results can be weakened because the artificial nature of their experiments makes the results too different from what might turn out in a natural setting.

One of the most important aspects of the experimental method is effectively planning the design of the experiment—not just how the data will be interpreted, but the entire experimental design.[27] One key concept involves independent and dependent variables. "A variable that the researcher manipulates is called an **independent variable**. A variable that is potentially influenced by the independent variable is called a **dependent variable**, because it is influenced by, and thus to some extent *depends on*, the independent variable."[28] If we were testing a new medication to reduce headaches for example, the drug itself would be the independent variable and the headaches, which hopefully will be reduced or eliminated by the drug, would be the dependent variable—the headaches would depend on the medication.

METHODOLOGIES IN A NUTSHELL

When trying to choose between quantitative and qualitative methods, it is important to take several things into consideration—to include your interests. Table 1 gives some good guidelines of what to take into consideration.

Table 1. Research Methodology Guidelines

Use this approach if:	Quantitative	Qualitative
1. You believe that:	There is an objective reality that can be measured	There are multiple possible realities constructed by different individuals
2. Your audience is:	Familiar with/supportive of quantitative studies	Familiar with/supportive of qualitative studies
3. Your research question is:	Confirmatory, predictive	Exploratory, interpretive
4. The available literature is:	Relatively large	Limited
5. Your research focus:	Covers a lot of breadth	Involves in-depth study
6. Your time available is:	Relatively short	Relatively long
7. Your ability/desire to work with people is:	Medium to low	High
8. Your desire for structure is:	High	Low
9. You have skills in the area(s) of:	Deductive reasoning and statistics	Inductive reasoning and attention to detail
10. Your writing skills are strong in the area of:	Technical, scientific writing	Literary, narrative writing

SOURCE: Leedy, Paul D, and Jeanne Ellis Ormrod, *Practical Research; Planning and Design* (Upper Saddle River, NJ: Merrill Prentice Hall, 2001), 112.

College students spend entire semesters studying the intricacies of the various research methods and this appendix has only briefly touched on a few of them. For more details, one good reference is Leedy & Ormrod's *Practical Research: Planning and Design*.

SUMMARY

The most important concept to take away from this appendix is that research must consist of more than just gathering information and spewing it back out again in a slightly different format. Research begins with the formulation of an appropriate research question. The research question, the data, and your interests influence the research methodology. In shorter research papers you may not be required to specifically state the methodology you used. However, it should be apparent from reading your paper that you aren't just presenting information, but are also drawing some new meaning from the information and actually contributing to the body of knowledge.

The real fun in life comes from total creative absorption in a task and not the external rewards for doing it.

— Michael Leboeuf

Notes

[1] Thomas S. Kuhn, *The Structure of Scientific Revolutions* (Chicago, IL: The University of Chicago Press, 1970), 38.

[2] J. J. Carr, *The Art of Science: A Practical Guide to Experiments, Observations and Handling Data* (San Diego, CA: HighText Publications, 1992), 85-87.

[3] Jacques Barzun and Henry F. Graff, *The Modern Researcher* (New York: Harcourt, Brace and World, 1985), 19.

[4] Paul D, Leedy and Jeanne Ellis Ormrod, *Practical Research; Planning and Design* (Upper Saddle River, NJ: Merrill Prentice Hall, 2001), 70.

[5] Marcia Watkins, *The Walden University Survival Guide* (NY: Love From the Sea Publishers, 1997), 56.

[6] Marvin Bassett, ed., *Air University Style Guide*, (Maxwell AFB, AL: Air University Press, 2001), 11.

[7] Bassett, 27.

[8] Bassett, 72.

[9] Bassett, 131.

[10] Goodfellow, Major Gerald V., "Citing an email." E-mail to Marcia Watkins. 1 April 2003.

[11] HCarolineH, HEricH, and HEmilyH, "How to Write a Clear Research Report," on-line, Internet, 8 October 2002, available from http://www.radix.net/%7Efornax/air/clearrep.html.

[12] HE. Robert SchulmanH, HC. Virginia CoxH, and HE. Anne SchulmanH, "How to Write a Scientific Research Report," on-line, Internet, 8 October 2002, available from http://www.radix.net/%7Efornax/air/scireport.html.

[13] Bassett, 86.

[14] Leedy & Ormrod, 147.

[15] Paul D. Leedy, *Practical Research; Planning and Design* (New York: MacMillan, 1989),125.

[16] Leedy, 128.

[17] Leedy, 133-134.

[18] Leedy, 134.

[19] Leedy & Ormrod, 149.

[20] Leedy & Ormrod, 149.

[21] Leedy & Ormrod, 151.

[22] Leedy & Ormrod, 151.

[23] Leedy & Ormrod, 153.

[24] Leedy & Ormrod, 154.

[25] Leedy & Ormrod, 155.

[26] Leedy, 141.

[27] Leedy & Ormrod, 230.

[28] Leedy & Ormrod, 233.

Let's face it—English is a crazy language

There is no egg in eggplant nor ham in hamburger; neither apple nor pine in pineapple.

Sweetmeats are candies while sweetbreads, which aren't sweet, are meat.

We take English for granted. But if we explore its paradoxes, we find that quicksand can work slowly, boxing rings are square and a guinea pig is neither from Guinea nor is it a pig.

And why is it that writers write but fingers don't fing, grocers don't groce and hammers don't ham?

If the plural of tooth is teeth, why isn't the plural of booth beeth?

One goose, 2 geese. So one moose, 2 meese? One index, 2 indices?

Doesn't it seem crazy that you can make amends but not one amend?

If you have a bunch of odds and ends and get rid of all but one of them, what do you call it?

If teachers taught, why didn't preachers praught?

If a vegetarian eats vegetables, what does a humanitarian eat?

Sometimes I think all the English speakers should be committed to an asylum for the verbally insane.

In what language do people recite at a play and play at a recital?

Ship by truck and send cargo by ship?

Have noses that run and feet that smell?

How can a slim chance and a fat chance be the same, while a wise man and a wise guy are opposites?

You have to marvel at the unique lunacy of a language in which your house can burn up as it burns down, in which you fill in a form by filling it out and in which an alarm goes off by going on.

English was invented by people, not computers, and it reflects the creativity of the human race (which, of course, isn't a race at all).

That is why, when the stars are out, they are visible, but when the lights are out, they are invisible.

Enuff? Gud!

APPENDIX 3

EFFECTIVE READING STRATEGIES

This appendix covers:

- Barriers to effective reading.

- Reading for study or research.

- Memory aids.

Why would we want to improve our reading skills? If we're reading for pleasure, we may enjoy reading slowly and savoring every word. However, if we need to read or review lengthy materials in the course of our jobs, our time is valuable. Improving reading speed and comprehension will help pinpoint the information we need in the limited time we have available. On a day-to-day basis, we must stay current with world events and read or review Air Force Instructions, operating instructions, plans, technical orders, and promotion study guides. Since we don't have time to spend our days reading, we have to learn to use our reading time more effectively.

This appendix provides some basic strategies for improving reading skills. First, we'll discuss some barriers to effective reading and techniques to overcome these barriers. Next, we outline a study and research a reading method to help you get the most from your time. Finally, we offer some memory improvement tips.

"Reading is to the mind what exercise is to the body."
– Joseph Addison

WHAT ARE SOME BARRIERS TO EFFECTIVE READING?

Some of us continue to use the habits we adopted when we first learned how to read. Most of us don't even realize we're still practicing them. These habits are keeping us from reading efficiently and effectively.

① **Reading at a fixed speed.** One habit we tend to overlook is reading different types of material at the same speed. Not all materials are equally difficult to read. In the same way, not all sections of a particular written work are of equal complexity.

To read efficiently, adjust your speed according to the difficulty of the text you're reading and your purpose for reading it. If you're reading for main ideas only, skim the material quickly. Shift speeds as needed for additional information. Slow down on the complex parts, and speed up on the easy ones. If you are somewhat familiar with the material, read faster. Read faster over broad overviews and unneeded details. Generally, you'll need to slow down while reading detailed technical material and material with unfamiliar concepts. If you are reading for study purposes, your overall rate will be slower.

② **Vocalizing, or subvocalizing, words.** Another common habit we typically don't think about is vocalization, or subvocalization. This occurs when we say the words we're reading aloud, or to ourselves. If we pronounce each word, we can't read any faster than we can speak.

To overcome this habit, learn to read faster than you can speak. One way to do this is to shift your focus from a small area of print to a larger one. We'll cover a couple of techniques under the next section.

Another way to help cut your dependence on saying the words is to think about the key words, ideas or images—picture what is happening. You have to go over the print fast enough to give your brain all the information quickly. Keep your mind clear so the picture or thought can enter it. Then, accept it without question. Try not to let other thoughts intrude.

③ **Reading words one at a time.** Reading individual words is very inefficient. To read faster, your eyes must move faster over the printed material. We can't continue reading one word at a time. However, just willing your eyes to move faster isn't enough. Many of us have been practicing this habit of slow eye movements for so long that it's too hard to give it up that easily.

One way to break this habit is to use your hand, or another object, as a pacing aid. Since your eyes tend to follow moving objects, your eye will follow your hand as it moves across the page. One method is to make one continuous movement with your hand across each line of text. For example:

The first operations conducted by airmen were designed to gain information superiority. Subsequently, air-to-air combat evolved as a means to deny information superiority to an adversary. Information operations from air and space and, today, in cyberspace, remain key elements of what our Service brings to the Nation, the joint force commander, and component and coalition forces.

AFDD 1, *Air Force Basic Doctrine*, 1 Sep 97

Use your hand
as a pacing aid

As your hand moves across the text, your eyes will follow it. You can change the speed as needed for the difficulty of the material you're reading.

Another method is to let your peripheral vision do some of the work. To use this technique, break up the page into groups of words, or columns. Instead of focusing your eyes on every word, begin by focusing on every second or third word. Practice moving your eyes smoothly from one group to the next, and let your peripheral vision pick up the words on each side of the break. For example, you could break up this sentence into groups of words like this:

For example, you / could break up / this sentence into / groups of words / like this.

To break the text up into columns, use the following pattern as an example for your eyes to follow. As you read each line of text, focus your eyes on the first dotted line. (This line represents the center of the first column of text.) Then jump to the second line and finally the third line. Let your peripheral vision pick up the words on either side of the line as you go along.

Thunderstorms contain the most severe weather hazards to flight. Many are accompanied by strong winds. severe icing and turbulence, frequent lightning, heavy rain, and hazardous windshear. If all of these are not enough, - consider the possibility of large hail, microbursts and even tornadoes. Thunderstorms are quite powerful.

Use your peripheral vision

AFH 11-203, Vol 1, *Weather for Aircrews*, 1 Mar 97

Practice is the key! You can increase your reading speed by spending a few minutes a day reading at a faster rate than normal.

④ **Rereading passages.** Another barrier to effective reading is rereading passages. This habit can slow your reading speed to a snail's pace. We can all probably relate to this one. Once we've read a sentence or passage, we realize we have no idea what we've just read. Why? Our thoughts have a tendency to wander.

To prevent reading passages over again, you need to increase your concentration. Try to isolate yourself from any outside distractions. Find a place away from phones, TVs and engaging conversations.

A simple way to keep from rereading is to cover up the material you've already read with an index card. You can also use one of the pacing tools discussed under the last section.

⑤ **Stumbling over unknown words or large numbers.** Don't let unfamiliar words or large numbers slow you down.

Even readers who have a large vocabulary will come across unfamiliar words. You can usually gain the meaning of a word by its use in the text. Keep in mind that your ability to read faster depends upon your ability to recognize words quickly. The more you read, the more your vocabulary will grow. Additionally, improving your vocabulary will improve your comprehension.

As for numbers, unless you need to remember specific data, just substitute "few" or "many" for the actual digits.

WHAT WILL HAPPEN TO MY COMPREHENSION?

When you first start practicing reading faster, you may understand very little of what you read. Be patient. Your eyes will get used to seeing the print at a rapid pace, and you'll understand more. Begin by reading faster for only 5-10 minutes at one time, and gradually increase your time. In the end, your comprehension and retention should be at least as good as it is now.

Research shows a proportional relationship between reading speed and comprehension. In most cases, an increase in comprehension follows an increase in reading speed. Comprehension decreases when speed decreases. Reading slowly, word-by-word, seems to inhibit understanding. (Virginia Tech and the University of Maryland.) Eliminating the barriers to effective reading can help.

> Practice is the key!

HOW SHOULD I APPROACH READING FOR STUDY AND RESEARCH PURPOSES?

SQR3 is one research reading method you can use. It stands for survey, question, read, recall, and review. F.P. Robinson coined the SQR3 acronym in a book entitled *Effective Study* (1946). University researchers continue to recommend this method to help you get the most from your reading time. It will help you separate the important information from the chaff.

① **Survey.** The first step is to survey the material to get the big picture. This quick preview allows you to focus your attention on the main ideas and identify the sections you want to read in detail. The purpose is to determine which portions of the text are most applicable to your task. Read the table of contents, any introductions, section headings, subheading, summaries and the bibliography. Skim the text in-between. Be sure to look at any figures, diagrams, charts, and highlighted areas.

> Skim for gist

② **Question.** Once you've gained a feel for the substance of the material, compose questions about the subject you want answered. First, ask yourself what you already know about the topic. Next, compose your questions. You can also turn section headings and subheadings into questions. For example:

- "How do I create or change a program element?"
- "What is the technical order change process?

For some materials, you'll want to use your critical thinking skills to interrogate the writer. The more you know about the author and his/her organization, the better you'll be able to evaluate what you read. Try to answer questions like these:

- What is the author's experience and credentials?
- What is the author's target audience?
- What is the author's purpose?
- What are the author's assumptions?
- What are the author's arguments?
- What evidence does the author use to support his/her arguments?
- What are the author's conclusions?
- What factors shaped the author's perspective?

③ **Read.** Now go back and read those sections you identified during your survey. Search for answers to your questions. Look for the ideas behind the words.

> Read for details

④ **Recall.** To help you retain the material, make a point to summarize the information you've read at appropriate intervals (end of paragraphs, sections, and chapters.) Your goal is not to remember everything you've read, just the important points. Recite these points silently or aloud. This will help improve your concentration. You can also jot down any important or useful points. Finally, determine what information you still need to obtain.

⑤ **Review.** This last step involves reviewing the information you've read. Skim a section or chapter immediately after you finish reading it. You can do this by skimming back over the material and by looking at any notes you made. Go back over all the questions you posed and see if you can answer them.

> Review for understanding

HOW CAN I REMEMBER MORE OF WHAT I READ?

If you need to improve your ability to remember information you've read, the following tips may help.

① **Improve your concentration.** Improving your reading environment can help. Minimize distractions. Choose a place away from visual and auditory distractions. Ensure your chair, desk, and lighting are favorable for reading. Establish a realistic goal for how much you intend to read in one sitting. Stop occasionally for short breaks.

② **Organize the information.** Arrange data or ideas in small groups that make sense to you. These groups will be easier to remember by association with the group.

③ **Make the information relevant.** Connect the new information with information you already know. Try to associate what you are reading with your job or your interests. Recalling the information you already know about a subject will make it easier to recall the new stuff.

④ **Learn Actively.** Researchers report "people remember 90 percent of what they do, 75 percent of what they see, and 20 percent of what they hear." (Dave Ellis, *Becoming A Master Student*, 8th ed., 1997.) Use all of your senses. Don't just speak aloud when recalling information you've read, get your entire body into the act. Get up and move around, as if you are practicing for a speech. Use visualization and picture the information you're reading. Write down the main points. These methods will help you to retain the material. Don't forget to relax. When we're relaxed, we think more clearly.

⑤ **Use your long-term memory.** Your short-term memory space is limited, so take advantage of your unlimited long-term memory. To commit information to your long-term memory, review the material several times. Change the order of the information you recite during your reviews. This takes advantage of your ability to remember best what you read last.

⑥ **Maintain a positive attitude.** If you think the information is going to help you in your job or personal life, it will be easier to remember. However, if you believe the information you're reading is too difficult to retain, it will be. Think positive!

⑦ **Use Mnemonics.** You probably remember this memory aid from your childhood. Remember "My Very Energetic Mother Just Served Us Nine Pizzas?" This catch phrase helped us remember the order of the planets: Mercury, Venus, Earth, Mars, Jupiter, Saturn, Uranus, Neptune, and Pluto. You probably also know this catch phrase to help remember General Officer rank order: "Be My Little General" (Brigadier General, Major General, Lieutenant General, and General.) You could use a catchword to remember the Principles of War. "MOSSMOUSE" = Maneuver, Objective, Security, Simplicity, Mass, Offensive, Unity of command, Surprise and Economy of force.

"I divide all readers into two classes: Those who read to remember and those who read to forget."

— William Phelps

SUMMARY

Improving effective reading skills means learning to break old reading habits that slow us down. The following tips will help. Be patient: Practice is the key!

- Adjust your reading speed to match the difficulty of the material.

- Learn to read faster than you can speak.

- Use pacing aids to help you read faster.

- Put your peripheral vision to work.

- Increase your concentration by removing distractions.

- Infer the meaning of unfamiliar words by their use in the text.

- Improve your vocabulary.

- Substitute "few" or "many" for numbers you don't need to remember.

Improving our study and research skills will help us make the most of our time. The following steps will help.

- Survey the material.

- Form questions you want to answer.

- Read relevant sections to obtain details and answer your questions.

- Recall information as you finish reading sections.

- Review the material.

Improving our ability to remember information will help us in all areas. The following tips may help.

- Plan your environment to optimize your ability to concentrate.

- Organize the information in small groups.

- Make the information relevant to you.

- Learn actively; use all of your senses.

- Review the information several times, changing your review patterns.

- Think positive.

- Use mnemonic catchwords or catch phrases.

> "An ounce of practice is worth more than tons of preaching."
> – Mohandas Gandhi

SOURCES

AFDD 1, *Air Force Basic Doctrine*, 1 September 1997.

AFH 11-203, Volume 1, *Weather for Aircrews*, 1 March 1997.

Ellis, Dave B., *Becoming a Master Student*, 8th ed., Houghton Mifflin Co., 1997.
Fowler, H. Ramsey, *The Little, Brown Handbook*, 2d ed., 1983.

Robinson, F. P., *Effective Study*, New York: Harper & Row, 1946.

Strategies for Improving Concentration and Memory, the Cook Counseling Center at Virginia Tech.

Suggestions for Improving Reading Speed, the Cook Counseling Center at Virginia Tech, RSSL at the University of Maryland.

USAFA Reading Program.

Wenick, Lillian P., *Speed Reading Naturally*, 2d ed., Englewood Cliffs, NJ: Prentiss Hall, 1990.

Some Actual Signs

In the front yard of a funeral home:
DRIVE CAREFULLY, WE'LL WAIT

On an electrician's truck:
LET US REMOVE YOUR SHORTS

Outside a radiator repair shop:
BEST PLACE IN TOWN TO TAKE A LEAK

In a nonsmoking area:
IF WE SEE YOU SMOKING, WE WILL ASSUME YOU ARE ON FIRE AND TAKE
APPROPRIATE ACTION

On a maternity room door:
PUSH, PUSH, PUSH

On a front door:
EVERYONE ON THE PREMISES IS A VEGETARIAN EXCEPT THE DOG

At an optometrist's office:
IF YOU DON'T SEE WHAT YOU'RE LOOKING FOR, YOU'VE COME TO THE RIGHT
PLACE

On a taxidermist's window:
WE REALLY KNOW OUR STUFF

On a butcher's window:
LET ME MEAT YOUR NEEDS

On a fence:
SALESMEN WELCOME. DOG FOOD IS EXPENSIVE

At a car dealership:
THE BEST WAY TO GET BACK ON YOUR FEET—MISS A CAR PAYMENT

Outside a muffler shop:
NO APPOINTMENT NECESSARY. WE'LL HEAR YOU COMING

In a dry cleaner's emporium:
DROP YOUR PANTS HERE

On a desk in a reception room:
WE SHOOT EVERY 3RD SALESMAN, AND THE 2ND ONE JUST LEFT

In a veterinarian's waiting room:
BE BACK IN 5 MINUTES. SIT! STAY!

At the electric company:
WE WOULD BE DELIGHTED IF YOU SEND IN YOUR BILL. HOWEVER, IF YOU DON'T,
YOU WILL BE

APPENDIX 4

BIBLIOGRAPHY AND OTHER REFERENCES

Barzun, Jacques and Henry F. Graff. *The Modern Researcher*. New York: Harcourt, Brace and World, 1985.

Bassett, Marvin, ed. *Air University Style Guide For Writers and Editors*, Maxwell AFB, AL: Air

Booth, Wayne C, Gregory C. Colomb, and Joseph M. Williams, *The Craft of Research*, Chicago, IL: University of Chicago Press, 1995.

Bronowski, J. *The Common Sense of Science*. Cambridge, MA: Harvard University Press, 1953.

Carr, J. J. *The Art of Science: A Practical Guide to Experiments, Observations and Handling Data*. San Diego, CA: HighText Publications, 1992.

Covey, Stephen R. *The 7 Habits of Highly Effective People*. New York: Simon and Schuster, 1989.

Cox, Caroline, Eric Schulman, and Emily Schulman, "How to Write a Clear Research Report," n.p. On-line. Internet, 8 October 2002. Available from http://members.verizon.net/~vze3fs8i/air/scireport.html

Doyle, Michael and David Straus, *How to Make Meetings Work*. New York: Berkley, 1976.

Kuhn, Thomas S. *The Structure of Scientific Revolutions*. Chicago, IL: The University of Chicago Press, 1970.

Leedy, Paul D. *Practical Research*. New York: MacMillan, 1989.

Paul, Richard and Linda Elder, *The Miniature Guide to Critical Thinking*. Dillon Beach, CA, The Foundation for Critical Thinking, 1999.

Ramage, John D., John C. Bean and June Johnson, *Writing Arguments* (concise edition). New York: Longman, 2001.

Schulman, E. Robert, C. Virginia Cox, and E. Anne Schulman, "How to Write a Scientific Research Report," np. On-line. Internet, 8 October 2002. http://members.verizon.net/~vze3fs8i/air/scireport.html

Smith, Perry M., *Rules and Tools for Leaders; A Down to Earth Guide for Effective Managing*. New York: Avery, 1998.

The Tongue and Quill (AFH 33-337). Maxwell AFB AL: Air Command and Staff College, 30 June 1997.

Watkins, Marcia *The Walden University Survival Guide*. Love From the Sea, 1997.

Wilkerson, D. & Kellogg, J. *Survey Assessment: A Practical Approach to Supporting Change*. Arlington, VA: Coopers and Lybrand, 1994.

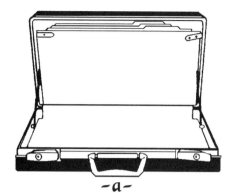

THE INDEX

Modified verb, 213
Modifier
    ambiguous, 97, 267
    dangling, 97, 267
    definition, 97
    misplaced, 97
Month, 312, 328

–n–

Nonconcurrence, 230
Nonverbal communication
    stage fright, 119
    visual aid, 126-130
Noun
    abstract, 268
    capitalization, 327
    collective, 268
    common, 320
    concrete, 268
    proper, 320
Number (see Grammar definition)
Number
    abbreviation, 332
    age, 332
    clock, 332
    compound modifier, 335
    date, 332
    dimension/size/temperature, 333
    document ID, 333
    executive correspondence, 331
    figure style, 332
    formal, 275, 334
    fraction, 335
    hyphenation, 333
    in general, 331
    latitude/longitude, 333
    literary, 331
    mathematical expression, 333
    measurement, 332
    military unit designator, 334
    million, 335
    money, 332
    nontechnical, 331
    percentage, 333
    plural, 335
    proportion, 333
    ratio, 333
    referred to as number, 333
    Roman numeral, 336
    rounded/indefinite, 335
    scientific/statistical, 331
    score, 333
    serial, 333
    spelling out, 334-335

    symbol, 333
    time, 332
    unit modifier, 333
    voting results, 333
    with noun, 327
    word style, 334-335

–o–

Object, 269
Official memorandum, 183-185
Organize and Outline
    bottom line first, 56
    cause and effect, 62
    classical, 58
    chronological, 59-60
    formats, 57
    logic, 61
    numerical, 58
    organizational patterns, 59-62
    outline, 58, 59, 61
    pattern, 59-62
    problem-solution, 61
    reason, 61
    spatial/geographic, 60
    outline formats, 57-58
    topical, 59
Organized communication, 7

–p–

Papers
    background, 215-216
    bullet background, 211-212
    point, 210
    position, 217-218
    staff summary sheet, 219-223
    talking, 209-210
Paragraphs
    arrangement/flow, 68-69
    dividing, 68-69, 297
    length, 68-69
    main point, 68
    topic sentence, 69
    transitional, 72-73
Parallel construction, 75
Paraphrase, 343
Parentheses, 299-300
Parenthetical expressions, 100
Participle, 270
Parts of speech, 269
Passive voice, 73-74, 269
Pattern
    cause and effect, 62
    outline, 58, 59, 61
    problem-solution, 61
    reason, 61
    spatial/geographical, 60
    chronological, 59-60

    topical, 59
Performance report, 225-235
Period, 301-302
Person, 269
Personal letter, 167-169
Phone (see Telephone)
Phonetic alphabet, 313
Phrase
    between subject/verb, 98
    buzzword, 240
    definition, 269
    simpler word/phrase, 81-87
    transitional, 72-73
Pictures, 126
Plagiarism, 269, 342
Plural (letter of alphabet),
Point paper, 210
Position paper, 217-218
Possessive, 266
Predicate, 269
Prefix, 295-296
Preposition, 269
Problem-solution
    description, 61
    limit, 200
    pro and con, 61
    solution, 61
    staff study report, 199-207
Proofreader marks, 102
Pronoun, 99, 269
    pronoun antecedent agreement, 99
Publications (AF)
    action officer role, 241
    contraction, 310
    editing, 242
    numbering, 302
    style, 242
    title, 242
    writing, 242
Punctuation
    apostrophe, 276-277
    asterisk, 278
    brackets, 279
    closed, 274-275
    colon, 280-281
    comma, 100, 282-286
    dash, 287-288
    display dot, 289-290
    ellipsis, 291
    exclamation mark, 292
    guidelines, 273-307
    hyphen, 293-297
    italics, 298
    open, 274-275
    parenthesis, 299-300
    period, 301-302
    question mark, 303-304
    quotation marks, 305-306

Some Actual Signs continue

In a Beauty Shop:
DYE NOW!

On the side of a garbage truck:
WE'VE GOT WHAT IT TAKES TO TAKE WHAT YOU'VE GOT. (BURGLARS PLEASE COPY.)

In a restaurant window:
DON'T STAND THERE AND BE HUNGRY, COME IN AND GET FED UP

Inside a bowling alley:
PLEASE BE QUIET. WE NEED TO HEAR A PIN DROP

In a cafeteria:
SHOES ARE REQUIRED TO EAT IN THE CAFETERIA. SOCKS CAN EAT ANY PLACE THEY WANT

Sign in a Laundromat:
AUTOMATIC WASHING MACHINES: PLEASE REMOVE ALL YOUR CLOTHES WHEN THE LIGHT GOES OUT

Sign in a London department store:
BARGAIN BASEMENT UPSTAIRS

In an office:
WOULD THE PERSON WHO TOOK THE STEP LADDER YESTERDAY PLEASE BRING IT BACK OR FURTHER STEPS WILL BE TAKEN

In an office:
AFTER TEA BREAK STAFF SHOULD EMPTY THE TEAPOT AND STAND UPSIDE DOWN ON THE DRAINING BOARD

Outside a secondhand shop:
WE EXCHANGE ANYTHING—BICYCLES, WASHING MACHINES, ETC. WHY NOT BRING YOUR WIFE ALONG AND GET A WONDERFUL BARGAIN?

Notice in health food shop window:
CLOSED DUE TO ILLNESS

Spotted in a safari park:
ELEPHANTS PLEASE STAY IN YOUR CAR

Seen during a conference:
FOR ANYONE WHO HAS CHILDREN AND DOESN'T KNOW IT, THERE IS A DAY CARE ON THE FIRST FLOOR

Notice in a field:
THE FARMER ALLOWS WALKERS TO CROSS THE FIELD FOR FREE, BUT THE BULL CHARGES

Message on a leaflet:
IF YOU CANNOT READ, THIS LEAFLET WILL TELL YOU HOW TO GET LESSONS

EXTRA

I know that
you believe you
understand what
you think I said,
but,
I am not sure
you realize that
what you heard
is not
what I meant!

Acknowledgements

The Tongue and Quill has been a valued Air Force resource for decades, and many people have contributed their talents to various editions over the years. I'd like to acknowledge the efforts of several personnel who overhauled chapters in this edition, including Lt Colonel Kimberly Demoret (Chapters 1, 2, 4, 5, and 11), Lt Colonel Janelle Costa (Chapter 7), Major Jane Palmisano (Chapters 3, 6, 8, 10, and 13), Major Keith Tonnies (Chapter 9), Major Marci Watkins (Appendix 2), and Major Kathy Dowdy (Appendix 3). Ms. Stephanie Rollins provided technical inputs for Chapter 4 and compiled its extensive list of Internet sources. Also, MSgt Douglas McCarty who provided the accomplishment-impact bullets for the Bullet Background Paper and the Performance Report. I would also like to thank Lt Col Demoret for her support during the planning, integration and editing processes.

– Sharon McBride, *Tongue and Quill* OPR

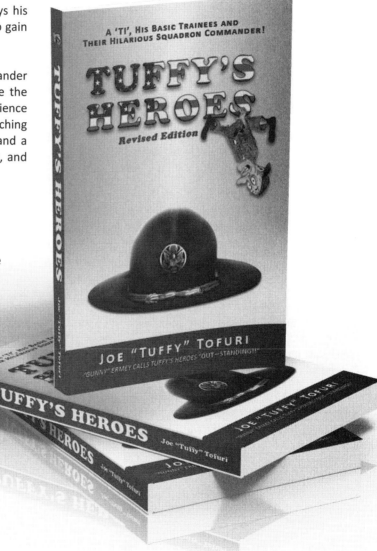

AIR FORCE
PROFESSIONAL DEVELOPMENT GUIDE

A complete USAFSE/PFE study guide in an attractive format for an affordable price

The Professional Development Guide (AFPAM 36-2241) is an Air Force publication from which the USAF Supervisory Examination and the Promotion Fitness Examination tests are developed. This book takes the information contained within the PDG and organizes it into an easy to read question and answer format to help you study for your exam.

You no longer have to rely on digging through the AFPAM, wading through paragraphs of filler and transitional text. This book gives you the important information right up front, organized by the same topics as the AFPAM.

- Over 2,300 questions and answers
- No multiple choice; study only the correct answers
- Charts, Forms, and Pictures help cement the information in your mind

PROFESSIONAL DEVELOPMENT GUIDE

STUDY AID
List Price $23.95

Published by:
Mentor Enterprises ™

What is the definition of unity of command?
Ensures concentration of effort for every objective under one responsible commander

What is DCAPES?
Deliberate and Crisis Action Planning and Execution Segment

What is the purpose of the annual training-needs survey?
It provides an opportunity for the supervisor to project training requirements for the upcoming fiscal year

What is the least severe disciplinary action?
An oral admonishment. It is often adequate to affect improvement or correction of work habits or behavior.

Study effectively. Develop yourself professionally. Get promoted.

Air Force Uniform Tools

Prepare for Your Next Inspection!

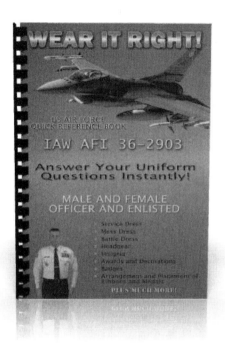

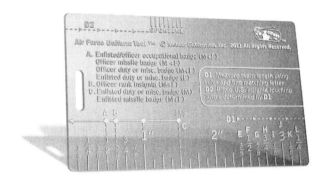

Are you tired of looking through AFI 36-2903 for answers to uniform and dress questions? Here is your solution.

No more searching through the bulky AFI just to get the basics. Grab Wear It Right and stick it in your pocket. This is a great, quick reference for Service Dress, Mess Dress, Formal Dress, Semi-Formal, Battle Dress, Maternity Uniforms, and Physical Fitness Uniforms.

The guide is organized by uniform type and lists general uniform requirements alphabetically.

From the creators of the Wear It Right! guidebooks, comes the greatest thing to happen to your next inspection. Everyone wants to look their best in their dress uniforms. Unfortunately, even the shiniest insignias won't impress your commander if they're in the wrong spot. That's why we created this all-in-one tool to help you position your insignia, medals, and more.

This handy tool is made from a durable translucent plastic, so you can see the measurements imposed on top of your uniform. All markings are gold-leafed to stand out boldly against dark fabrics. Combined with the ruler across that bottom edge and you have yourself the Swiss Army Knife for Wear and Appearance that fits right in your wallet.